Evidence based practice in Neuro-oncology

Supriya Mallick • Prashanth Giridhar
Goura K. Rath
Editors

Evidence based practice in Neuro-oncology

Springer

Editors
Supriya Mallick
Department of Radiation Oncology
National Cancer Institute, AIIMS
Haryana
India

Prashanth Giridhar
Department of Radiation Oncology
National Cancer Institute, AIIMS
Haryana
India

Goura K. Rath
Department of Radiation Oncology, All
India Institute of Medical Sciences
National Cancer Institute, All India Institute
of Medical Sciences, Jhajjar, Haryana
New Delhi
India

ISBN 978-981-16-2658-6 ISBN 978-981-16-2659-3 (eBook)
https://doi.org/10.1007/978-981-16-2659-3

This Springer imprint is published by the registered company Springer Nature Singapore Pte Ltd.
The registered company address is: 152 Beach Road, #21-01/04 Gateway East, Singapore 189721, Singapore

Foreword

The editors of *Evidence Based Practice in Neuro-oncology* need to be commended for compiling a succinct and informative book. The text integrates the radiological, pathological, and clinical aspects of management of central nervous system tumors. Optimal utilization of surgery, radiation, chemotherapy, and newer targeted therapy as a treatment in brain tumors requires a clear understanding of a range of issues, and this book provides information vividly to the clinician involved in brain tumor management, particularly those who are in training and new to practice. The book is divided into mainly four parts—the first part includes anatomy relevant to oncology, epidemiology, and pathology of CNS tumors, the second part includes clinical examination and imaging relevant to neuro-oncology, the third part includes basics relating to radiotherapy, and the last part includes a description of individual tumors including clinical management and follow-up. The chapter on neuro-anatomy has been well compiled and necessary for every neuro-oncologist—surgeon, radiation or medical oncologist. The chapters on pathology explain nicely the molecular classifications relevant to the oncologist keeping in line the WHO updates. The chapters on individual tumors are crisp and point wise. Future directions explained in chapters provide an insight to the reader regarding direction of research and help keep the reader up to date with current research. Careful attention is given to the concepts that are crucial in understanding modern treatment modalities. The chapters on neuro-cognitive deficits and endocrine effects make the reader aware of long-term issues for patients of brain tumors and will help in better management. The information in this book is presented in a logical and straightforward manner, thus offering an enjoyable learning experience.

During my long association with prestigious All India Institute of Medical Sciences, New Delhi, I got the opportunity to work with Prof. Rath who is well known for his contribution to the fraternity of Oncology. I feel proud of Dr. Supriya Mallick and Dr. Prashanth G whom I myself introduced to the subject of Neuro-Oncology. In fact, most of the authors are well known in their respective fields and have presented a beautiful teamwork. It is a great honor and a privilege for me to write a foreword for this book. I am confident that many clinicians would significantly benefit from the information provided by the authors.

P. K. Julka
All India Institute of Medical Sciences
New Delhi, India

Contents

Part III Management of Brain Tumours

About the Editors

Supriya Mallick completed his training in radiation oncology from All India Institute of Medical Sciences, New Delhi. He is now working as assistant professor, Department of Radiation Oncology, National Cancer Institute-India (NCI-India). He has more than 100 peer-reviewed publications and has delivered oral and poster presentations in numerous international meetings. He has published numerous original articles and meta-analysis of various tumors of the brain and spine. He is also conducting various trials in the field of neuro-oncology. He has also contributed chapters to many textbooks on radiation oncology.

Prashanth Giridhar is Radiation Oncologist by training and is currently working as Senior Research Associate at National Cancer Institute, Jhajjar, AIIMS, New Delhi. Dr. Prashanth has done his MBBS degree from JIPMER and then completed his postgraduation in AIIMS, New Delhi. He is a very refined researcher and has many publications in journals of international repute to his credit.

Goura K. Rath completed his training in radiation oncology from All India Institute of Medical Sciences, New Delhi. He is currently chief of B.R. Ambedkar Institute-Rotary Cancer Hospital, All India Institute of Medical Sciences, New Delhi. He has more than 250 peer-reviewed publications and is the recipient of Doctor K.M. Rai Memorial Oration award. He is also one of the editors of the textbook *Textbook of Radiation Oncology: Principles and Practice*, Elsevier, 2003, and is also the author of many chapters in several textbooks in the field of radiation oncology.

Part I

Epidemiology, Neuro-Anatomy and Neuro-Pathology for a Neuro-Oncologist

Epidemiology and Demography of Brain Tumors

1

Pritee A. Patil and Prashanth Giridhar

1.1 Introduction

- Brain and nervous system tumors account for 1.6% of all new cancer cases and 2.5% of cancer-related deaths globally [1].
- Approximately 70% of CNS tumors are nonmalignant and *30% are malignant.*
- According to global burden of disease study (GBD) [2], the incidence of CNS tumors in 2016 was 330,000 (95% UI: 299,000–349,000) and mortality was 227,000 (95% UI: 205,000–241,000). The age-standardized incidence rates were 4.3 per 100,000 per year (95% UI: 4.17–4.90) in 2016 which have increased globally by 17.3% (95% UI: 11.4–26.9) from 1990 to 2016.
- Incidence rates vary across the population of varying *sociodemographic index (SDI)* with highest age-standardized incidence rate seen in highest quantile of SDI (6.91 [5.71–7.53]) [2].
- The incidence of brain tumors also varies as per *geographical location*. The highest incidence is seen in East Asia followed by Western Europe and South Asia [2]. Incidence of these tumors in India ranged from 5 to 10 per 100,000 populations with an increasing trend [3].
- The overall incidence of brain tumors is higher in females compared to males (25.84 compared to 20.82 per 10,000) but *malignant tumors occur more commonly in males* [4].

P. A. Patil (✉)
Department of Radiation Oncology, National Cancer Institute, AIIMS, New Delhi, India

P. Giridhar
Department of Radiation Oncology, National Cancer Institute, AIIMS, Haryana, India

S. Mallick et al. (eds.), *Evidence based practice in Neuro-oncology*,
https://doi.org/10.1007/978-981-16-2659-3_1

1.2 Risk Factors

- Multiple environmental and behavioral risk factors have been evaluated for the development of these tumors.
- Exposure to ionizing radiation has been associated with increased risk of meningeal tumors and glial tumors.

1.3 Risk Factors Common for Brain Tumors

Glioblastoma	*Radiation exposure*: Rare *Use of cell phone*: Cell phones do not emit ionizing radiation and hence unlikely to cause cancer. However, some studies show higher incidence of benign brain tumors like meningioma, vestibular schwannomas, as well as glioma with cell phone use *Syndromic or associated with family history* Neurofibromatosis type 1 also known as von Recklinghausen disease: Other than neurofibroma, it is also associated with increased incidence of gliomas, schwannomas, meningioma Li-Fraumeni syndrome shows increased incidence of gliomas, breast, adrenal cancer, sarcoma Gorlin syndrome (basal cell nevus syndrome) Turcot syndrome Cowden syndrome
Medulloblastoma	Small numbers have inherited disorder with germline mutation. E.g., Gorlin syndrome, Turcot syndrome, li-Fraumeni syndrome
Pituitary adenoma	Syndromic association is seen in approximately 5% of the patients The genes responsible include (1) multiple endocrine neoplasia I (MEN I), (2) cyclin-dependent kinase inhibitor 1 B (CDKN), (3) protein kinase, cAMP-dependent regulatory 1A (PRKAR)—Carney complex (CNC)—Characterized by endocrine overactivity, schwannomas, skin pigmentation, and myxoma, (4) aryl hydrocarbon receptor-interacting protein (AIP) No causal relationship has been identified with smoking, alcoholism, or trauma
Pilocytic astrocytoma	Associated with neurofibromatosis 1

1.4 Pediatric Brain Tumors

- Brain tumors are the most common solid tumors in pediatric age group, with average annual age-adjusted incidence rate of 5.74 per 100,000 populations (both malignant and nonmalignant tumors) among males and females of 0–14 years of age [4].
- Glial cell tumors are the most common childhood brain tumors, pilocytic astrocytoma being the commonest among all glial tumors representing nearly 17.8% of all CNS tumors in 0–14 years of age followed by glioma NOS 14% and embryonal tumors 13.1% [4, 5].
- Relative frequency of various pediatric tumors is represented in Fig. 1.1.

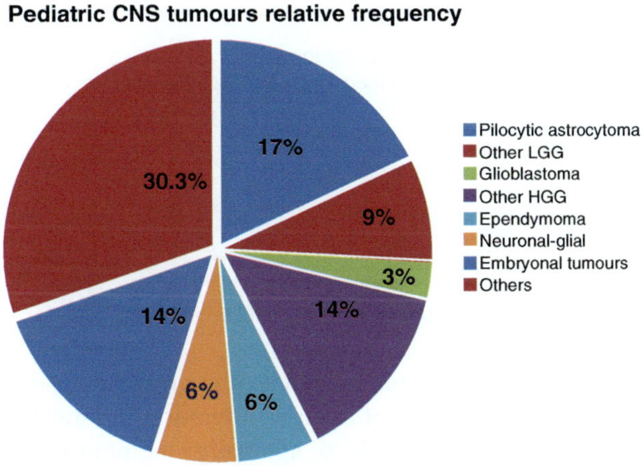

Fig. 1.1 Relative frequency of various pediatric tumors

1.5 Genetic Syndromes Associated with Pediatric Brain Tumors

Syndrome	Gene (chromosome locus)	CNS tumors	Clinical findings
Li-Fraumeni syndrome	TP53 (17p13)	Astrocytoma; cPNET; medulloblastoma, choroid plexus tumor	
Familial adenomatous polyposis (Turcot syndrome)	APC (5q21)	Medulloblastoma	Multiple adenomatous colonic polyps
Cowden syndrome	PTEN (10q23)	Dysplastic cerebellar gangliocytoma	Trichilemmoma, thyroid disorders, gastrointestinal polyps, macrocephaly
Tuberous sclerosis	TSC1 (9q34) TSC2 (16p13)	Sub-ependymal giant cell astrocytoma (SEGA)	Neurodevelopmental defects, renal angiomyolipoma, cardiac rhabdomyoma, ash-leaf spots, shagreen patches
von Hippel–Lindau syndrome	VHL (3p25)	Hemangioblastoma	Clear cell RCC, pheochromocytoma, pancreatic neuroendocrine tumor
Neurofibromatosis 1	NF1 (17q11)	Optic pathway glioma and other astrocytoma	Café au lait spots, neurofibroma, scoliosis, behavioral issues
Neurofibromatosis 2	NF2 (22q12)	Bilateral vestibular schwannomas, peripheral schwannomas, meningioma, astrocytoma, spinal ependymoma	Café au lait spots, neurofibroma

Syndrome	Gene (chromosome locus)	CNS tumors	Clinical findings
Goltz-Gorlin syndrome	PTCH (9q31)	Medulloblastoma	Odontogenic keratocysts, palmar and/or plantar pits, and ectopic calcifications of the falx cerebri

1.6 Brain Tumor in Adults

- The most common malignant tumor in adults is glioblastoma (14.6% of all tumors and 48.3% of malignant tumors) and most common nonmalignant tumor is meningioma (53.3% of nonmalignant tumors) (Fig. 1.2) [4].
- Median age at diagnosis for all primary brain tumors is 60 years, though median age varies according to histologic types.

1.7 Survival Outcomes

- CNS tumors poses significant mortality, globally it accounted for 227,000 death (205,000–241,000) with age-standardized death rates of 3.24 per 100,000 person-year (2.91–3.43) [2].

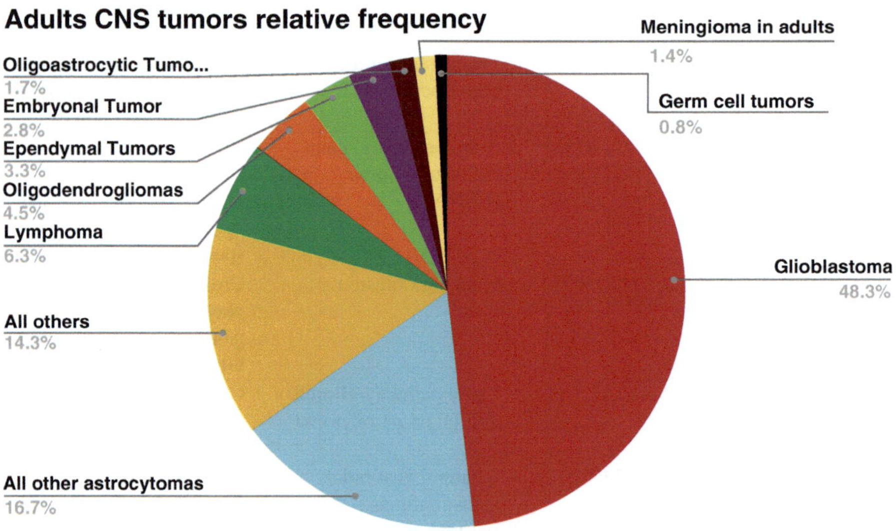

Fig. 1.2 Relative frequency of various adult CNS tumors. *In the below pie diagram, note that the block marked in yellow refers to Meningioma in adults 1.4% and block marked in black refers to Germ cell tumors 0.8%

- Overall, average annual mortality rate of malignant brain tumor is 4.42 per 100,000 and an average 15,944 deaths per year caused by malignant brain and other CNS tumors in USA [4].
- Average annual age-adjusted mortality rate for adults more than 40 year is 9.10 per 100,000 and 0.72 per 100,000 in children aged 0–14 years [4].
- Survival depends on the type of histology and on the age at diagnosis, as survival decreases with increasing age. Children and young adults have better survival outcomes for most histologies.
- Overall, 5-year survival in adolescents and young adults is 98.1%, children 97.2%, and adults 90.1%. The 5-year relative survival for malignant tumor was 35.8% and for nonmalignant tumors was 91.5% [4].
- According to different histology in malignant tumors, the 5-year relative survival rate is lowest for glioblastoma (6.8%). Among benign CNS tumor, the relative 5-year survival was highest for nerve sheath tumor (99.3%) [4].

References

1. Bray F, Ferlay J, Soerjomataram I, Siegel RL, et al. Global cancer statistics 2018: GLOBOCAN estimates of incidence and mortality worldwide for 36 cancers in 185 countries. CA Cancer J Clin. 2018;68:394–424.
2. Patel AP, Fisher JL, Nichols E, et al. Global, regional, and national burden of brain and other CNS cancer, 1990–2016: a systematic analysis for the Global Burden of Disease Study 2016. Lancet Neurol. 2019;18:376–93.
3. Dasgupta A, Gupta T, Jalali R. Indian data on central nervous tumors: a summary of published work. South Asian J Cancer. 2016;5(3):147–53.
4. Ostrom QT, Cioffi G, Gittleman H, et al. CBTRUS statistical report: primary brain and other central nervous system tumors diagnosed in the United States in 2012–2016. Neuro-Oncology. 2019;21(S5):1–100.
5. Johnson KJ, Cullen J, Barnholtz-Sloan JS, et al. Childhood brain tumor epidemiology: a brain tumor epidemiology consortium review. Cancer Epidemiol Biomark Prev. 2014;23(12):2716–36.

Neuro-Anatomy for Oncologist

2

Doris George Yohannan, Reba Babu Alex,
Aswathy Maria Oommen,
and Chandrasekharan Kesavadas

2.1 Introduction

Neuro-anatomy involves the study of the nervous system, which consists of the central nervous system (CNS) and peripheral nervous system (PNS). This chapter confines itself mainly to the study of the CNS, which is formed of the brain and spinal cord. The brain includes the cerebrum, diencephalon, cerebellum and brainstem (midbrain, pons and medulla oblongata) (Fig. 2.1a, b). The spinal cord is a downward extension of the CNS. The cranial nerves emerge from the brain and the spinal nerves from the spinal cord (Fig. 2.1c). An in-depth anatomical study of these structures is beyond the scope of this chapter and may not be necessary for oncologists. To understand the different signs and symptoms produced due to lesions in different parts of the nervous system, a sound knowledge of anatomy, its relations to adjacent structures and the functional importance of each part are absolutely essential. In this chapter, each part has been described in brief, trying to highlight the relevant anatomical details and relations of each part, so that the oncologist finds it easier to correlate and analyse the various lesions seen.

D. G. Yohannan (✉) · R. B. Alex · A. M. Oommen
Department of Anatomy, Government Medical College
Thiruvananthapuram, India

C. Kesavadas
Department of Imaging Sciences and Interventional Radiology
Sree Chitra Tirunal Institute of Medical Sciences and Technology
Thiruvananthapuram, India

S. Mallick et al. (eds.), *Evidence based practice in Neuro-oncology*,
https://doi.org/10.1007/978-981-16-2659-3_2

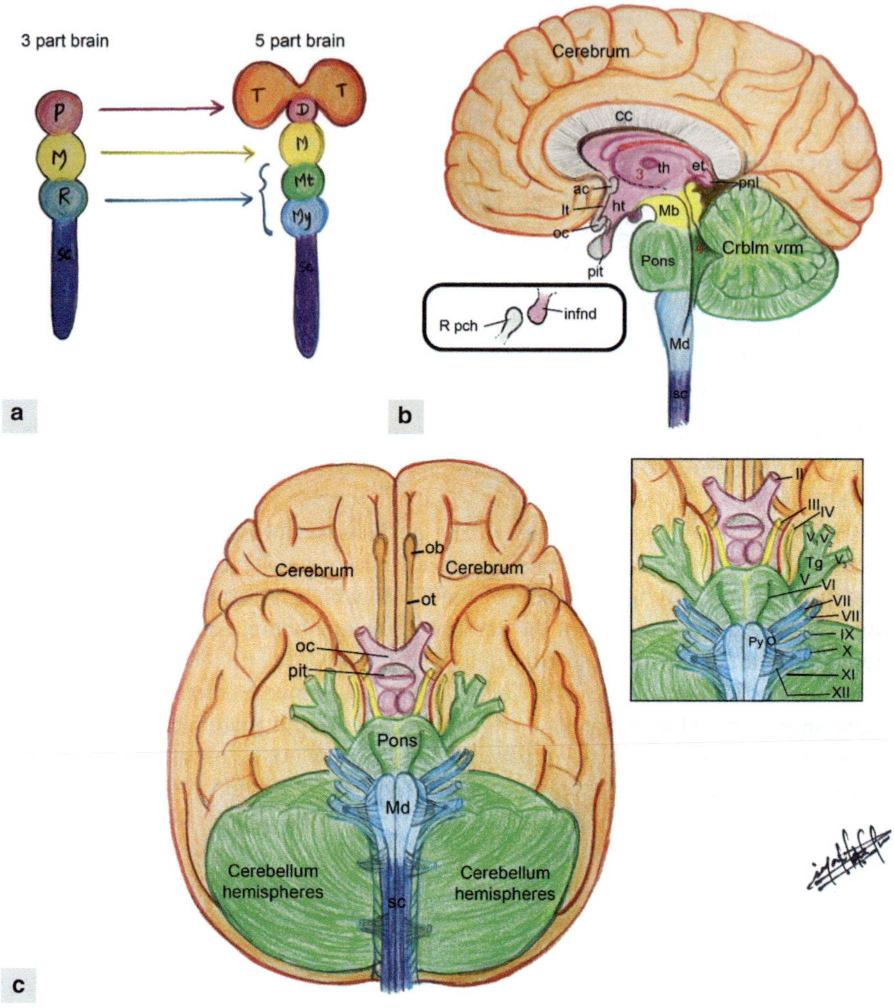

Fig. 2.1 Development of brain and its derivatives. (**a**) The 3-part brain with prosencephalon (P), mesencephalon (M) and rhombencephalon (R) and the 5-part brain with telencephalon (T), diencephalon (D); mesencephalon (M); metencephalon (Mt), myelencephalon (My) and the spinal cord (sc). (**b**) Midline section of brain (colour codes, corresponding to its embryologic origins as in **a**). Spinal cord (sc); medulla (Md); cerebellum vermis (Crblm vrm); fourth ventricle (4). The cerebral aqueduct runs through the midbrain (Mb). The diencephalon (pink) has three components: the part ventral to the hypothalamic sulcus (dashed black curve) is the hypothalamus (ht), the part above the sulcus is the thalamus (th) and the dorsal part is the epithalamus (et), with its main component—pineal gland (pnl). The cavity of the diencephalon is the midline third ventricle (3). (inset) Note the pituitary (pit) with its developmental sources—Rathke's pouch (R pch) and infundibulum (infnd) and its relation to optic chiasma (oc). Lamina terminalis (lt); anterior commissure (ac); corpus callosum (cc). Correlate with MRI in Fig. 2.3c. (**c**) *Ventral view of brain* with colour codes as in **a**. *ob* olfactory bulb, *ot* olfactory tract, The cranial nerves are shown in the inset in Roman numerals, *Tg*,trigeminal ganglion; *Py* pyramid, *O* olive (see Fig. 2.2b also)

2.2 The 'Environments' of the Brain

2.2.1 The Bony Framework

- The brain is enclosed within the upper skull—*neuro-cranium.*
- It is protected above, in front, behind and the sides by a flattened curved bone plate—the vault, formed by the parts of parietal, frontal, temporal and occipital bones.
- Below, the brain is related to the cranial fossa (skull base).
- The cranial fossa can be subdivided to three levelled 'steps', like that of a staircase (Fig. 2.2a, b). Most anteriorly, there lies an anterior cranial fossa (ACF), lodging the *inferior frontal lobes of the cerebrum.* Below the ACF, there are *orbits* laterally and *nasal cavity* in midline. It is important to know that the *frontal air sinuses* are also related to the anteromedial roof of the orbit; and the *ethmoid air sinuses* near the midline. This relation can cause *direct malignant invasion* of these paranasal sinus tumours which may require a multidisciplinary anterior craniofacial approach and resection for its management [1]. The ridges

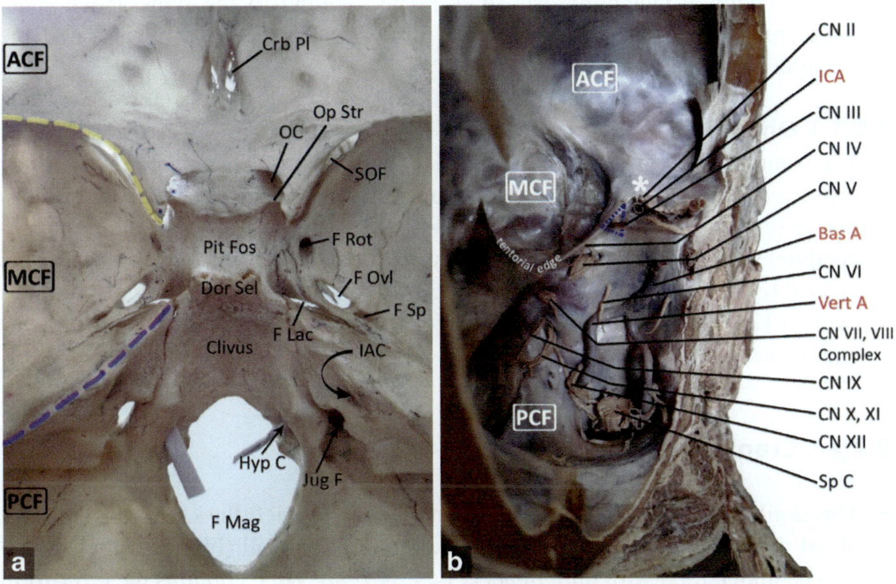

Fig. 2.2 Skull base (interior view). (**a**) Dry skull (view from top) and (**b**). dura intact (parasagittal section, view from posterior and right). *ACF* anterior cranial fossa, *MCF* middle cranial fossa, *PCF* posterior cranial fossa. (**a**) *Pit Fos* pituitary fossa, *Dor Sel* dorsum sellae, *Crb Pl* cribriform plate, *OC* optic canal, *Opt Str* optic strut, *SOF* superior orbital fissure, *F Rot* foramen rotundum, *F Ovl* foramen ovale, *F Sp* foramen spinosum, *F Lac* foramen lacerum, *IAC* internal acoustic canal, *Jug F* jugular foramen, a folded paper probe passes through, *Hyp C* hypoglossal canal, *F Mag* foramen magnum. (**b**) Cranial nerves are numbered in Roman numerals. Blood vessels are shown in red. *ICA* internal carotid artery, *Bas A* basilar artery, *Vert A* vertebral artery; blue dotted triangle: oculomotor trigone, white asterisk: falciform ligament, *Sp C* spinal cord

formed by the posterior border of lesser wings of sphenoid form the posterior boundary of the ACF (Fig. 2.2a).

- The 'butterfly'-shaped middle cranial fossa (MCF) (Fig. 2.2a) lodges the anterior half of the inferior temporal lobes in its 'wings' and pituitary gland (in sella turcica) and parasellar structures related to its 'central body'. The dorsum sella, in the midline, and the superior borders of the petrous temporal bone on two sides (where tentorium cerebelli attaches) form the posterior limit of the MCF.
- Behind it, there is the posterior cranial fossa (PCF) (Fig. 2.2a) with a conspicuously large foramen magnum through which the lower medulla continues as the spinal cord, with its surrounding meninges, the vertebral arteries, the accessory nerve and a few anterior ligaments related to the cranio-vertebral junction. The anterolateral aspects of the PCF have the posterior surface of petrous bone, which faces the cerebellum and the pons, a relation important while considering a CP angle tumour. Behind the foramen magnum on both sides lies the shallow cerebellar fossae which occupies the cerebellar hemispheres. The cerebellar fossae are limited above by the transverse sinuses lodged in the posterior attachment of tentorium cerebelli. In front of the foramen magnum there lies the clivus—a bony backward slope, formed by the fusion of the sphenoid and occipital—the spheno-occipital synchondrosis. It is related behind to the pons above (with intervening prepontine cistern, basilar artery, basilar venous plexus) and the medulla below (Fig. 2.3c).

2.2.1.1 Oncological Significance

- The commonest benign tumours related to the skull base are *meningiomas*, *sellar neoplasms* and *schwannomas*. Malignant tumours arising here are *chordomas* that arise from the notochordal remnant (seen usually in the clivus at the spheno-occipital synchondrosis; recall from development, notochord was in *midline* and *ventral* to the neural tube) and *chondrosarcomas* (from embryonic cartilaginous matrix) [2]. Skull base is also a site for *metastasis*, commonly from prostate and breast.

2.2.2 Cranial Nerves

- The skull base is studded with foramina through which the cranial nerves and blood vessels pass. They are best learned in the dry skull (Fig. 2.2a) but should also be understood with dura intact (Fig. 2.2b). The ACF contains the cribriform plate close to midline, which is made 'sieve-like' by 15–20 slender olfactory nerve rootlets (CN I).
- The optic nerve (CN II) runs through the narrow optic canal, from the orbital apex to the MCF. The optic canal is bounded medially by a surgically important bony ridge, the *optic strut* (Fig. 2.2a), which separates from the laterally lying superior orbital fissure (SOF). In the specimen, note a surgically important *falciform ligament* superior to it (Fig. 2.2b).

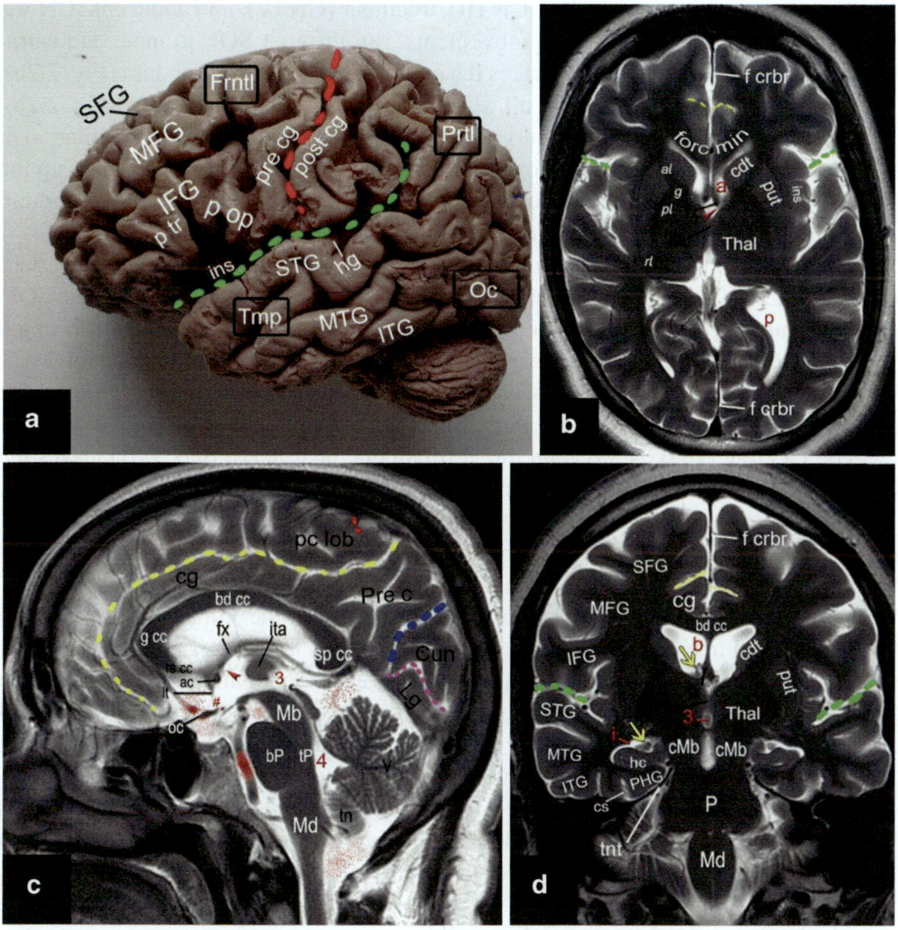

Fig. 2.3 (**a**) Lateral cerebral surface showing frontal (Frntl), parietal (Prtl), occipital (Oc) and temporal (Tmp) lobes. (**b**) Axial MRI (T2w). (**c**) Sagittal MRI (T2w). (**d**) Coronal MRI (T2w). (**a**) Dashed lines show important sulci. Red: central sulcus; green: Sylvian fissure; yellow: cingulate sulcus; blue: parieto-occipital sulcus; pink: calcarine sulcus. *SFG, MFG, IFG* superior, middle and inferior frontal gyri, *STG, MTG, ITG* superior, middle and inferior temporal gyri; p tr: pars triangularis, *p op* pars opercularis, *pre cg* pre-central gyrus, *post cg* post-central gyrus, *ins* insula, *hg* Heschl's gyrus, (**b**) *f crbr* falx cerebri, *forc min* forceps minor, *al, g, pl, rl* anterior limb, genu, posterior limb and retro lentiform parts of internal capsule, *fx* fornix, *cdt* caudate nucleus, *put* putamen, *ita* interthalamic adhesion, *red arrow head*: foramen of Monro, (red) b, a, p, i: body, anterior, posterior and inferior horns of lateral ventricle. (**c**) *Pc lob* paracentral lobule, *Pre c* pre cuneus, *Cun* cuneus, *Lg* lingual gyrus, *rs cc, g cc, bd cc and sp cc* rostrum, genu, body and splenium of corpus callosum, *Mb* midbrain, *bP and tP* basilar and tegmental pons, *Md*: medulla, *3* third ventricle and its supra optic recess (#), above the optic chiasma (oc) and infundibular recess (asterisk); ac: anterior commissure, *lt* lamina terminalis, *4* fourth ventricle, *V* vermis of cerebellum, *tn* tonsil of cerebellum, *'Red particles'* indicate the CSF-filled basal cisterns: chiasmatic around chiasm; interpeduncular and quadrigeminal, in front and behind midbrain; prepontine and cerebellomedullary (see Fig. 2.4 also). (**d**) Yellow arrows: choroid plexus; *cs* collateral sulcus, *PHG*: parahippocampal gyrus, *hc* hippocampus, *cMb* crus cerebri of midbrain, *tnt* tentorium

- The cranial nerve oculomotor (CN III), trochlear (CN IV), and abducens (CN VI) that primarily control the eye movements, run through SOF to innervate extra-ocular muscles. The CN III pierces the oculomotor trigone in the dura (Fig. 2.2b), and the CN IV, the thinnest of all, slips underneath the tentorial edge (Fig. 2.2b), endangering it in skull base surgeries. The course of abducens is peculiar, as it emerges from the brainstem lower than CN III and IV, bends and climbs up the clivus, pierces the dura in the clival slope (Fig. 2.2b) and runs through a subdural (Dorello's) canal to bend again and reach the SOF.
- The trigeminal nerve (CN V) divisions—ophthalmic nerve (CN V_1), maxillary nerve (CN V_2) and mandibular nerve (CN V_3) (Fig. 2.1b) pass through the SOF, the foramen rotundum and the foramen ovale, respectively, the three forming a 'crescent of foramina' at the root of the greater wing of sphenoid in MCF (Fig. 2.2a), though the sensory root pierces the dura at the *trigeminal pore* more posteriorly (Fig. 2.2b) and the ganglion is within the Meckel's cave.
- The facial nerve (CN VII) and the vestibulocochlear nerve (CN VIII) leaves the PCF through the internal auditory canal (IAC) found at the posterior surface of the petrous temporal bone.
- The glossopharyngeal (CN IX), the vagal rootlets (CN X) and the spinal accessory nerve (CN XI) exit the PCF through the jugular foramen. Of these nerves, the spinal accessory, arising from the upper five segments of the spinal cord, ascends up the foramen magnum and leaves through the jugular foramen (Fig. 2.2b).
- The hypoglossal nerve (CN XII) leaves the lower most part of PCF through the hypoglossal canal, located inferomedial to the jugular foramen, in the occipital condyle (Fig. 2.2a).
- The vascular relations of foramina are also important. The foramen lacerum (lacerated foramen) is in fact a dried skull artefact, formed as the petrous apex is lateral to sphenoid (Fig. 2.2a). It is related above to the C3 segment (lacerum segment) of internal carotid artery (ICA), just after its exit from the petrous course [3]. Then it continues as the cavernous (C4) and the clinoid (C5) parts, which pierces the dura mater (Fig. 2.2b).
- The vertebral arteries (VA) enter the skull through the foramen magnum (Fig. 2.2b). The two ICAs and two VAs supply the brain. The foramen spinosum is the site of entry of the middle meningeal artery (MMA) which occupies the plane outside the dura. The jugular foramen is also the exit point of the sigmoid sinus and the inferior petrosal sinuses, where it is related posterior to jugular bulb. The sigmoid sinus leaves the jugular foramen as the internal jugular vein (IJV). Figure 2.4 depicts various cranial nerves on MR imaging.

2.2.2.1 Oncological Significance

- *Falciform ligament* is a surgically important landmark, and this fold may need to be incised for mobilizing the optic nerve during surgery.
- Due to peculiar course of CN VI, even a mild central herniation of the brainstem (downward displacement due to elevated intracranial pressure) can cause a potential traction on the CN VI causing a lateral rectus palsy and diplopia/squint

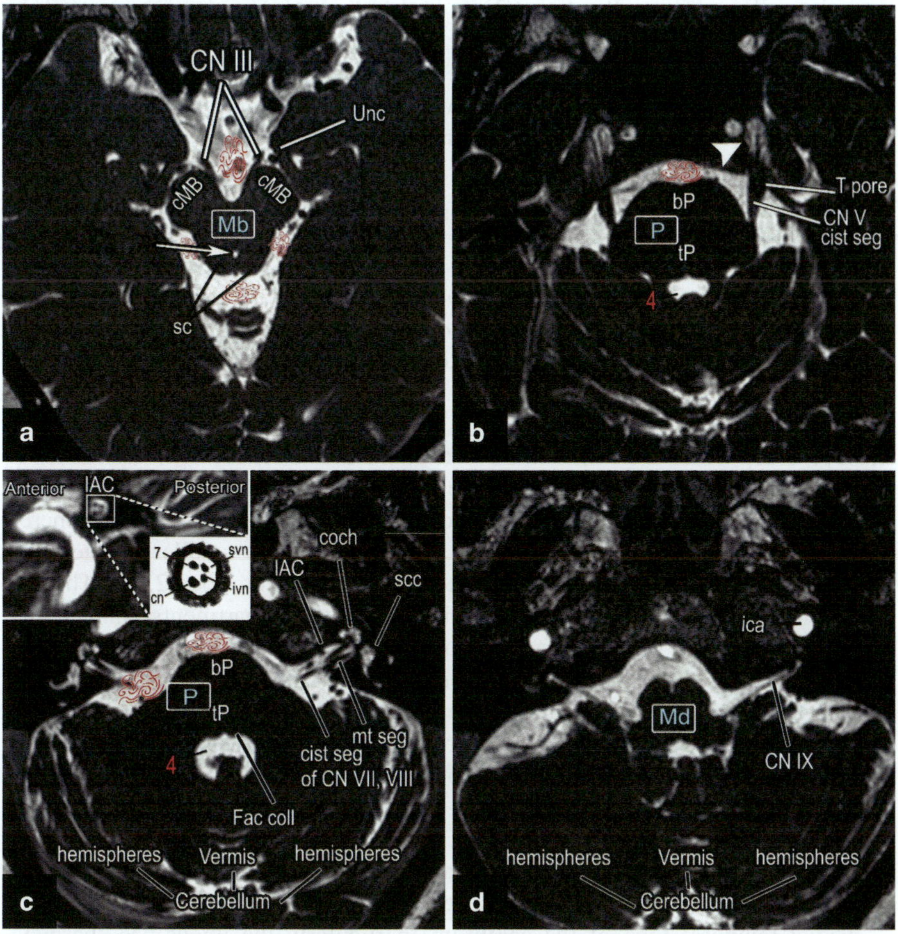

Fig. 2.4 (**a**–**d**) MRI (CISS sequences) axial sections in a cranio-caudal order to show cranial nerves and related anatomy. (**a**) *Upper midbrain (MB).* CN III exiting medial to crus cerebri (cMB). Uncus (Unc) of medial temporal lobe is related anterolateral to the crus. *Arrow* aqueduct, *sc* superior colliculus, Red particles indicate the sites of peri-mesencephalic—interpeduncular (in front), ambient (by sides) and quadrigeminal cisterns (behind the midbrain). (**b**) *Upper pons (P),* with basilar part (bP) anteriorly and tegmental part (tP) posteriorly. 4: fourth ventricle. CN V exits from lateral pontine surface, cisternal segment (cist seg) lies in the lateral aspect of prepontine cistern (red particles), enters the trigeminal pore (T pore). White arrowhead: Meckel's cave (see Fig. 2.2b also). (**c**) *Lower pons,* with CN VII and VIII complex lying in the cerebellopontine angle cistern (red particles, laterally)—the cisternal segment (cis seg) and then the internal auditory canal (IAC)—the meatal segment (mt seg). The inner ear apparatus can be seen laterally with cochlea (coch) anteriorly and semi-circular canals (scc) posteriorly. The cistern in front of pons is the pre-pontine cistern. Inset: The nerves occupying the IAC is seen in an oblique plane MRI. A drawing of the same shows *7* seventh nerve, *cn* cochlear nerve, *svn* superior vestibular nerve, *ivn* inferior vestibular nerve. (**d**) Upper medulla (Md), with CN IX heading to leave the jugular foramen, which lies posterior to the internal carotid artery (ica)

unilaterally or bilaterally. A neoplasm in the clivus (say, a chordoma) can compress the CN VI, quite early.
- The knowledge of the skull base and these foramina should aid oncologists to correlate a case of craniofacial pain with signs of cranial nerve involvement at the foramen level to a possible skull base metastasis [4].

2.2.3 The Meninges

- The meninges covers the CNS. It can be classified as *pachymeninges* (containing the outer tough dura mater) and *leptomeninges* (containing the inner thinner, arachnoid mater and pia mater). The disposition of meninges is different around the brain and the spinal cord.

2.2.3.1 Dura Mater Around the Brain and Dural Folds

- The most important dural folds are the *falx cerebri* (Fig. 2.3b), a 'sickle shaped' partition in the mid-sagittal plane, lodging between the cerebral hemispheres; and *tentorium cerebelli*, a 'tent like fold', separating a smaller infratentorial compartment from a larger supratentorial compartment (Fig. 2.3d). The two folds incompletely divide the intracranial space into compartments. An increased pressure in any of the compartments may cause a 'brain herniation' syndrome.
- A *diaphragma sellae* is a small horizontally oriented fold above pituitary gland, with an opening in it that allows the passage of the pituitary stalk. The diaphragma and its opening may have roles in directing the growth of pituitary adenomas and thus to compress the optic neural pathways above it [5–7].

Oncological Significance

- A large tumour in the supratentorial compartment can increase the intra-compartmental pressure, thereby 'squeezing' the softer brain tissue, mainly the medial temporal lobe of the cerebrum (particularly the uncus), to 'herniate' beyond the tentorial edge, a phenomenon called as the *trans-tentorial herniation/uncal herniation*. Such an uncal herniation on a side can compress the closely lying ipsilateral third nerve and the ipsilateral crus cerebri of midbrain (see Fig. 2.4a), causing corresponding focal neurological deficits (in this case, an ipsilateral third nerve palsy with a dilated, unresponsive pupil; a contralateral hemiplegia, as the crus cerebri contains pyramidal tract fibres that will cross to opposite side).
- In some cases, the uncus may push the midbrain all the way, that the crus cerebri of the contralateral side is compressed by the opposite side tentorial notch creating an ipsilateral hemiplegia respective to the side of the tumour—the Kernohan–Woltman phenomenon; a false localizing sign in the setting of neoplastic lesions or more commonly, trauma [8].
- The increased pressure in one half of the intracranial space can cause a *subfalcine herniation* where the cingulate gyrus can herniate below the lower free edge of the falx cerebri entrapping the anterior cerebral artery.

- An increased intracranial pressure within the infratentorial compartment can cause the tonsils of the cerebellum (Fig. 2.3c) (a small subdivision of the inferior cerebellar hemisphere) to herniate down through the foramen magnum—*tonsillar herniation*. This can compress the medulla (with the vital centres within it) and cause a cardiorespiratory collapse.

2.2.3.2 The Pia, Arachnoid and Subarachnoid Space

- Deep to the meningeal layer of the dura lies the arachnoid mater, a web-like, flimsy layer apposed onto the dura.
- In this plane, from arachnoidal cap cells arise meningiomas, one of the common tumours related to the brain.
- The pia mater invests the brain parenchyma throughout the sulci, gyri and fissures. This creates *an actual subarachnoid space*, filled with cerebrospinal fluid (CSF).

2.3 Spinal Cord

- Spinal cord-related tumours may arise from within the substance of the cord (*intramedullary*[1]) or outside the cord (*extramedullary*) (Fig. 2.5a).
- The spinal cord is enclosed in the vertebral column and lies within the spinal canal. The cord is related to vertebral bodies anteriorly, the pedicles anterolaterally, the laminae posterolaterally and the spinous processes (palpable clinically) posteriorly (Fig. 2.5a). Vertebral metastases are common in posterior vertebral body with accompanying pedicular destruction [9].
- The spinal cord is cylindrical, around 45-cm long and extends as a continuation of the medulla oblongata (from the upper border of C1 spine) extending down till a conical termination—the conus medullaris (at L1 vertebra in the adult). In the foetus, the lower limit is at the S2 vertebral level; in the new-born, the lower limit ascends to level of L3 spine, and by adulthood reaches the L1 vertebra; a phenomenon due to a normal developmental lag of growth of spinal cord with respect to the bony spinal column.
- The spinal cord can be thought to be a 'stack' that contains 31 spinal segments. From each spinal segment arises a spinal nerve, through a dorsal and ventral root (Fig. 2.5b). Due to the developmental disparity of cord with respect to the vertebral column, mentioned above, the spinal segments *will be at the level of numerically corresponding vertebrae* in the *upper cervical cord only* and *will have a disparity with the vertebral relation downward the cord.*
- The spinal cord is surrounded by the meninges. However, the dura lies separated from the inside of the spinal canal, separated by the low pressure, loose areolar tissue filled epidural space/*extradural* (cf. cranium, where there is no actual

[1] Note that spinal 'medulla' is an older anatomical term, but retained in these medical lexicon, intra 'medullary', conus 'medullaris', etc. It should not be confused with medulla oblongata, a component of the brainstem.

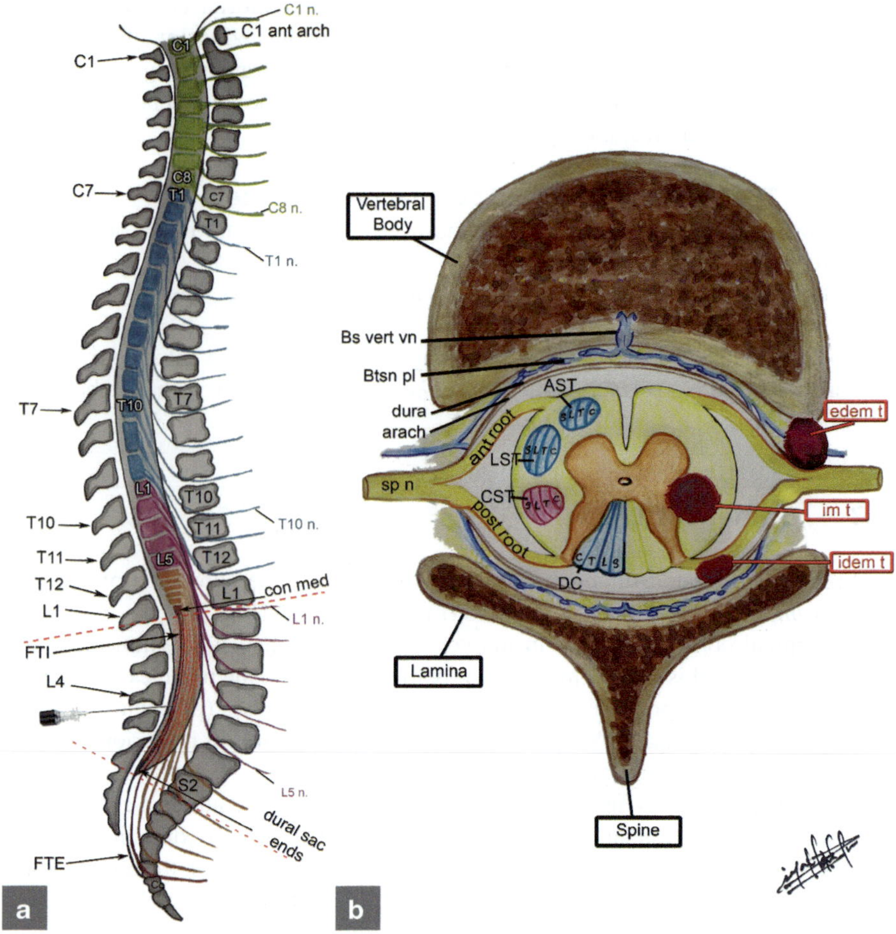

Fig. 2.5 (**a**) Diagram showing relations of vertebral column and its contents. Note the disparity of levels of spinal segments (white writings) and that of vertebral body/spine. Ant arch: anterior arch; Con med: conus medullaris; FTI: filum terminale internum; n.: nerve; FTE: filum terminale externum. Red dashed lines indicate the lower level of spinal cord and dural sac. *'Red particles'* below L1 vertebra indicate the location of the CSF filled *lumbar cistern* where the cauda equina and FTI lie and where a lumbar puncture can be safely done below L4 spine. (**b**) Transverse section showing spinal foramen and its contents. *Bs vert vn* Basi vertebral veins, *Btsn pl* Bateson's plexus, *arach* arachnoid layer, *ant root* anterior root, *post root* posterior root, *sp n* spinal nerve, *AST and LST* anterior and lateral spinothalamic tract, *CST* cortico spinal tract, DC: dorsal column pathway. Note the lamination of important tracts and how a tumour can compress some tracts preferentially. *C* cervical, *T* thoracic, *L* lumbar, *S* sacral, *Edem t*: Extra dural extra medullary; *im t* intra medullary tumour, *idem t* intra dural extra medullary tumour

epidural space). Tumour in extradural space is an important cause of neoplastic spinal cord compression [10] and can result from metastasis to vertebrae nearby. The dural sac and arachnoid however extend till the S2 vertebra, and hence, it is evident that below the conus medullaris, there is a CSF-filled lumbar cistern, which contains only the dangling nerve roots—the cauda equina and the filum terminale (FT). FT is the pial extension below the conus medullaris, the upper 3/4th lies within the dural sac (FT internum) and the lower 1/4th (FT externum) tethers the dural sac to the coccyx (Fig. 2.5a).

- The ventral root of the spinal nerve contains purely motor and the dorsal root purely sensory fibres. The spinal nerves exit the intervertebral foramen in a systematic manner. The *first seven cervical nerves* exit *above* the numerically corresponding vertebrae, the *C8 nerve* exits *below C7 vertebra (as there are seven cervical vertebrae and eight cervical nerves/segments*), and the rest of the spinal nerves exit *below* the corresponding numbered vertebra. For example, the L4 nerve will leave the spinal canal through the L4–L5 intervertebral foramen. Due to this, the lumbar and the sacral nerve roots dangle down, almost vertically, below the conus medullaris within the lumbar cistern (Fig. 2.5a). These roots form the cauda equina (similar to the hairs of a horse tail) (Fig. 2.5a).
- The spinal cord contains grey matter in the centre and white matter surrounding it. The white matter has the ascending and descending tracts. The major tracts that need to be understood are the descending *corticospinal tract* (the motor fibres controlling distal muscles) and the ascending *dorsal column pathway* (carrying sensations of vibration and conscious proprioception). Both these tracts are *located on the same side of innervation*. The ascending *spinothalamic pathway* (carrying pain, crude touch, pressure and temperature) are *located on the opposite side of innervation*, as these fibres cross soon after entering the spinal cord. More important is to understand the somatotopic arrangement/lamination of these tracts. The arrangement of the fibres in the *spinothalamic tract* and *corticospinal* is from medial to lateral, *cervical*, *thoracic*, *lumbar* and *sacral* (Fig. 2.5b). The somatotopy of dorsal column is just the reverse.

2.3.1 Oncological Significance

- The disparity of vertebral level and spinal cord (as explained above) is clinically significant. The *lower cervical segments* will be corresponding to *one vertebral level higher* (e.g., *C7 spinal segment will be at the C6 vertebral level)*. The *upper thoracic segments* will be *two vertebral levels higher*. The *lower thoracic segments* will be *two to three vertebral levels higher*. Hence, a sensory loss below a circumferential level at the umbilicus (representing T10 dermatome) should alert the evaluation of T7 or T8 vertebrae. The *lumbar segments* will be at the *T10 to T12 vertebral levels*. The *sacral spinal segments* (within the conus medullaris) will be at the *L1 vertebral level*. These are important as the lower thoracic and upper lumbar vertebrae are commoner sites for metastases.

- As the *corticospinal tract* (the motor fibres controlling distal muscles) and the ascending *dorsal column pathway* contain sacral fibres most superficially, an *extramedullary* mass lesion can cause *early sacral sensory loss* and *weakness* (UMN spastic type) *of the legs*. In contrast, an *intramedullary* tumour (e.g., astrocytoma, ependymoma), arising from the substance of the spinal cord, usually causes a characteristic 'sacral sparing' of sensations, causing motor involvement or UMN signs of lower limbs to appear later due to the somatotopy of the aforesaid tracts. The extramedullary tumours can be further subdivided into extradural or intradural depending on whether it is outside or inside the dural sac.
- The *intradural tumours* are the most common (70–80%) and commonly benign, the common being meningiomas, schwannomas (from nerve sheaths) and neurofibromas (commonly from dorsal root) [11]. These tumours can compress nerve roots and cause radicular pain early on, in addition to the cord compression that can ensue later. *Extradural tumours* are generally malignant and usually metastatic as mentioned above (Fig. 2.5b).

2.4 Brainstem

- The brainstem is an upward continuation of the spinal cord, above C1 vertebra. It contains three components stacked from below upwards—the medulla oblongata, the pons and the midbrain (Figs. 2.1b, c and 2.3c).
- The brainstem contains in its core a *reticular formation* that includes a 'cranial' component that is important to maintain consciousness and a 'caudal' component that is more important for visceral, motor and other reflex functions. The cranial nerves in brainstem can be conveniently (though, not precise) remembered with the mnemonic 4-4-2 for medulla, pons and midbrain. The lower four cranial nerves (CN IX, X, XI and XII) arise from medulla, the next four (CN V, VI, VII and VIII) arise mainly from pons and the next two cranial nerves (CN III and IV) arise from midbrain. It is helpful if the brainstem is learned based on these fundamental principles as it will be helpful to correlate lesions in the brainstem grossly with functions and clinical presentations.
- The topography of the brainstem with respect to the ventricular space also needs to be understood. The lower part of the medulla, the 'closed' medulla, contains a central canal similar to the spinal cord. The upper medulla becomes 'open' as the central canal opens up to the fourth ventricle, with the upper medulla and the pons, in front of the fourth ventricle. At the upper end, the fourth ventricle leads to the aqueduct of Sylvius, which runs through the midbrain.

2.4.1 Oncologic Significance

- Tumours in the brainstem are usually *gliomas* which can destroy the cranial nerve nuclear masses mentioned above or their fascicles, depending upon its location. Hence the clinical presentations depend on the specific location and whether it affects unilaterally or bilaterally. The lateral gaze centre (PPRF, in pons) can be involved which can create a lateral gaze palsy. The ascending or descending tract involvement can create paralysis or hemianesthesia. These tracts can be visualized in diffusion tensor imaging and studies have proved that it can be a sensitive investigation to detect tract displacement, distortion or invasion in brainstem tumours and also may be promising in surgical planning [12, 13]. There are nuclei and fibres involving cerebellar circuitry within the brainstem (note red nucleus in midbrain, pontine nuclei and pontocerebellar fibres in pons and olivary nuclei in medulla). When the brainstem is involved in a disease a presentation simulating cerebellar involvement (nystagmus, ataxia, intention tremors, dysmetria) can occur and can be false localizing signs.

2.5 Cerebellum

- The cerebellum is located in the posterior cranial fossa, behind the fourth ventricle, behind the pons and the medulla (Figs. 2.1 and 2.3c). Cerebellum is snugly fit in a relatively unyielding space with an inferior (occipital) surface facing the squamous parts of the occipital bone, a lateral (petrous) surface facing the petrous temporal bone and a superior (tentorial surface) facing tentorium cerebelli above. The cerebellum is connected by three cerebellar peduncles (superior, middle and inferior) to the brainstem components in a logical sequence—to the midbrain, pons and medulla. Anatomically, cerebellum has a flocculonodular lobe (in front of the posterolateral fissure), an anterior lobe (in front of the primary fissure) and a posterior or middle lobe (between the two fissures).
- Functionally, the flocculonodular lobe is particularly concerned with vestibular connections and coordination of eye movements. The cerebellar vermis is concerned with trunk and proximal limb coordination.
- The paramedian cerebellum is concerned with distal limb coordination and lateral cerebellum is concerned with motor planning. One cerebellar hemisphere receives the inputs of the 'intended movement' from the *contralateral* cerebral cortex, through the cortico-ponto-cerebellar pathway. It also receives unconscious proprioceptive inputs from the *ipsilateral* body, about the 'actual movements' occurred, through spinocerebellar pathways.
- It is very important to understand a neuroanatomical localization principle that *one side cerebellar hemisphere* deals with coordinating the movement of the *ipsilateral limbs*.

2.5.1 Oncological Significance

- A lesion in the vermis is likely to cause truncal ataxic symptoms, unsteady 'drunken' gait, vestibular symptoms and ocular movement incoordination (nystagmus).

- A tumour involving the right cerebellar hemisphere can cause right appendicular cerebellar ataxia (a characteristic uncoordinated movement of the limbs, since coordination of muscle movements is a function of the cerebellum). Patients may present with uncoordinated movements, dysmetria and walking difficulty. This is in contrast with the general neuroanatomical localization principle of the cerebrum where the cerebral cortex of one side controls the activity of the *contralateral* side.
- The common tumours with respect to the cerebellum and fourth ventricle are *medulloblastoma*, *ependymoma* and *metastases*. These usually obstruct the CSF circulation in the fourth ventricle and cause hydrocephalus early on and a presentation of cerebellar involvement follows later. Medulloblastoma arises exclusively in the cerebellum, the site of origin being most commonly at the inferior medullary velum. In children it arises in the vermis and in adults involvement of the cerebellar hemispheres is commoner.

2.6 Diencephalon: The Thalamus, Hypothalamus and Epithalamus

Diencephalon being in midline is best learned by studying a midline sagittal section of the brain. In such a view, the three parts of the diencephalon that can be seen are: (1) the large nuclear mass called the thalamus, (2) the epithalamus above and behind and (3) the hypothalamus below and in front.

2.6.1 Thalamus

- The part above and behind the hypothalamic sulcus is the thalamus. It is an important relay station in the brain where almost all of the sensory tracts (except olfaction) are relayed. Not only the ascending tracts but the cerebellar and basal ganglia outputs also reach the thalamus for processing, before being projected to the cerebral cortex, concerned with motor function.
- The view seen in a median sagittal section of the brain will be the medial surface of the thalamus, i.e. the wall of third ventricle. It should also be understood that the superior surface of thalamus is the floor of the lateral ventricle (see Fig. 2.6d). The thalami lie deep, almost completely hidden from the intact brain surface, making thalamic tumours challenging for the neurosurgeon. They are shaped like two eggs lying very close to each other (Fig. 2.6d).
- The two thalami often remain very close, even bridging each other, creating a massa intermedia[2] in up to 70% of individuals and is composed of thalamic grey matter. This creates a 'bullet-hole' in cast models of ventricular system (see Fig. 2.6c). Its functional significance is largely unknown [14] though some con-

[2] Also called interthalamic adhesion.

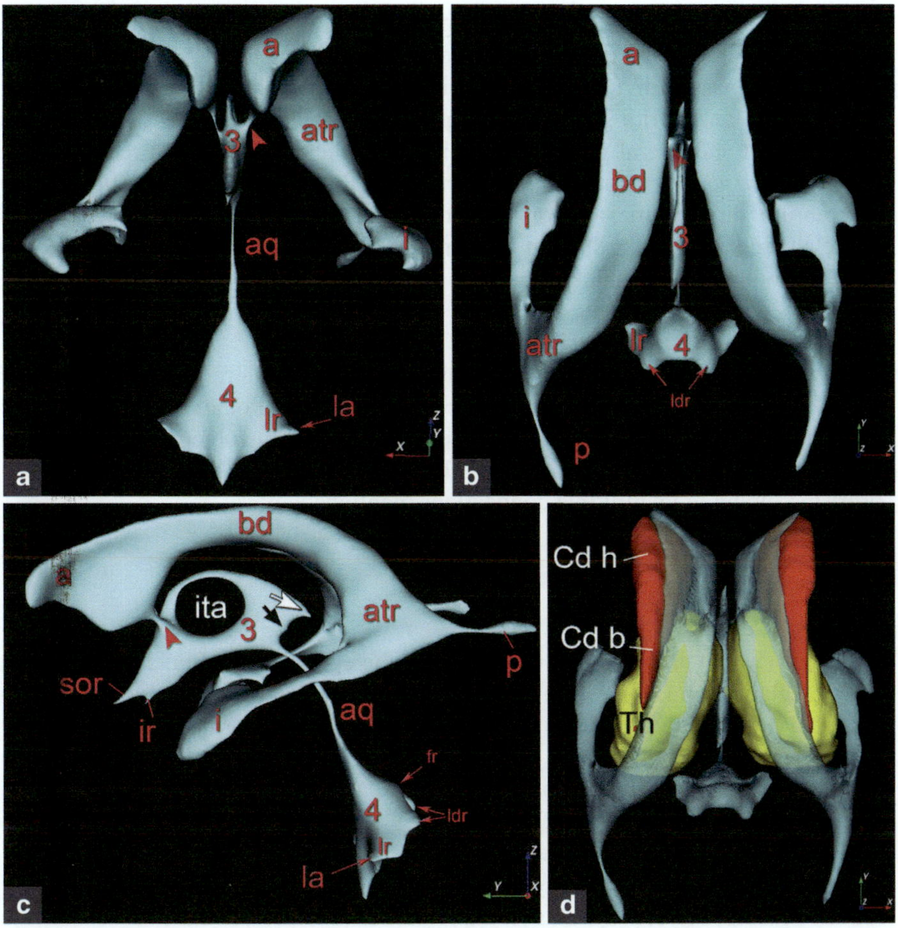

Fig. 2.6 3D reconstruction (from MRI images) of the ventricular system seen in (**a**). anterior; (**b**). top; (**c**) left side views. Lateral ventricle with its body (bd), anterior (a), posterior (p) and inferior (i) horns with the atrium (atr). The foramen of Monro (red arrowhead) communicates the lateral ventricles to the third ventricle (3). Third ventricle recesses seen anteriorly are supraoptic recess (sor), infundibular recess (ir), the small pineal recess (black arrow) and the larger suprapineal recess/habenular recess (white arrow). The inter-thalamic adhesion (ita) is visualized as a 'hole' in the third ventricle reconstruction image. The aqueduct communicates the third ventricle to the fourth. The fourth ventricle has its spout like lateral recesses (lr) with lateral apertures (la) at its tip. The dorsal recesses are the fastigial recess (fr) and the lateral dorsal recesses (ldr). (**d**) *Top view* shows the relation of the head (hc) and body (bc) of caudate nucleus and thalamus (Th) to the third and lateral ventricles (made translucent). Note the head of the caudate bulging to the anterior horn and the lateral ventricle wrapping around the thalamus

sider it likely that bilateral thalamic glioma evolve near central thalamic area, spreading through interthalamic adhesion [11].

- This large mass is divided into smaller nuclear groups by an internal medullary lamina into an anterior group, medial group and a lateral group. Posteriorly, there lies medial and lateral geniculate bodies (MGB and LGB) which receive *auditory* and *visual* sensory signals and project to the corresponding cortices in the cerebrum through the auditory and optic radiation. The ventral lateral (VL) nucleus receives its input from the cerebellum and basal ganglia and project to the premotor areas of the frontal lobe.

2.6.1.1 Oncological Significance

A tumour invading thalamus can generally cause contralateral general cutaneous sensory involvements and loss in proprioception. Reports of other rare presentations of thalamic tumours like movement disorders [15], motor involvement, hemianopia [16] and behavioural and other frontal lobe involvement [17, 18] signs may not be surprising owing to its multiple connections mentioned above.

2.6.2 Epithalamus

- Posteriorly and above, to the thalamus lies the epithalamus (Fig. 2.1b), with the pineal gland being its main component. The pineal gland may involve melatonin-related day-light cycles. It is one of the sites that can be normally calcified (seen in CT images; hence, pineal gland shift is a useful landmark to assess midline shift at the brainstem-diencephalon level).

2.6.2.1 Oncological Significance

Pineal gland tumours or pinealomas, though rare, can obstruct the aqueduct of Sylvius causing a hydrocephalus. It can compress the pretectal area rostral to the midbrain which contains gaze centres for vertical gaze (cf. lateral gaze centres in the pontine tegmentum) and result in an up gaze palsy, loss of pupillary reflexes as in Parinaud syndrome [19].

2.6.3 Hypothalamus

- The region below and in front of the hypothalamic sulcus is the hypothalamus (Fig. 2.1b). It is located on both sides of the inferior aspect of the third ventricle which is almost like a narrow midline slit (Fig. 2.6a). It is one of the highest regulatory centres for *autonomic*, *limbic* and *endocrine* control at the brain as well as controlling many homeostatic functions like day-night circadian rhythm, temperature regulation, hunger, thirst, sexual desire, etc.

- *Autonomic fibres* from hypothalamus descend through the posterolateral aspect of the brainstem and in the spinal cord mediating the sympathetic and parasympathetic control. The hypothalamic nuclei and *limbic circuits* (mentioned in cerebral cortex) are reciprocally interconnected and probably reflects how the emotions (limbic circuits) are strongly linked to autonomics. It *regulates the pituitary gland* by producing *releasing* and *inhibiting hormones*, that transit through the *hypothalamo-hypophyseal portal system*, that control the trophic hormone secretions of anterior pituitary. The supraoptic and paraventricular nuclei in the hypothalamus synthesize the hormones oxytocin and antidiuretic hormone which will be transported through the *hypothalamo-hypophyseal tract* to be stored in the posterior pituitary gland.
- It also contains specific areas that controls primitive attributes that are needed for *homeostatic functions* like circadian rhythm, appetite, thirst, thermoregulation [10]. For example, the suprachiasmatic nucleus in the anterior hypothalamus is the 'master clock' that controls circadian rhythm. Thermoregulation ensures optimum temperature in the body, critical for body function. Anterior hypothalamus has sensors to detect increased heat and activate body mechanisms that dissipate heat.

2.6.3.1 Oncological Significance

- Tumours of anterior hypothalamus can cause *insomnia*, and lesions of posterior hypothalamus can cause *hypersomnia.* Similarly, lateral hypothalamic lesions cause decreased appetite and a *decrease in body weight*; while medial hypothalamic lesions can affect inhibition of appetite and cause *obesity*. Lateral hypothalamic lesions can cause a decrease in the person's water intake.
- A lesion in the anterior hypothalamus can cause the body to become hyperthermic.

2.7 Pituitary Gland

- The pituitary gland is an amalgamation of the Rathke's pouch (from the roof of the primitive pharynx) and the evagination from the lower diencephalon (Fig. 2.1b).
- The cell rests of the Rathke's pouch are thought to be one of the origins of craniopharyngioma and can present in the suprasellar area in children. The part of the pituitary contributed by the Rathke's pouch is the anterior pituitary (or pars distalis/adenohypophysis).
- The posterior pituitary (pars nervosa/neurohypophysis) is in fact the axonal endings of the supraoptic and paraventricular nuclei of the hypothalamus and secretes oxytocin and vasopressin (antidiuretic hormone/arginine vasopressin). Pituitary tumours are relatively common of all intracranial neoplasms (~15%) and are almost always benign [20].

2.8 Cerebrum: Sulci, Gyri, Functional Areas

- The cerebrum lies in the supratentorial compartment. The surfaces of the cerebral hemispheres are remarkably folded by the presence of sulci (fissures) and gyri (bumps).
- The cerebral hemispheres comprise *frontal*, *parietal*, *temporal* and *occipital* lobes, named after the cranial bones to which they are related. Within these lobes lies some areas known as 'eloquent cortices' which sub-serve important functions. Lesions in them will cause a loss of those functions, and so it is important to know these areas when considering surgical approaches to resect tumours near these regions [21]. The *insula*, seen in the depth of the Sylvian fissure (Fig. 2.3a), and the *limbic lobe* are also considered as lobes by many authors. The limbic areas are evolutionarily primitive and particularly important in memory, emotions and olfaction.
- The frontal and parietal lobes are separated by the obliquely oriented *central sulcus* (Fig. 2.3a). In front of it lies the *pre central gyrus*, which is the *primary motor cortex* classified as the Brodmann's area (BA) 4. This area functionally represents the cell bodies of the *upper motor neurons (UMN)*, whose axons will form the *corticospinal tract*. Similar to UMN fibres of the cranial nerves, the *corticobulbar*[3] *fibres* will descend to the cranial nerve nuclei and will primarily innervate head and neck. The representation of the entire body on the pre-central gyrus is called the *motor homunculus* and is in such a way that the *legs* and perineal areas are on the medial surface of the gyrus. The *arms* are higher on the superolateral surface, and the *head* is represented in the lower most part (Fig. 2.7a, b).
- While a tumour nearby is resected, the corticospinal tract has to be preserved to the best extent possible. Behind the central sulcus is the *post central gyrus*, which is the *primary sensory cortex* (BA 3,1,2). The entire general sensations of pain, touch and temperature are projected to this area, and are similarly represented by a sensory homunculus. The pre- and post-central gyri thus run parallel, separated by the central sulcus.
- The areas in front of the primary motor cortex are *premotor areas*, which are involved in planning of the motor programs for movement. A lesion in this site can cause a deficit in motor planning. Behind the primary sensory cortex lies the *sensory association cortices* which is concerned with higher order processing of sensory information.
- The *Sylvian fissure* is prominent and lies separating the temporal lobe below from the frontoparietal lobes above (Fig. 2.3a). Just above the Sylvian fissure lies the *anterior* and *ascending rami* of the Sylvian fissure creating a pars orbitalis, pars triangularis and the pars opercularis. These are in fact subdivisions of the inferior frontal gyrus. The pars opercularis (BA44) and triangularis (BA45)

[3] 'Bulb'—an older anatomical term indicating medulla oblongata; used in clinical vocabulary in terms like bulbar palsy, pseudobulbar palsy. The term bulb also interchangeably indicates brainstem as in terms like corticobulbar neurons.

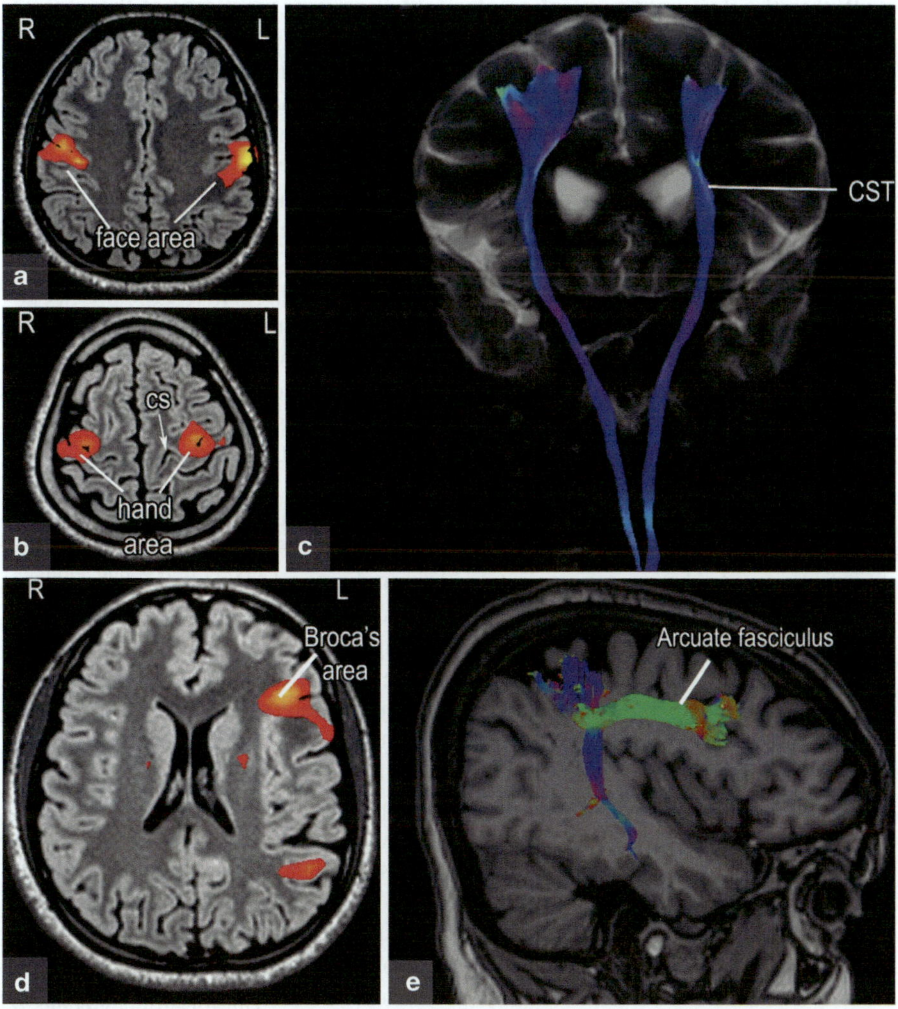

Fig. 2.7 Advanced neuroimaging reveals functional neuroanatomy. (**a**, **b**) Functional MRI (fMRI) images reveal cortical areas of the motor homunculus responsible for the motor activity of the face (lower down in the pre-central gyrus) and the hand (higher in the pre-central gyrus). cs: central sulcus. (**c**) Tractography demonstrates the corticospinal tract (CST) descending from the motor cortex through the internal capsule and brainstem. (**d**) fMRI co-registered on FLAIR demonstrates Broca's area in the left inferior frontal area. Note that the sides of MRI are according to radiological convention. (**e**) Tractography co-registered on T1-weighted image reveals the arcuate fasciculus connecting the Broca's area to the Wernicke's area and also to the parietal regions

(Fig. 2.3a) on the *dominant hemisphere* (in most people, the left) is the *motor speech area*, the *Broca's area* (Fig. 2.7d). This cortex represents the area which produces the neural programs to express language. Lesions in this region will cause a Broca's aphasia/motor aphasia with markedly reduced word output.

- Within the *dominant hemisphere* itself, the posterior most area of the superior temporal gyrus, in the lower bank of the Sylvian fissure, there lies the *sensory speech area*, the *Wernicke's area* (BA 22). This area is responsible for comprehension of language. Patients with lesions in this area will have Wernicke's aphasia with severe affliction to understand words and usually will not be able to follow commands, will have normal fluency but a nonsensical speech.
- Besides these, more cortical and subcortical areas maybe involved in the language function. An important related association tract is the arcuate fasciculus which is a fibre bundle connecting the Wernicke's area to the Broca's area (Fig. 2.7e). If the Sylvian fissure is opened up, in the superior surface of the superior temporal gyrus, there lies a transversely oriented finger—like gyri that run from superficial surface to the depth (Fig. 2.3a). This is the Heschl's gyri (BA 41), which is the *primary auditory cortex*. The surrounding areas have the auditory association cortices that are needed for the processing of the auditory stimuli. Also note the proximity of the auditory cortices to the Wernicke's area, which concerns with the processing of words heard.
- The parieto-occipital sulcus is best seen on the medial surface of cerebrum and demarcates the parietal lobe from the occipital lobe. The calcarine sulcus seems to converge to the parieto-occipital sulcus and cut out a 'wedge' in the medial surface of brain called cuneus, and the area anterior to it is the precuneus. Below the calcarine sulcus lies the 'tongue like' lingual gyrus (Fig. 2.3c). The *primary visual cortex* [11, 13, 14] occupies the areas on the 'banks' of the calcarine sulcus. The upper bank is part of the cuneus, where the contralateral inferior quadrant of visual field is represented. The lower bank, lingual gyrus represents the contralateral upper quadrant of visual field (Fig. 2.3c).
- On the tentorial surface of the temporal lobe, medial to the collateral sulcus lies the parahippocampal gyrus, the anterior end of which forms the uncus. The name 'parahippocampal' ('near the hippocampus') indicates that above it lies the hippocampus, which is the phylogenetically older cerebral cortex (archicortex) that got hidden from the outside by an 'in rolling' of the cortex [10]. The hippocampus occupies the floor of the inferior horn of lateral ventricle (Fig. 2.3d) and is seen in a coronal section in the medial temporal lobe (Fig. 2.3d). Hippocampus is involved in memory consolidation and other limbic functions.

2.8.1 Oncological Significance

- Preoperative diffusion tensor imaging and tractography to map the corticospinal tract and arcuate fasciculus to know the proximity and location of a tumour with respect to it is promising to preserve as much of motor and language function respectively during surgery for brain tumours (see Fig. 2.7c, e) [22].
- A lesion affecting the cuneus will cause a contralateral inferior quadrantanopia—'pie on the floor' and lesion in the lingual gyrus can cause a 'pie in the sky'—contralateral superior quadrantanopia.

- Tumours in the cerebral hemispheres can cause focal neurological deficit and can also commonly cause focal or generalized seizures. Lesions in certain lobes may have more specific features. A tumour in the *frontal lobe* can affect the *motor functions* or create *motor aphasia* syndromes (if in the dominant hemisphere). In addition to this, the frontal lobe is also the seat of intellect, cognition, personality, judgement, drive and initiative. These 'silent' areas (cf. eloquent cortices) are important, and hence, it is not uncommon for a person developing a frontal lobe tumour to present with behavioural and psychiatric manifestations with severe intellectual involvement, lack of initiative, abulia, akinetic mutism and the patient may seem euphoric or indifferent.
- Tumours in the *temporal lobe* may cause seizures. Tumours in the hippocampal region may cause characteristic odours as part of their fits (due to its limbic connections). *Parietal lobe* tumours may create astereognosis (inability to identify the form by touch alone), cortical sensory deficits (two-point discrimination, tactile localization), sensory aphasic syndromes and also often an inferior quadrantanopia if optic radiation fibres traversing the parietal lobe are affected. A Gerstmann's syndrome (acalculia, agraphia, right left disorientation and finger agnosia) may be found in a person with a lesion in the dominant inferior parietal lobe [23].
- An *occipital lobe* tumour may present with contralateral quadrantanopia or a hemianopia, and there may usually be macular sparing due to the wide representation of the macular visual area in the cortex.

2.9 White Matter of Cerebrum

- These are classified into three main types: association, commissural and projection fibres.
- Association fibres extend between gyri of the same cerebral hemisphere and can be short or long. The short association fibres connect one gyrus with an adjacent gyrus, whereas a long association fibre bundle extends between two areas lying further apart, may even be in two lobes of a hemisphere. The named association fibre bundles are superior longitudinal fasciculus, inferior longitudinal fasciculus, uncinate fasciculus, fronto-occipital fasciculus and cingulum.
- Commissural fibres connect similar cortical areas of both hemispheres. The minor ones are: anterior (Fig. 2.1b), posterior, habenular (traverse through the upper and lower laminae of the stalk of the pineal gland, respectively) and hippocampal commissures.
- The corpus callosum, the largest commissure, is seen on a median sagittal section as a 'C'-shaped structure (Fig. 2.3). From the tapered anterior end, rostrum, it takes a 'bend' at the genu and continues as the body, isthmus and splenium, from anterior to posterior. The rostrum is seen in the floor and the genu forms the anterior wall, of the anterior horn of the lateral ventricle. The body of the corpus callosum forms the roof of the anterior horn and body of the lateral ventricle. Laterally and posteriorly, the fibres of the body and splenium forms the tapetum,

which forms the roof and lateral wall of the posterior horn and then continues as the roof of the inferior horn of the lateral ventricle. From a 3D perspective, the genu radiates out anteriorly into the frontal lobes as *forceps minor*, whereas fibres from the splenium extend to the occipital lobes as *forceps major*. Laterally the fibres of the corpus callosum spread out to all parts of the cerebral hemispheres, connecting corresponding areas of both hemispheres. A glioblastoma multiforme that spreads along the fibres of corpus callosum forming the classic 'butterfly' tumour/glioma recaps this arrangement.
- Projection fibres project from the cerebral cortex to subcortical structures and vice versa. The most important projection fibres are internal capsule (IC) and fornix.
- The IC has an anterior limb (lateral to caudate head), genu (bend), posterior limb (lateral to thalamus), retro-lentiform (behind the lentiform nucleus) and sub-lentiform parts (inferior to the lentiform nucleus, seen below that plane) (Fig. 2.3b).
- IC is a major pathway of ascending and descending fibres, the motor fibres innervating the contralateral muscles of head and neck (corticobulbar fibres) occupy the genu; and the ones innervating the contralateral half of the body (corticospinal fibres) occupy the posterior limb (Fig. 2.3b). If these fibres of one side are involved (commonly, as in a case of a deep-seated tumour), it will result in contralateral hemiparesis or hemiplegia with UMN facial palsy. In addition to this, there are descending corticopontine fibres (see Cerebellum) involved in cerebellar circuitry and the thalamic radiations (see Thalamus).
- Among the thalamic radiations, of particular importance are the superior thalamic radiation in the posterior limb projecting ascending general sensory fibres to the parietal lobe; and the posterior thalamic radiation, in the retro-lentiform part, which is the optic radiation. Hence, contralateral hemi-anaesthesia and/or contralateral homonymous hemianopia can also be seen when these parts of IC are afflicted. The sub-lentiform part occupies the auditory radiation.
- The axons from the hippocampus continue as the fornix which passes below the splenium of corpus callosum as the crura of fornix join as the body and split again as the columns of fornix. As it contains fibres from the hippocampus, it has a role in limbic circuits, specifically in memory (see below).

2.10 Basal Ganglia

- The basal ganglia are deep nuclear masses which are located in the deeper cerebral hemispheres. It includes corpus striatum, substantia nigra and subthalamic nucleus. Conventionally, they are considered to modulate movement.
- The substantia nigra in the *midbrain* contains dopaminergic neurons that project to the striatum as the nigrostriatal pathway. These neurons are degenerated in Parkinson's disease, affecting the function of the striatum causing the movement manifestations of the disease. The subthalamic nucleus, seen in the *diencephalon,* is involved in the indirect loop of basal ganglia, important in abolishing unintended movements.

2.10.1 Oncological Significance

- Invasion of a deep cerebral tumour on basal ganglia can cause movement disorders like choreo-athetosis, dystonias and tremors [15]. A parkinsonian presentation is unusual but has been reported to occur in mesencephalic tumours or in midbrain compression by nearby tumours [15, 24].

2.11 Ventricles of the Brain

- There are four CSF-filled cavities within the brain: two lateral ventricles, a third ventricle, and a fourth ventricle. Developmentally, they are derived from the cavity of the neural tube.
- Lateral ventricles are C-shaped cavities situated within each cerebral hemisphere. It somewhat wraps around the thalamus (see Fig. 2.6d). The lateral ventricle has a body and three horns: anterior, posterior and inferior. The body is roofed by corpus callosum and in its floor, lie body of caudate nucleus laterally and thalamus medially. The lateral ventricle extends into the frontal lobe as the anterior horn, into the occipital lobe as the posterior horn and into the temporal lobe as the inferior horn (Fig. 2.6). The atrium (trigone) is where the body, posterior horn and inferior horn meet. In the floor of the anterior horn lies the bulging head of caudate nucleus. The hippocampus lies in the floor of inferior horn and in its roof lies tail of caudate nucleus. The inferior horn ends slightly short of the amygdala. The lateral ventricle communicates with the third ventricle, through interventricular foramen (of Monro). The foramen is situated in the anterior part of medial wall of lateral ventricle bounded anteriorly by the column of fornix and posteriorly by anterior end of thalamus. The fornix should be spared while approaching a tumour in its vicinity [25–27], to avoid memory deficits, as it is an important pathway in limbic circuits (see above). Between the fornix above and thalamus below lies a slit called the choroid fissure through which the choroid plexus from the foramen of Monro in front to the body, atrium and the inferior horn.
- The third ventricle is a slit-like cavity in the midline (Fig. 2.6) between two thalami and hypothalami. It communicates with the lateral ventricles via interventricular foramen (Monro) and with the fourth ventricle via cerebral aqueduct. The anterior limit is the thin lamina terminalis, which separates the subarachnoid cistern in front of it from the third ventricle itself (see in Fig. 2.3c); hence, an important neurosurgical corridor is used to approach a third ventricle tumour or to fenestrate the third ventricle into the subarachnoid space [28]. Below, it is related to the optic chiasm and the tuber cinereum of hypothalamus. The third ventricle possesses a few recesses. They are the supraoptic recess between the lamina terminalis and optic chiasm and the infundibular recess that directs into the pituitary stalk (see Figs. 2.3c and 2.6a). The site between the infundibular recess and mammillary body is also a corridor for an endoscopic ventriculostomy to treat an obstructive hydrocephalus. The roof is formed by tela choroidea of third ventricle which invaginates the ependyma on either side of midline to

form choroid plexus of third ventricle. The blood vessels to choroid plexus of lateral and third ventricles are derived from anterior and posterior choroidal branches of internal carotid and basilar arteries.

- The fourth ventricle is a tent-shaped cavity of hindbrain. Superiorly, it communicates with third ventricle through cerebral aqueduct and inferiorly with the central canal of spinal cord. It has a roof, directed posteriorly projecting into cerebellum. The apex of the tent is the fastigial point where there extends a median fastigial recess (median dorsal). The floor, which is rhomboid in shape formed by the dorsal surface of pons and cranial part of medulla. Upper half of the roof is formed by superior cerebellar peduncles and superior medullary velum (a sheet of white matter) connecting the two peduncles. Lower half of the roof is ependymal and has a median aperture (*foramen of Magendie*). The floor has important cranial nerve nuclei creating distinct features on its surface: nucleus of VI, with the looping facial nerve fascicles, forms the facial colliculus close to the midline; vestibular nuclei are found in the vestibular area laterally; the hypoglossal nucleus and the dorsal nucleus of vagus found in the hypoglossal and vagal triangles medially. Area postrema/chemoreceptor trigger zone which is devoid of blood–brain barrier is situated close to the lower lateral margin of the fourth ventricle floor. The cavity of fourth ventricle extends laterally as two spout-like lateral recesses, which have the lateral apertures (*foramina of Luschka*) at its extreme angles. Through foramina of Magendie and Luschka, the fourth ventricle communicates with the cisterna magna and cerebellopontine angle cisterns and hence to the subarachnoid space. This is an important 'exit' of the fourth ventricle and can be the site of a fourth ventricle outlet obstruction (FVOO) which may cause an obstructive hydrocephalus [29]. The choroid plexus of fourth ventricle is contributed by the posterior inferior cerebellar arteries (PICA) which enter through the lower half of roof and is the most important artery of the fourth ventricle.
- Total volume of CSF is approximately 150 ml of which 125 ml is in the cranium and 25 ml in the spine. The circulation of cerebrospinal fluid (CSF) begins with its secretion from choroid plexus. Through interventricular foramen, CSF reaches third ventricle from lateral ventricles. It then passes through cerebral aqueduct into fourth ventricle and via foramen of Luschka and Magendie, it reaches subarachnoid (cerebellomedullary and pontine) cisterns. From there, it flows superiorly through tentorial notch to the inferior surface of cerebrum. The flow is aided by pulsations of vessels in choroid plexus and cilia on ependymal cells. It then moves superiorly over the lateral aspect of cerebral hemispheres, assisted by the pulsations of cerebral arteries. The major site of absorption is arachnoid villi projecting into the dural venous sinuses. Thus, the CSF gets absorbed into venous sinuses. Some gets directly absorbed into veins of subarachnoid space and some enters perineural lymph of cranial and spinal nerves. Since the production of CSF is constant, the rate of absorption through arachnoid villi controls the cerebrospinal fluid pressure.

2.11.1 Oncological Significance

- Lateral ventricle tumours can be primary, originating from the wall and remaining within the ventricle (ependymomas or choroid plexus papillomas) or can be secondary, which are paraventricular arising from anatomically adjacent structures (gliomas).
- Third ventricle tumours can be primary, arising from the ventricular walls—colloid cysts, commonly seen at the roof of the third ventricle, thalamic or hypothalamic gliomas, in the lateral walls and ependymomas. They can also be secondary paraventricular and mostly from below—sellar and suprasellar tumours.
- The most common tumour of the fourth ventricle is ependymoma in the adult and medulloblastoma in children (mentioned in cerebellum). Ependymomas originate from the ependymal lining of the ventricles and is most seen with relation to fourth ventricle. These may fill the fourth ventricle, enlarging it, compressing the foramina of Luschka and Magendie and may exit the ventricle to the cerebellopontine angle cisterns, cisterna magna or even to the foramen magnum below it to enter the cervical region.
- A rise in cerebrospinal fluid pressure can occur when there is either an increased production of fluid (as in tumours of choroid plexus) or an obstruction in the cerebrospinal fluid circulation (as in tumours blocking interventricular foramen, or cerebral aqueduct; by a colloid cyst of third ventricle or ependymomas or medulloblastoma in cerebellar vermis) or a decrease in its absorption (as in inflammatory exudates or thrombosis of venous sinuses).

Acknowledgements The authors would like to thank the staff members of Department of Anatomy, Government Medical College, Thiruvananthapuram. We also thank Dr. John Joseph, DMRT, DNB, Assistant Professor, Radiation Oncology, Regional Cancer Centre, Trivandrum; Dr. Sreejith MD, MS, MCh, Fellowship in Paediatric Neurosurgery, Associate Consultant Neurosurgeon, KIMS Hospital, Trivandrum; Dr. Nivedita Nair, MS DNB, FICO, Senior Resident, Sankara Nethralaya, Chennai, for their opinions and contributions for improving the content of this chapter. We thank Ms. Riya Mariam George, second year medical student who has hand drawn many of the figures. We also thank Dr. Nithin K. Raju (postgraduate student in Anatomy), Ms. Josline Elsa Joseph (research fellow, SCTIMST, Trivandrum) and Ms. Mythri V (research scholar, IIIT, Hyderabad) for preparing the figures.

References

1. Varshney S, Bist S, Gupta N, Singh R, Bhagat S. Anterior craniofacial resection—for paranasal sinus tumors involving anterior skull base. Indian J Otolaryngol Head Neck Surg. 2010;62(2):103–7.
2. Almefty K, Pravdenkova S, Colli BO, Al-Mefty O, Gokden M. Chordoma and chondrosarcoma: similar, but quite different, skull base tumors. Cancer. 2007;110(11):2467–77.

3. Bouthillier A, Van Loveren HR, Keller JT. Segments of the internal carotid artery: a new classification. Neurosurgery. 1996;38(3):425–33.
4. Greenberg HS, Deck MD, Vikram B, Chu FC, Posner JB. Metastasis to the base of the skull clinical findings in 43 patients. Neurology. 1981;31(5):530.
5. Campero A, Martins C, Yasuda A, Rhoton AL Jr. Microsurgical anatomy of the diaphragma sellae and its role in directing the pattern of growth of pituitary adenomas. Neurosurgery. 2008;62(3):717–23.
6. Won H-S, Han S-H, Oh C-S, Lee J-I, Chung I-H, Kim SHJS. Topographic variations of the optic chiasm and the foramen diaphragma sellae. Surg Radiol Anat. 2010;32(7):653–7.
7. Yohannan DG, Krishnapillai R, Suresh R, Ramnarayan S. A morphometric study of the foramen of diaphragma sellae and delineation of its relation to optic neural pathways through computer aided superimposition. Anat Res Int. 2015;2015:618042.
8. Zhang C, DeSouza R, Kho J, Vundavalli S, Critchley G. Kernohan–Woltman notch phenomenon: a review article. Br J Neurosurg. 2017;31(2):159–66.
9. Algra PR, Heimans JJ, Valk J, Nauta J, Lachniet M, Van Kooten B. Do metastases in vertebrae begin in the body or the pedicles? Imaging study in 45 patients. AJR Am J Roentgenol. 1992;158(6):1275–9.
10. Blumenfeld H. Neuroanatomy through clinical cases. Sunderland, MA: Sinauer Associates; 2010.
11. Ganewatte E, Al-Zahrani Y, Purohit B, Kollias S. A case report on 1-year follow-up of bilateral thalamic glioma. Neuroradiol J. 2015;28(6):584–6.
12. Helton KJ, Phillips N, Khan R, Boop F, Sanford R, Zou P, et al. Diffusion tensor imaging of tract involvement in children with pontine tumors. AJNR Am J Neuroradiol. 2006;27(4):786–93.
13. Chen X, Weigel D, Ganslandt O, Buchfelder M, Nimsky C. Diffusion tensor imaging and white matter tractography in patients with brainstem lesions. Acta Neurochir. 2007;149(11):1117–31.
14. Damle NR, Ikuta T, John M, Peters BD, DeRosse P, Malhotra AK, et al. Relationship among interthalamic adhesion size, thalamic anatomy and neuropsychological functions in healthy volunteers. Brain Struct Funct. 2017;222(5):2183–92.
15. Krauss JK, Nobbe F, Wakhloo AK, Mohadjer M, Vach W, Mundinger F. Movement disorders in astrocytomas of the basal ganglia and the thalamus. J Neurol Neurosurg Psychiatry. 1992;55(12):1162–7.
16. Beks J, Bouma G, Journee H. Tumours of the thalamic region. Acta Neurochir. 1987;85(3–4):125–7.
17. Nagaratnam N, Ting A, Jolley D. Thalamic tumour presenting as frontal lobe dysfunction. Int J Clin Pract. 2001;55(7):492–3.
18. Kwon N, Kim HS. Bilateral thalamic glioma in a young woman: a case report. Brain Neurorehabil. 2019;12(2)
19. Moffie D, de Visser BO, Stefanko S. Parinaud's syndrome. J Neurol Sci. 1983;58(2):175–83.
20. Melmed S. Pathogenesis of pituitary tumors. Nat Rev Endocrinol. 2011;7(5):257.
21. Kim SS, McCutcheon IE, Suki D, Weinberg JS, Sawaya R, Lang FF, et al. Awake craniotomy for brain tumors near eloquent cortex: correlation of intraoperative cortical mapping with neurological outcomes in 309 consecutive patients. Neurosurgery. 2009;64(5):836–46.
22. Maruyama K, Koga T, Kamada K, Ota T, Itoh D, Ino K, et al. Arcuate fasciculus tractography integrated into gamma knife surgery. J Neurosurg. 2009;111(3):520–6.
23. Ardila A, Concha M, Rosselli M. Angular gyrus syndrome revisited: Acalculia, finger agnosia, right-left disorientation and semantic aphasia. Aphasiology. 2000;14(7):743–54.
24. Pohle T, Krauss JK. Parkinsonism in children resulting from mesencephalic tumors. Mov Disord. 1999;14(5):842–6.
25. Kehler U, Regelsberger J, Gliemroth J. The mechanism of fornix lesions in 3rd ventriculostomy. Minim Invasive Neurosurg. 2003;46(04):202–4.
26. Azab WA, Salaheddin W, Alsheikh TM, Nasim K, Nasr MM. Colloid cysts posterior and anterior to the foramen of Monro: anatomical features and implications for endoscopic excision. Surg Neurol Int. 2014;5:124.

27. Gaffan E, Gaffan D, Hodges JR. Amnesia following damage to the left fornix and to other sites: a comparative study. Brain. 1991;114(3):1297–313.
28. Tubbs RS, Nguyen HS, Loukas M, Cohen-Gadol AA. Anatomic study of the lamina terminalis: neurosurgical relevance in approaching lesions within and around the third ventricle. Childs Nerv Syst. 2012;28(8):1149–56.
29. Ishi Y, Asaoka K, Kobayashi H, Motegi H, Sugiyama T, Yokoyama Y, et al. Idiopathic fourth ventricle outlet obstruction successfully treated by endoscopic third ventriculostomy: a case report. Springerplus. 2015;4(1):565.

Pathology, Molecular Biology and Classification of Gliomas

3

Suvendu Purkait

3.1 Evolving Concept of Classification of Glioma and Integrated Diagnosis

Gliomas are the most common primary CNS tumors. In 1926, Bailey and Cushing set the foundation for classifying gliomas based on the morphology. This histology-driven system was the cornerstone of the evolution of the subsequent classification systems for almost seven to eight decades [1, 2]. The glioma cases were histologically classified based on their microscopic resemblance to different glial cell types and designated as astrocytomas (astrocytic differentiation), oligodendrogliomas (oligodendroglial differentiation), or oligoastrocytoma (mixed histology). In addition to the morphological differentiation, the World Health Organization (WHO) classification of tumors of the CNS tumors adopted the *St. Anne-Mayo grading* approach for assigning a malignancy grade depending upon different histological parameters including cellularity, cellular atypia, mitosis, necrosis, and endothelial proliferation. The grading system provided a better idea to determine the management and predict the clinical outcome across the tumor subtypes [3].

Traditionally, gliomas were also subdivided into two major groups: "diffuse" and "circumscribed" gliomas. The astrocytic and oligodendroglial tumors which had a high cell migration with diffuse infiltration into adjacent brain parenchyma were designated as diffusely infiltrating/diffuse glioma. Due to its infiltrating nature, complete resection of these tumors was very difficult, and they were all considered potentially malignant. This group comprised of the following tumor entities: diffuse astrocytoma (WHO grade II), anaplastic astrocytoma (WHO grade III), glioblastoma (WHO grade IV), oligodendroglioma (WHO grade II), and anaplastic Oligodendroglioma (WHO grade III). On the other hand, gliomas with a pushing

S. Purkait (✉)
Department of Pathology, All India Institute of Medical Sciences, Bhubaneshwar, India

S. Mallick et al. (eds.), *Evidence based practice in Neuro-oncology*,
https://doi.org/10.1007/978-981-16-2659-3_3

noninfiltrative margin were considered as circumscribed glioma (e.g., pilocytic astrocytoma, pleomorphic xanthoastrocytoma). The more recent 2003 and 2007 WHO classifications were also essentially based on the histomorphological parameters substituted by some of the immunohistochemistry-based key molecular markers [3].

Although most of the histomorphology-based diagnoses were user friendly and robust, there was a significant inter- and intra-observer variation in the diagnosis and grading of gliomas. Also, there was substantial heterogeneity in terms of biological behavior within some of the histologically defined tumor groups. These raised the possibility of detrimental consequences for patients [4]. Further, in recent decades, a plethora of high throughput studies have increased the knowledge of the genetic basis of glioma exponentially and revealed many biomarkers with potential clinical implications [5]. Hence, there was a need to integrate at least some of the important molecular parameters into the histology-based diagnostic scheme to make it more objectivized and informative. To address this challenge, the International Society of Neuropathologists (ISN) convened an expert group of neuropathologists from different parts of the world in Haarlem, the Netherlands. This group recommended incorporating histological and molecular information to define tumor entities as narrowly as possible to create homogeneous tumor groups. They suggested a four-layered standardized reporting system of integrated tumor classification for providing more detailed, clinically relevant information. The layers comprised of information regarding molecular genetic alteration (layer 4), WHO grade (layer 3), a histological impression (layer 2), and an integrated diagnosis including molecular and morphological data (layer 1) [6]. The incorporation of well-established molecular biomarkers with histological parameters in the 2016 WHO classification led to a paradigm shift in the glioma classification and nomenclature. Mutation status of isocitrate dehydrogenase gene (IDH) and co-deletion/loss of heterozygosity (LOH) for the short arm of chromosome 1 and the long arm of chromosome 19 (1p, 19q co-deletion) appeared to be essential for glioma classifications. Other markers that also assumed importance are alpha-thalassemia/mental retardation syndrome X (*ATRX*) gene mutation, *TP53* gene mutation, H3 Histone Family Member 3A (H3F3A) mutation, CDKN2A homozygous deletion, EGFR amplification, etc. [7, 8]. Unfortunately, the molecular testing was not possible in each case in every laboratory owing to technical issues related to the tumor tissue and the infrastructural heterogeneity throughout the world. Further, large-scale studies have also raised the possibility of sub-classifying or refining some of the tumor entities with additional molecular markers. Hence, the Consortium to Inform Molecular and Practical Approaches to CNS Tumor Taxonomy (cIMPACT-NOW) was established to evaluate and recommend changes to future central nervous system (CNS) tumor classifications in between WHO editions. The committee has published seven updates to date where they addressed specific questions related to molecular diagnosis or diagnostic entities. As per the most recent recommendation, it is preferable to use Arabic number for grading instead of Roman numbers [9, 10]. However, as per the latest WHO classification of CNS tumor 2016 Roman numbers for grading have been used.

3.2 Diffuse Gliomas

This subtype of glioma, as described above, is conventionally diagnosed based on their diffusely infiltrative growth pattern. The tumor cells spread into the adjacent CNS tissue and show clustering around the vessels, individual tumor cells around entrapped neuronal cells, or subpial spread, which are known as secondary structures. The astrocytic and oligodendroglial tumors show respective phenotypic differentiation. It is important to note that all these tumors are potentially malignant and can progress to a higher grade. The grading of almost all the gliomas is still based on the histological parameters [8, 11]. The mutation status of the *IDH* gene and 1p/19q co-deletions status is an integral part of diffuse glioma classification. Although molecular testing is integrated into the WHO classification system, the tumor histology is a valuable guide to choose subsequent molecular markers, especially in recourse limited settings.

The following histological entities are described under diffuse glioma—(a) diffuse astrocytoma (DA), IDH mutant, WHO grade II; (b) anaplastic astrocytoma (AA), IDH mutant, WHO grade III; (c) diffuse astrocytoma (DA), IDH wild type, WHO grade II; (d) anaplastic astrocytoma (AA), IDH wild type, WHO grade III; (e) glioblastoma (GBM), IDH mutant, WHO grade IV; (f) glioblastoma (GBM), IDH wild type, WHO grade IV: oligodendroglioma (OG), IDH mutant, 1p/19q co-deleted, WHO grade II; anaplastic oligodendroglioma (AO), IDH mutant, 1p/19q co-deleted, WHO grade III. The key markers in glioma are summarized in Table 3.1.

When the required molecular data are not available and the diagnosis is based on the histology, the tumor is designated with not otherwise specified (NOS). The details of NOS category are discussed at the end of the chapter [12, 13].

3.2.1 Diffuse Astrocytomas (DA), IDH Mutant (WHO Grade II) and Anaplastic Astrocytoma (AA), IDH Mutant (WHO Grade III)

DA and AA usually affect patient of third to fourth decade with cerebral hemisphere being the common location. By definition, both these tumor entities are IDH mutant, and the differences between these two are barely histomorphological. The molecular profile of both tumors is similar. Large-scale studies could not find any prognostic difference between IDH mutant grade II and III astrocytic tumors [14]. Further, the morphological parameters, including cellular atypia and mitotic activity, which are used to differentiate these two tumors, have uncertain clinical significance [14–16]. Hence, these two entities are discussed together. As per the recent recommendation it is preferable to use Astrocytoma, WHO grade II or III insted of Diffuse and Anaplastic astrocytoma [17].

3.2.1.1 Microscopy

Diffuse astrocytoma (DA): Histologically, DA cases are moderately cellular and show diffuse infiltration in the adjacent brain parenchyma. The tumor cells are

Table 3.1 Key genetic markers of diffuse gliomas

Genetic alteration	Molecular mechanism	Frequency in glioma subtypes	Commonly used methods of detection	Biomarker type	Comments
IDH mutation	• Point mutations of the citric acid cycle enzyme IDH • Mutant enzyme protein leads to production of oncometabolites 2-hydroxyglutarate which inhibits demethylases • Increase gene methylation particularly at CpG site—G-CIMP phenotype	• All oligodendroglial tumors • Approximately 80-90% of DA and AA—classified as DA/AA, IDH mutant • Approximately10% of GBM—classified as GBM, IDH mutant	• IHC for IDH-1 (R132H)—the most common mutation • Sequencing—can detect all mutation variants	• Diagnostic—essential marker for glioma classification • Prognostic—mutants have better prognosis	• IHC is highly sensitive and specific • Sequencing is required only in limited number of cases with noncanonical mutations
1p/19q co-deletion	• The mechanism is poorly understood • Deletion of long arm of ch 1 and short arm of ch 19 is associated with less tumor-associated microtubule which helps to interconnect the tumor cell	• Characteristic of OG and AO and present in all cases • Not present in any other CNS tumor	• Fluorescence in situ hybridization (FISH)	• Diagnostic—essential marker for diagnosis of oligodendroglial tumors	• Microdeletions can be missed in FISH which can be detected by single-nucleotide polymorphism chromosomal microarrays
ATRX mutation	• A chromatin remodeling protein crucial for genomic stability and genomic repair • Mutation leads with loss of protein expression which is associated with alternative lengthening of telomeres	• Almost all cases IDH mutant DA & AA • Approximately 70% of IDH mutant GBM • Not seen in oligodendroglioma • Rare in IDH wild-type GBM	• IHC—loss of protein expression	• Diagnostic—present mainly in IDH mutant astrocytic tumors	• Mutually exclusive with 1p/19q co-deletion

TERT promoter mutation`	• Activating point mutation associated with upregulated expression of TERT, enzyme responsible for maintaining telomere length	• High frequency (>90%) in molecular oligodendrogliomas (1p/19q co-deleted, IDH mutated) • Common in IDH wild-type GBM (50–75%) • Rare in IDH mutant astrocytoma and GBMs	• Sequencing	• Prognostic—associated with poor prognosis in GBMs and other astrocytic tumors	• Mutually exclusive with ATRX mutation
H3F3A K27M	• Point mutations at one allele of the histone H3F3A gene • Substitution of Lysine 27-to-methionine • Downregulation of the suppressive histone mark H3K27me3	• Diffuse midline glioma, H3K27M mutant—all cases	• IHC analysis for K27M mutant protein and H3K27me3 • Sequencing	• Diagnostic—essential for diagnosis of diffuse midline glioma, H3K27M mutant	• For IHC analysis, combined assessment of both the markers is preferable
TP53 mutation	• One of the major tumor suppressor gene. Mutation associated with DNA repair defect • In gliomas, missense mutations are common	• Usually co-exist with ATRX mutation. Predominantly seen in IDH mutant grade II/III astrocytomas and GBMs (80%) • Rare in IDH wild type GBMs and oligodendroglial tumors	Immunohistochemistry Sequencing	Diagnostic—marker of astrocytic lineage	• IHC is not sensitive or specific—result must be interpreted in context with morphology and other findings
CDKN2A/B homozygous deletion	• Important negative regulator of cell cycle progression • Homozygous deletion associated with uncontrolled cell proliferation	• Present is all adult diffuse glioma with variable frequency • Higher grade > lower grade	• FISH • PCR	• Prognostic—poor prognosis in IDH mutant astrocytic tumor	• IDH mutant grade II/III astrocytic tumor with this genetic alteration now designated as grade IV

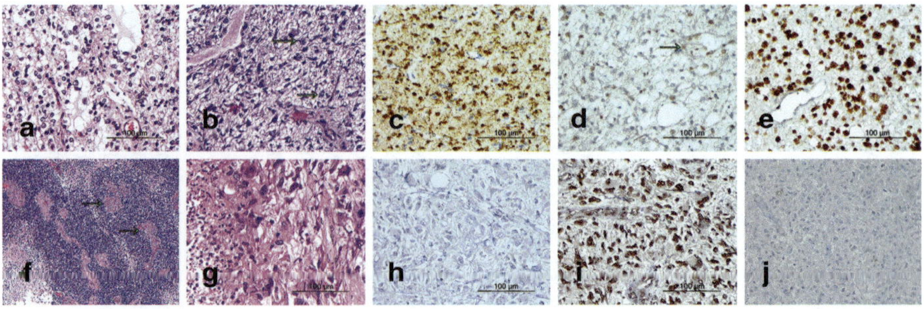

Fig. 3.1 (**a**) Photomicrographs of a representative case of diffuse astrocytoma (DA), showing moderate cellularity and microcystic changes. (**b**) Photomicrographs of a representative case of an anaplastic astrocytoma (AA), showing high cellularity and mitotic activity (arrow). (**c**) Photomicrograph showing IDH-1 immunopositivity in the tumor cells (feature of IDH mutant DA and AA). (**d**) Photomicrograph showing loss of ATRX expression in the tumor cells [endothelial cells (arrow) and inflammatory cells are positive] (feature of IDH mutant DA and AA). (**e**) Photomicrograph showing strong immunopositivity for p53 (feature of IDH mutant DA and AA). (**f**) Photomicrographs of a representative case of glioblastoma showing high cellularity with endothelial proliferation (arrow) and necrosis (star). (**g**) Photomicrograph showing significant cellular pleomorphism and presence of mitotic activity in a case of glioblastoma. (**h**–**j**) Immunohistochemical profile of a case of IDH wild-type GBM showing immunonegativity for IDH-1 (**h**), retained ATRX expression (**i**), and negativity for p53 (**j**)

highly differentiated and show astrocytic morphology (fibrillary and/or protoplasmic and/or gemistocytes) with variable microcystic changes. If gemistocytes constitute more than 20% of the tumor cell population, it is designated as *gemistocytic astrocytoma*. There is a moderate degree of nuclear atypia. However, mitotic activity is usually absent. Microvascular proliferation or necrosis is characteristically absent in these cases (Fig. 3.1) [8].

Anaplastic astrocytoma: These diffusely infiltrating astrocytomas are more cellular as compared to their grade II counterpart. Additionally, AA cases show more prominent nuclear atypia and high mitotic activity. The WHO 2016 classification system has not provided any cut-off for the mitotic activity for diagnosing AA and recommended to assess the mitotic count with respect to the size of the biopsy specimen [8]. Many neuropathologists use a cut-off of ≥2 mitoses in the whole specimen or single mitosis in a small biopsy [18]. Presence of necrosis or endothelial proliferation is not consistent with the diagnosis of AA (Fig. 3.1).

3.2.1.2 Molecular Profile

As described above, both the IDH mutant astrocytomas (grade II and III) have a similar molecular profile. These tumors show a mutation of the *IDH1* or *IDH2* gene. A missense mutation in the codon 132 of the *IDH1* gene, namely R132H, constitutes the majority of IDH mutations (approximately >90%). More importantly, this particular mutation can also be detected by immunohistochemistry using a specific primary antibody against the mutant protein. However, it should be kept in mind that this antibody is not able to identify other IDH mutations (IDH1 R132C, IDH1 R132G, IDH1 R132S, IDH1 R132L, IDH2), which are also known as non-canonical mutations. Hence, cases that are negative for IDH1 R132H immunohistochemistry should be screened for other uncommon IDH mutations, preferably by sequencing-based molecular techniques [19–21].

Almost all cases of IDH mutant DA/AA invariably shows mutation in the *ATRX* and/or *TP53* gene which is also described by some authors as "astrocytic genotype." The mutation in the *ATRX* gene can be detected by immunohistochemical analysis. On immunohistochemistry, the ATRX protein, unlike IDH1, is expressed in the normal brain as well as wild-type cases. The mutated cases usually show loss of expression in the tumor cells. Most of the TP53 mutant cases have a strong nuclear expression of p53 protein on immunohistochemistry. However, immunohistochemistry is less reliable for the detection of TP53 mutation status [15, 18, 22, 23].

Interestingly, TP 53 and ATRX mutations are mutually exclusive with 1p/19q co-deletion. The IDH mutated and 1p/19q co-deleted tumors, which classically show oligodendroglial morphology, are usually negative for p53 immunostaining and also retain nuclear ATRX expression. Hence, cIMPACT update 2 has recommended that the diagnosis of DA/AA, IDH mutant, can be made without 1p/19q testing in cases of astrocytic morphology along with loss of ATRX expression and/or diffuse, strong immunoreactivity for p53. Hence, the majority of these cases can be diagnosed based on the immunohistochemical findings only [8, 18, 23].

The IDH mutant astrocytic tumors (grade II/III) can further be stratified based on the *CDKN2A/B* gene status. Homozygous deletion of *CDKN2A/B* gene emerged as an independent prognostic marker in IDH mutant grade II and III diffusely infiltrating astrocytic gliomas and associated with poor prognosis [24, 25]. Hence, cases of IDH mutant grade II/III astrocytic tumors with CDKN2A homozygous deletion is upgraded and now designated as astrocytoma, IDH mutant, grade IV which is synonymous with GBM, IDH mutated, WHO grade IV [9].

3.2.2 Diffuse Astrocytoma, IDH Wild Type (WHO Grade II) and Anaplastic Astrocytoma, IDH Wild Type (WHO Grade III)

3.2.2.1 Histology

Although regarded as distinct categories by WHO 2016 classification system, these IDH wild-type tumors are morphologically indistinguishable from their IDH mutant counterpart.

3.2.2.2 Molecular Profile

A substantial amount of published literature on IDH wild-type grade II/III astrocytic tumor revealed that most of these cases, especially in the adult age group, have prognostic outcomes comparable to IDH wild-type glioblastomas. On the other hand, some of these tumors can also mimic other glial and glioneuronal tumors, including pediatric low-grade glioma, pilocytic astrocytoma, pleomorphic xanthoastrocytoma, which are also IDH wild type but have a more favorable prognosis [26–28]. Based on the available literature, the working committee of cIMPACT reached a consensus that cases of IDH wild-type grade II/III astrocytic tumors which have EGFR amplification or combined whole chromosome 7 gain and 10 loss or TERT promoter mutation have an aggressive clinical course equivalent to IDH wild-type glioblastoma. However, as these cases do not have the characteristic morphological features of glioblastoma, they are designated as "diffuse astrocytic glioma, IDH-wild type, with molecular features of glioblastoma, WHO grade IV" [29]. For the purpose of simplification, in the recent update the tumor is designated as GBM, IDH wild type [17].

3.2.3 Glioblastoma, IDH Wild Type, WHO Grade IV

It is the most common malignant primary CNS tumor. Approximately 90% of cases of GBM are IHD wild type. The affect relatively older population (usually >50 years) with cerebral hemisphere being the commoner site. These tumors usually arise de novo without any identifiable precursor lesion and were previously designated as primary GBM [8].

3.2.3.1 Microscopy

These tumors are highly cellular, and like all other diffuse gliomas, there is infiltration into the adjacent brain parenchyma. The tumor cells have marked cellular atypia and pleomorphism. The cellular composition may be variable and includes small cell, gemistocytic cell, granular, multinucleated giant cell, etc. Often, different cell types may be present in a single tumor with marked histological diversity. Previously the term "multiforme" was used for the same reason. Mitotic activity is brisk. The presence of endothelial proliferation and/or necrosis is characteristic. Often the necrosis is pseudopalisading type (Fig. 3.1).

Apart from the classical morphological features described above, three distinct histological variants of GBM have also been described, namely giant cell GBM, gliosarcoma, and epithelioid GBM. Histologically, the giant cell GBM cases show a significant proportion of multinucleated giant cells and syncytial cells. On the other hand, gliosarcomas are biphasic tumors with gliomatous (glial) and sarcomatous (mesenchymal) areas on histology. The epithelioid GBMs predominantly comprise epithelioid cells with few rhabdoid cells [8].

3.2.3.2 Molecular Profile

Characteristically, these cases do not show IDH mutation and, hence, negative for IDH-1 immunohistochemistry. Notably, IDH-1 immunohistochemistry does not completely exclude the possibility of mutation, especially in younger patients and patients with a prior history of lower-grade glioma. Hence, sequencing for IDH-1/2 is essential to rule out non-canonical IDH mutations. However, IDH mutation is rare in patients aged more than 54 years (<1%) [30]. These cases can be diagnosed as IDH wild type based on the IDH-1 immuno-negativity only, even without gene sequencing results [8].

A plethora of genetic and epigenetic alterations have been reported in IDH wild-type GBM over the years. However, only a few of them appeared to be clinically relevant. A point mutation in the TERT promoter region (74–83%) is one of the prevalent genetic alterations which carries a poor prognosis [31, 32]. The other critical genetic changes are CDKN2A homozygous deletion (37–58%) and EGFR amplification (25–50%) [32, 33]. These genetic alterations are also regarded as a molecular signature of IDH wild-type GBMs as they are rare in other astrocytic tumors. Hence, IDH wild-type lower-grade astrocytomas with any of these alterations are regarded as molecular GBM/GBM. However, apart from IDH, none of the genetic alteration is essential for classifying GBM as per WHO 2016. MGMT promoter methylation status is one of the predictive biomarkers that can provide critical inputs regarding the tumor response to alkylating agents. It is also associated with a good prognosis [32, 34].

These tumors rarely show ATRX mutation (loss of nuclear expression of ATRX protein on immunohistochemistry). The frequency of TP-53 is also very low (25–37%) [34, 35].

Apart from the morphology, the molecular profile of histological variants GBMs differs to a certain extent. The giant cell variant has a higher frequency of TP53 mutation, while the TERT promoter mutation frequency is significantly lower. Further, both giant cell GBM and gliosarcoma cases rarely show EGFR amplification [36]. Approximately half of the cases of epithelioid GBMs harbor BRAF V600E mutation [37].

The molecular profile of the pediatric GBM cases is also distinctly different from its adult counterpart. These cases are usually IDH wild types. There is a high frequency of mutations in the H3.3/ATRX/DAXX chromatin remodeling pathway [38]. The majority of the K27M mutant tumors, which usually occur in midline location, are now considered separate entities. These tumors rarely show TERT promoter mutation, EGFR amplification, CDKN2A deletion, or PTEN mutations [38–40].

3.2.4 Glioblastoma, IDH Mutant Type

These cases are morphologically indistinguishable from their IDH wild-type counterpart. However, the demographic profile, molecular spectrum, and clinical outcome of IDH mutant GBMs are entirely different. The age at presentation is significantly lower than the IDH wild-type cases (usually <50 years). These cases are thought to arise from the other lower grade diffuse gliomas through malignant progression. IDH mutant GBM carries a significantly better prognosis as compared to its wild-type counterpart [41]. As per the consensus of the cIMPACT working committee, the term "glioblastoma" should be reserved for the IDH wild-type cases. The IDH mutant cases are designated as astrocytoma, IDH mutant, WHO grade 4, and clubbed together with other IDH mutant grade 2/3 astrocytic tumors harboring CDKN2A/B homozygous deletion [9].

3.2.4.1 Molecular Profile

The molecular pattern of this group of tumors predominantly overlaps with other IDH mutant astrocytic tumors. In addition to IDH mutation, they have a high frequency of TP53 (approximately 80%) and ATRX mutation (approximately 70%) [42, 43]. These cases characteristically show a G-CIMP phenotype and have a higher incidence of MGMT promoter methylation. TERT promoter mutation frequency is significantly lower (5–28%) compared to IDH wild type cases [31, 32] Usually, IDH mutant GBMs do not show EGFR amplification or PTEN mutation [42].

3.2.5 Diffuse Midline Glioma (DMG), H3K27M Mutant

This is a newly introduced entity in the WHO 2016 classification based on the location and characteristic molecular genetic alteration, irrespective of histological grade. These tumors are located in the midline location (thalamus, brain stem, and spinal cord) and harbor a specific mutation (K27M) in the H3F3A, a histone coding gene. The subset of tumor which is located in the pons was previously termed as diffuse intrinsic pontine glioma (DIPG). DMG cases are predominantly seen in children. However, sporadic cases are also reported in adults. Because of its poor clinical outcome independent of tumor location, K27M mutant diffuse midline gliomas are considered as WHO grade IV [8, 44]. It is noteworthy that K27M mutation can be present in small subsets of posterior fossa ependymoma, ganglioglioma, and pilocytic astrocytoma cases [45–47]. Nevertheless, all these are circumscribed glioma and hence cannot be included in this category [12, 13].

3.2.5.1 Microscopy

Infiltration of the adjacent brain structure is a typical feature and necessary for diagnosis. It can show astrocytic or oligodendroglial morphology, the former being far more common (Fig. 3.2). The cases may show the presence of mitotic activity, necrosis, or endothelial proliferation. However, these are not essential for diagnosis [12, 13].

3.2.5.2 Molecular Profile

The K27M is a missense mutation in the *H3F3A* gene associated with substitution of lysin with methionine in the 27th position. This mutation leads to a decrease in the expression of H3K27me3 protein. The mutant protein (K27M) and H3K27me3 both can be detected by immunohistochemistry. However, these two immunohistochemistry should be interpreted together and with great caution for assessing mutational status [48]. Typically, DMG cases are IDH wild type. However, the frequency of TP53 mutation (47–68%), PDGFRA amplification (28–50%), and ATRX mutation (15–46%) is relatively higher as compared to the classical IDH wild-type GBMs. TERT promoter mutation and EGFR amplifications are rare in these tumors [49–51].

3.2.6 Diffuse glioma H3.3 G34 – Mutant

The working committee 3 of cIMPACT- NOW has recommended individualizing this entity into the upcoming WHO classification for CNS tumors in update 6. Like

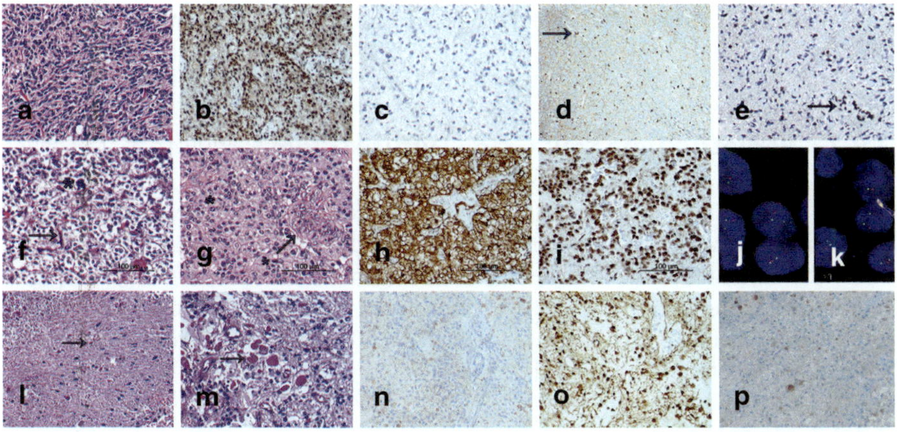

Fig. 3.2 (**a**–**e**) Photomicrograph of a thalamic (midline) diffuse glioma with high cellularity (**a**). The tumor was immunopositive for K27M mutant specific antibody (**b**). Staining for IDH-1 was negative (**c**). There was loss of expression of ATRX and H3K27me3 in the tumor cells (internal control positive—arrow). (**f**) Photomicrographs of a representative case of oligodendroglioma showing clear cell morphology with arborizing thin-walled vascular channels (arrow). There was focus of calcification (star). (**g**) Photomicrographs of a representative case of anaplastic oligodendroglioma. In addition to clear cells morphology, there were endothelial proliferation (arrow) and high mitosis (star). (**h**–**k**) Classical immunohistochemical/molecular profile of oligodendroglial tumors. Photomicrographs showing immunoreactivity for IDH-1 (**h**) and retained positivity for ATRX (**i**). FISH showing codeletion for 1p (**j**) and 19q (**k**) [test (orange)—single signal; control (green)—two signals]. (**l**–**m**) Photomicrographs of a representative case of pilocytic astrocytoma (PA) showing compact (**l**) and loosely textured (**m**) areas. The compact areas had Rosenthal fibers (arrow) while the loosely textured areas had eosinophilic granular bodies (arrow). (**n**–**p**) Classical immunohistochemical profile of PA. The tumor cells showing immunonegativity for IDH1, retained ATRX expression and negativity for p53

the DMG, H3K27M mutant, it is also considered a tumor entity defined by its characteristic molecular alterations in the histone H 3.3 G34 [17]. These tumors are usually having a hemispheric location and commonly seen in pediatric and young adult patients. The prognosis is significantly shorter than IDH mutant grade IV tumors and graded as WHO grade IV [17, 40].

3.2.6.1 Microscopy

The histomorphological features are similar to GBMs/grade IV astrocytomas with the presence of cellular atypia, mitotic activity, necrosis, and/or endothelial proliferation. A small cell morphology mimicking CNS embryonal tumor has also been described [52].

3.2.6.2 Molecular profile

Two distinct missense mutations are described, namely H3.3 G34R and H3.3 G34V have been documented (replacement of glycine with arginine or valine, respectively). Although sequencing is the gold standard for detection, specific primary antibodies against the mutant protein are available commercially. Hence, immunohistochemistry can be used as an alternate method for routine practice [53]. These tumors are usually lacking IDH mutation. However, the frequency of ATRX and TP53 mutations are very high (approximately 90%) [52].

3.2.7 Pediatric-Type Low-Grade Diffuse Glioma

Diffuse low-grade gliomas are rare in the pediatric age group. The genomic profile and clinical outcome of these tumors are largely different from their adult counterpart. These tumors contain a distinct set of genetic alterations with therapeutic implications. The WHO 2016 classification system did not consider this group of tumors as a separate entity. However, the cIMPACT update 4 has provided a detailed insight into this tumor entity [54].

3.2.7.1 Microscopy

These tumors are morphologically similar to the adult grade II astrocytic and oligodendroglial tumors. The cellular morphology can also mimic other circumscribed gliomas and often tricky to diagnose, especially in small biopsy specimens. Conventionally, the mitotic activity, necrosis, endothelial proliferation and necrosis are not seen.

3.2.7.2 Molecular Profile

Typically, these tumors are IDH and K27M wild type. They also do not show other genetic alterations commonly presents in adult diffuse gliomas, including TERT mutation, EGFR amplification. Instead, NGS-based studies documented sets of novel genetic alterations including MYB alteration, MYBL1 alteration, FGFR1 TKD-duplication, FGFR-1 mutation, BRAF V600E mutation, MAPK pathway alterations. All these genetic alterations carry prognostic implications. Hence, cIMPACT recommended an integrated diagnosis of these cases with the listed genetic alterations [54–56].

3.2.8 Oligodendroglioma (OG), IDH Mutant, and 1p/19q Co-Deleted

3.2.8.1 Anaplastic Oligodendroglioma (AO), IDH Mutant, and 1p/19q Co-Deleted

These two diffusely infiltrative gliomas have characteristic morphological features. The peak incidence of these tumors is in the third to fifth decade. The age at presentation of AO is slightly older than OG cases. The grade II and grade III tumors differ histologically only in their malignant grades. However, the molecular profile of both these tumors is almost similar. Further, the prognostic significance of this histological grading is not consistent [22, 57]. As per WHO 2016 classification system, the presence of IDH mutation and 1p/19q co-deletion are prerequisite molecular features for the diagnosis of any oligodendroglial tumor [8].

3.2.8.2 Microscopy

Oligodendroglioma Like the other diffuse gliomas, OG cases also show diffuse infiltration in the adjacent brain parenchyma with the presence of secondary structures. The tumor comprises of a monomorphic round cell arranged in a sheet-like pattern with intervening thin wall arborizing vascular network. The cells have a round nucleus with perinuclear cytoplasmic clearing and distinct cell borders, characteristically seen in the formalin-fixed paraffin-embedded section. This appearance is often described as a fried egg or honeycomb appearance. Many cases show the presence of intra-tumoral calcification. Occasional mitosis can be seen (<6/10 HPF). Endothelial proliferation and necrosis are characteristically absent (Fig. 3.2) [8].

Anaplastic Oligodendroglioma The tumor is more cellular and shows a significant degree of nuclear atypia. However, the characteristic vascular pattern and individual cellular morphology are similar to OG cases. The mitotic activity is brisk (>6/10 HPF). Endothelial proliferation and necrosis are usually present. The presence of high mitotic activity or endothelial proliferation is required for the diagnosis of AO. Necrosis and calcification are seen in most of the cases (Fig. 3.2) [8].

3.2.8.3 Molecular Profile

All cases of OG and AO characteristically have IDH mutation and 1p/19q co-deletion. These molecular alterations are essential for diagnosis. It must be kept in mind that if a tumor with astrocytic histology demonstrates molecular features of oligodendroglial tumors, it should be designated as OG or AO [8]. Almost all the cases harbor TERT promoter mutation [58]. Other commonly encountered molecular alterations in OG and AO cases are CIC (~60%) and FUBP-1 (~30%) mutation [26]. In contrast to grade II/III astrocytic tumors, these tumors do not harbor ATRX and TP53 mutations. Hence, all the cases of OG/AO have retained nuclear immune-reactivity of ATRX and negative for p53 immunohistochemistry. In other words, combining these immunohistochemical markers with the morphology can predict the status of 1p/19q. In the absence of a molecular testing facility for 1p/19q, these immunohistochemical markers are helpful for diagnosing OG/AO in a resource-limited setting [59].

3.2.9 Oligoastrocytoma

These tumors morphologically show both astrocytic and oligodendroglial differentiation. However, large-scale studies have shown that oligoastrocytoma cases have molecular profile characteristics of either astrocytoma (IDH, ATRX, and TP53mutant) or oligodendroglioma (IDH mutant and 1p/19q co-deleted). More importantly, the cases with astrocytic and oligodendroglial molecular features have the similar clinical outcome as the astrocytic and oligodendroglial tumors of the same grades [15, 16, 60]. Hence, it is now considered only as a morphological

Fig. 3.3 Flowchart for the molecular diagnosis of diffuse gliomas (non-midline) on the basis of the WHO and cIMPACT NOW recommendation [Gr-grade, IHC- immunohistochemistry, Seq.- Sequencing, Del - Deleton, Mut- Mutation, WT- wild type, Amp - Amplification. ($,*- occurance is extremely rare)

entity without much clinical significance. The diagnosis of oligoastrocytoma is strongly discouraged by the WHO. However, in the absence of any molecular data, it may be designated as oligoastrocytoma, not otherwise specified (see below) [8].

3.3 Diffuse Glioma: Approach to Diagnosis

The approach to diagnosis of different entities enumerated above as per the WHO 2016 classification and various consensus update published by cIMPACT NOW is depicted in Fig. 3.2. The molecular classification schema of gliomas is provided in Fig. 3.3.

3.4 Other Astrocytic Tumors

These astrocytic tumors are relatively slow-growing and have a circumscribed/non-diffuse margin. Even if focal infiltration is present at the tumor–brain interface, the secondary structures are characteristically absent. As per the 2016 WHO classification, four histological entities are included in this group: Pilocytic astrocytoma (WHO grade I), sub-ependymal giant cell astrocytoma (SEGA) (WHO grade I), pleomorphic xanthoastrocytoma (PXA) (WHO grade II), and anaplastic PXA (WHO grade III) [8].

3.4.1 Pilocytic Astrocytoma (PA), WHO Grade1

It is the most common glial tumor in children and adolescent age groups. Although any part of the CNS can be involved, occurrence in the cerebellum, optic nerve, and optic chiasma, are more frequent [8].

3.4.1.1 Microscopy

These tumors are usually well-circumscribed and have low cellularity. Characteristically PA cases have biphasic histology with compact and loosely textured areas. The compact areas consist of bipolar piloid cells having long, thin, and fibrillary cytoplasmic processes. These areas also contain bright eosinophilic Rosenthal fibers. The loose areas usually show multipolar cells and microcystic areas with the presence of eosinophilic granular bodies. There may be degenerative cellular atypia and vascular proliferations. Features of malignancies, including mitosis, necrosis, and endothelial proliferation, are usually absent [8, 61].

3.4.1.2 Molecular Profile

On molecular genetic background, PA is a single pathway disease associated with the activation of the mitogen-activated protein kinases (MAPK) pathway. The most common genetic alteration being a tandem duplication in chromosome 7q34, leading to the formation of an oncogenic fusion gene KIAA1549:BRAF. This alteration is seen in approximately 70% of cases of PA and reported mainly in the cerebellar tumor [62, 63]. The other alterations include BRAF mutations, FGFR1 mutation/duplication, NTRK fusion, etc. [64]. However, these molecular features are not essential for the diagnosis of PA.

3.4.1.3 Pilocytic Astrocytoma with Anaplasia

Rarely cases of PA can show features of anaplasia in addition to their classic histology. To date, no fixed criteria exist to define anaplasia in PA. However, most of the previous studies have used the following histological features: increase mitotic activity, necrosis, and significant cellular atypia. PA with anaplasia has a poor clinical outcome and shorter survival as compared to classic cases [65]. The survival outcome of these cases is better than GBM and corresponds to WHO grade II/III astrocytomas. These cases commonly show CDKN2A/B deletion (17–85%) and ATRX mutation (45–57%) apart from MAPK pathway alterations [66, 67].

3.5 Not Otherwise Specified (NOS) and Not Elsewhere Classified (NEC) Categories

3.5.1 Not Otherwise Specified (NOS)

The diagnosis of different diffuse gliomas requires analysis of one or more molecular alterations in addition to the histology. If these diagnostic tests are not/cannot be pre-confirmed due to technical, infrastructural, or any other issue, these tumors are designated as NOS. For example, the diagnosis of oligodendroglioma requires analysis of IDH mutation and 1p/19q co-deletion. If a tumor has oligodendroglial morphology, but the analysis of one or more of these markers is not available, then the diagnosis of oligodendroglioma, NOS, is rendered.

Another typical example of the NOS category is oligoastrocytoma in the absence of required molecular testing for sub-classifying it into either astrocytic or oligodendroglial group [12, 13].

3.5.2 Not Elsewhere Classified (NEC)

This category indicates that the necessary molecular tests are performed, but the results do not fit into any particular WHO diagnosis. These cases are mostly related to a mismatch between the histological and molecular features and may include a newer entity which is still not incorporated in the WHO classification system [12, 13].

References

1. Bailey OT. Genesis of the Percival Bailey-Cushing classification of gliomas. Pediatr Neurosci. 1985;12(4–5):261–5.
2. Ferguson S, Lesniak MS. Percival Bailey and the classification of brain tumors. Neurosurg Focus. 2005;18(4):e7.
3. Louis DN, Ohgaki H, Wiestler OD, Cavenee WK, editors. WHO classification of tumours of the central nervous system. Lyon: IARC; 2007.
4. van den Bent MJ. Interobserver variation of the histopathological diagnosis in clinical trials on glioma: a clinician's perspective. Acta Neuropathol. 2010;120(3):297–304.
5. Cancer Genome Atlas Research Network, Brat DJ, Verhaak RG, Aldape KD, Yung WK, Salama SR, et al. Comprehensive, integrative genomic analysis of difuse lower-grade gliomas. N Engl J Med. 2015;372:2481–98.
6. Louis DN, Perry A, Burger P, Ellison DW, Reifenberger G, von Deimling A, et al. International society of neuropathology--Haarlem consensus guidelines for nervous system tumor classification and grading. Brain Pathol. 2014;24(5):429–35.
7. Kristensen BW, Priesterbach-Ackley LP, Petersen JK, Wesseling P. Molecular pathology of tumors of the central nervous system. Ann Oncol. 2019;30(8):1265–78.
8. Louis DN, Ohgaki H, Wiestler OD, et al. World Health Organization histological classification of tumours of the central nervous system. Lyon: International Agency for Research on Cancer (IARC); 2016.
9. Brat DJ, Aldape K, Colman H, et al. cIMPACT-NOW update 5: recommended grading criteria and terminologies for IDH-mutant astrocytomas. Acta Neuropathol. 2020;139(3):603–8.

10. Louis DN, Ellison DW, Brat DJ, et al. cIMPACT-NOW: a practical summary of diagnostic points from round 1 updates. Brain Pathol. 2019;29(4):469–72.
11. Perry A, Wesseling P. Histologic classification of gliomas. Handb Clin Neurol. 2016;134:71–95.
12. Louis DN, Giannini C, Capper D, et al. cIMPACT-NOW update 2: diagnostic clarifications for diffuse midline glioma, H3 K27M-mutant and diffuse astrocytoma/anaplastic astrocytoma. IDH-mutant Acta Neuropathol. 2018;135(4):639–42.
13. Louis DN, Wesseling P, Paulus W, et al. cIMPACT-NOW update 1: not otherwise specified (NOS) and not elsewhere classified (NEC). Acta Neuropathol. 2018;135(3):481–4.
14. Olar A, Wani KM, Alfaro-Munoz KD, et al. IDH mutation status and role of WHO grade and mitotic index in overall survival in grade II-III diffuse gliomas. Acta Neuropathol. 2015;129(4):585–96.
15. Reuss DE, Mamatjan Y, Schrimpf D, et al. IDH mutant diffuse and anaplastic astrocytomas have similar age at presentation and little difference in survival: a grading problem for WHO. Acta Neuropathol. 2015;129(6):867–73.
16. Reuss DE, Sahm F, Schrimpf D, et al. ATRX and IDH1-R132H immunohistochemistry with subsequent copy number analysis and IDH sequencing as a basis for an "integrated" diagnostic approach for adult astrocytoma, oligodendroglioma and glioblastoma. Acta Neuropathol. 2015;129(1):133–46.
17. Louis DN, Wesseling P, Aldape K, et al. cIMPACT-NOW update 6: new entity and diagnostic principle recommendations of the cIMPACT-Utrecht meeting on future CNS tumor classification and grading. Brain Pathol. 2020;30(4):844–56.
18. Onizuka H, Masui K, Komori T. Diffuse gliomas to date and beyond 2016 WHO classification of tumours of the central nervous system. Int J Clin Oncol. 2020;25(6):997–1003.
19. Agarwal S, Sharma MC, Jha P, et al. Comparative study of IDH1 mutations in gliomas by immunohistochemistry and DNA sequencing. Neuro-Oncology. 2013;15(6):718–26.
20. Mellai M, Piazzi A, Caldera V, Monzeglio O, Cassoni P, Valente G, et al. IDH1 and IDH2 mutations, immunohistochemistry and associations in a series of brain tumors. J Neuro-Oncol. 2011;105(2):345–57.
21. Turkalp Z, Karamchandani J, Das S. IDH mutation in glioma: new insights and promises for the future. JAMA Neurol. 2014;71(10):1319–25.
22. Suzuki H, Aoki K, Chiba K, et al. Mutational landscape and clonal architecture in grade II and III gliomas. Nat Genet. 2015;47(5):458–68.
23. Tanboon J, Williams EA, Louis DN. The diagnostic use of immunohistochemical surrogates for signature molecular genetic alterations in gliomas. J Neuropathol Exp Neurol. 2016;75(1):4–18.
24. Appay R, Dehais C, Maurage CA, et al. CDKN2A homozygous deletion is a strong adverse prognosis factor in diffuse malignant IDH-mutant gliomas. Neuro-Oncology. 2019;21(12):1519–28.
25. Korshunov A, Casalini B, Chavez L, et al. Integrated molecular characterization of IDH-mutant glioblastomas. Neuropathol Appl Neurobiol. 2019;45(2):108–18.
26. Brat DJ, Verhaak RG, et al. Cancer genome atlas research network. Comprehensive, integrative genomic analysis of diffuse lower-grade gliomas. N Engl J Med. 2015;372(26):2481–98.
27. Ceccarelli M, Barthel FP, Malta TM, et al. Molecular profiling reveals biologically discrete subsets and pathways of progression in diffuse glioma. Cell. 2016;164(3):550–63.
28. Stichel D, Ebrahimi A, Reuss D, et al. Distribution of EGFR amplification, combined chromosome 7 gain and chromosome 10 loss, and TERT promoter mutation in brain tumors and their potential for the reclassification of IDHwt astrocytoma to glioblastoma. Acta Neuropathol. 2018;136(5):793–803.
29. Brat DJ, Aldape K, Colman H, et al. cIMPACT-NOW update 3: recommended diagnostic criteria for "diffuse astrocytic glioma, IDH-wildtype, with molecular features of glioblastoma, WHO grade IV". Acta Neuropathol. 2018;136(5):805–10.
30. Chen L, Voronovich Z, Clark K, et al. Predicting the likelihood of an isocitrate dehydrogenase 1 or 2 mutation in diagnoses of infiltrative glioma. Neuro-Oncology. 2014;16(11):1478–83.

31. Killela PJ, Reitman ZJ, Jiao Y, et al. TERT promoter mutations occur frequently in gliomas and a subset of tumors derived from cells with low rates of self-renewal. Proc Natl Acad Sci U S A. 2013;110(15):6021–6.
32. Purkait S, Mallick S, Sharma V, et al. Prognostic stratification of GBMs using combinatorial assessment of IDH1 mutation, MGMT promoter methylation, and tert mutation status: experience from a tertiary care center in India. Transl Oncol. 2016;9(4):371–6.
33. Umehara T, Arita H, Yoshioka E, et al. Distribution differences in prognostic copy number alteration profiles in IDH-wild-type glioblastoma cause survival discrepancies across cohorts [published correction appears in Acta Neuropathol Commun. 2019;7(1):131]. Acta Neuropathol Commun. 2019;7(1):99.
34. Weller M, Tabatabai G, Kästner B, et al. MGMT promoter methylation is a strong prognostic biomarker for benefit from dose-intensified temozolomide rechallenge in progressive glioblastoma: the DIRECTOR trial. Clin Cancer Res. 2015;21(9):2057–64.
35. Liu XY, Gerges N, Korshunov A, et al. Frequent ATRX mutations and loss of expression in adult diffuse astrocytic tumors carrying IDH1/IDH2 and TP53 mutations. Acta Neuropathol. 2012;124(5):615–25.
36. Oh JE, Ohta T, Nonoguchi N, et al. Genetic Alterations in Gliosarcoma and Giant Cell Glioblastoma. Brain Pathol. 2016;26(4):517–22.
37. Broniscer A, Tatevossian RG, Sabin ND, et al. Clinical, radiological, histological and molecular characteristics of paediatric epithelioid glioblastoma. Neuropathol Appl Neurobiol. 2014;40(3):327–36.
38. Schwartzentruber J, Korshunov A, Liu XY, et al. Driver mutations in histone H3.3 and chromatin remodelling genes in paediatric glioblastoma [published correction appears in nature. 2012 Apr 5;484(7392):130]. Nature. 2012;482(7384):226–31.
39. Brennan CW, Verhaak RG, McKenna A, et al. The somatic genomic landscape of glioblastoma [published correction appears in cell. 2014 Apr 24;157(3):753]. Cell. 2013;155(2):462–77.
40. Sturm D, Witt H, Hovestadt V, et al. Hotspot mutations in H3F3A and IDH1 define distinct epigenetic and biological subgroups of glioblastoma. Cancer Cell. 2012;22(4):425–37.
41. Yan H, Parsons DW, Jin G, et al. IDH1 and IDH2 mutations in gliomas. N Engl J Med. 2009;360(8):765–73.
42. Ohgaki H, Kleihues P. The definition of primary and secondary glioblastoma. Clin Cancer Res. 2013;19(4):764–72.
43. Purkait S, Miller CA, Kumar A, et al. ATRX in diffuse gliomas with its mosaic/heterogeneous expression in a subset. Brain Pathol. 2017;27(2):138–45.
44. Karremann M, Gielen GH, Hoffmann M, et al. Diffuse high-grade gliomas with H3 K27M mutations carry a dismal prognosis independent of tumor location. Neuro-Oncology. 2018;20(1):123–31.
45. Gessi M, Capper D, Sahm F, et al. Evidence of H3 K27M mutations in posterior fossa ependymomas. Acta Neuropathol. 2016;132(4):635–7.
46. Hochart A, Escande F, Rocourt N, et al. Long survival in a child with a mutated K27M-H3.3 pilocytic astrocytoma. Ann Clin Transl Neurol. 2015;2(4):439–43.
47. Kleinschmidt-DeMasters BK, Donson A, Foreman NK, Dorris K. H3 K27M mutation in gangliogliomas can be associated with poor prognosis. Brain Pathol. 2017;27(6):846–50.
48. Bechet D, Gielen GG, Korshunov A, et al. Specific detection of methionine 27 mutation in histone 3 variants (H3K27M) in fixed tissue from high-grade astrocytomas. Acta Neuropathol. 2014;128(5):733–41.
49. Buczkowicz P, Hoeman C, Rakopoulos P, et al. Genomic analysis of diffuse intrinsic pontine gliomas identifies three molecular subgroups and recurrent activating ACVR1 mutations. Nat Genet. 2014;46(5):451–6.
50. Solomon DA, Wood MD, Tihan T, et al. Diffuse midline gliomas with histone H3-K27M mutation: a series of 47 cases assessing the spectrum of morphologic variation and associated genetic alterations. Brain Pathol. 2016;26(5):569–80.
51. Wu G, Diaz AK, Paugh BS, et al. The genomic landscape of diffuse intrinsic pontine glioma and pediatric non-brainstem high-grade glioma. Nat Genet. 2014;46(5):444–50.

52. Picart T, Barritault M, Poncet D, et al. Characteristics of diffuse hemispheric gliomas, H3 G34-mutant in adults. Neurooncol Adv. 2021;3(1):vdab061.
53. Haque F, Varlet P, Puntonet J, et al. Evaluation of a novel antibody to define histone 3.3 G34R mutant brain tumours. Acta Neuropathol Commun. 2017;5(1):45.
54. Ellison DW, Hawkins C, Jones DTW, et al. cIMPACT-NOW update 4: diffuse gliomas characterized by MYB, MYBL1, or FGFR1 alterations or BRAFV600E mutation. Acta Neuropathol. 2019;137(4):683–7.
55. Jones DTW, Kieran MW, Bouffet E, et al. Pediatric low-grade gliomas: next biologically driven steps [published correction appears in Neuro Oncol. 2018 Jan 22;20(2):293]. Neuro-Oncology. 2018;20(2):160–73.
56. Sturm D, Pfister SM, Jones DTW. Pediatric gliomas: current concepts on diagnosis, biology, and clinical management. J Clin Oncol. 2017;35(21):2370–7.
57. Killela PJ, Pirozzi CJ, Healy P, et al. Mutations in IDH1, IDH2, and in the TERT promoter define clinically distinct subgroups of adult malignant gliomas. Oncotarget. 2014;5(6):1515–25.
58. Arita H, Narita Y, Fukushima S, et al. Upregulating mutations in the TERT promoter commonly occur in adult malignant gliomas and are strongly associated with total 1p19q loss. Acta Neuropathol. 2013;126(2):267–76.
59. Rajeswarie RT, Rao S, Nandeesh BN, Yasha TC, Santosh V. A simple algorithmic approach using histology and immunohistochemistry for the current classification of adult diffuse glioma in a resource-limited set-up. J Clin Pathol. 2018;71(4):323–9.
60. Sahm F, Reuss D, Koelsche C, et al. Farewell to oligoastrocytoma: in situ molecular genetics favor classification as either oligodendroglioma or astrocytoma. Acta Neuropathol. 2014;128(4):551–9.
61. Pujadas E, Chen L, Rodriguez FJ. Pathologic and molecular aspects of anaplasia in circumscribed gliomas and glioneuronal tumors. Brain Tumor Pathol. 2019;36(2):40–51.
62. Bar EE, Lin A, Tihan T, Burger PC, Eberhart CG. Frequent gains at chromosome 7q34 involving BRAF in pilocytic astrocytoma. J Neuropathol Exp Neurol. 2008;67(9):878–87.
63. Jones DT, Kocialkowski S, Liu L, et al. Tandem duplication producing a novel oncogenic BRAF fusion gene defines the majority of pilocytic astrocytomas. Cancer Res. 2008;68(21):8673–7.
64. Jones DT, Hutter B, Jäger N, et al. Recurrent somatic alterations of FGFR1 and NTRK2 in pilocytic astrocytoma. Nat Genet. 2013;45(8):927–32.
65. Rodriguez FJ, Scheithauer BW, Burger PC, Jenkins S, Giannini C. Anaplasia in pilocytic astrocytoma predicts aggressive behavior. Am J Surg Pathol. 2010;34(2):147–60.
66. Reinhardt A, Stichel D, Schrimpf D, et al. Anaplastic astrocytoma with piloid features, a novel molecular class of IDH wildtype glioma with recurrent MAPK pathway, CDKN2A/B and ATRX alterations. Acta Neuropathol. 2018;136(2):273–91.
67. Rodriguez FJ, Brosnan-Cashman JA, Allen SJ, et al. Alternative lengthening of telomeres, ATRX loss and H3-K27M mutations in histologically defined pilocytic astrocytoma with anaplasia. Brain Pathol. 2019;29(1):126–40.

Pathology and Molecular Pathology of Ependymoma

4

Aruna Nambirajan, Agrima Sharma,
and Mehar Chand Sharma

4.1 Introduction

Ependymomas are a heterogenous group of glial neoplasms that morphologically and ultrastructurally recapitulate the ependymal cells lining the ventricles and central canal of the spinal cord. They account for ~2% of all brain tumors and 6–7% of gliomas [1]. They occur across all age groups; median age at diagnosis irrespective of location and subtype is 43 years [1]. They are more commonly diagnosed in males as compared to females (male: female ratio of 1.32:1).

4.2 Pathology

The World Health Organization classification (revised fourth edition) classifies ependymal tumors [2] as follows:

- Subependymoma, WHO grade 1
- Myxopapillary ependymoma, WHO grade 1
- Ependymoma, WHO grade 2
- Anaplastic ependymoma, WHO grade 3
- Ependymoma, *RELA* fusion-positive

A. Nambirajan (✉) · A. Sharma · M. C. Sharma
Department of Pathology, AIIMS, New Delhi, India

S. Mallick et al. (eds.), *Evidence based practice in Neuro-oncology*,
https://doi.org/10.1007/978-981-16-2659-3_4

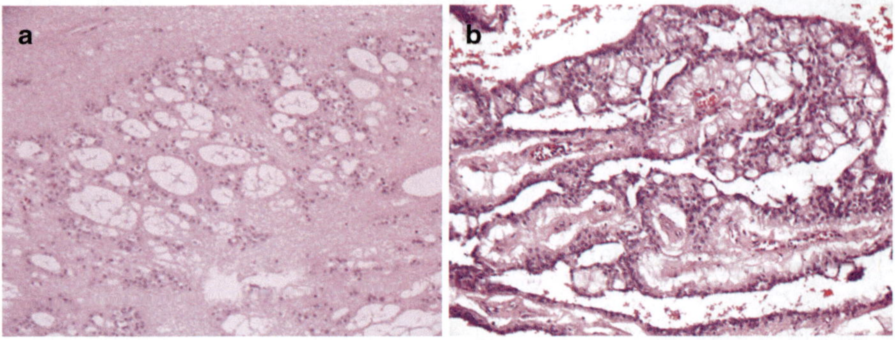

Fig. 4.1 Histology of subependymoma and myxopapillary ependymoma. Hematoxylin and eosin stained sections show classical histomorphological features of a subependymoma—hypocellular tumor, microcystic change, and nodular aggregates of bland tumor cells (**a**); and a myxopapillary ependymoma—papillary architecture with myxoid material in the perivascular regions (**b**)

4.2.1 Subependymoma

Subependymomas account for ~8% of ependymal tumors.

- These tumors are hypocellular and show a nodular architecture on light microscopy.
- The tumor cells are bland, embedded in an abundant fibrillary matrix with frequent microcystic change and are devoid of significant nuclear pleomorphism or mitotic activity (Fig. 4.1a).

4.2.2 Myxopapillary Ependymoma

Myxopapillary ependymomas account for 9–13% of ependymal tumors.

- These tumors show a papillary to solid architecture composed of bland cuboidal to elongated tumor cells radially arranged around fibrovascular cores (Fig. 4.1b).
- Alcian blue-positive mucoid material is prominent with formation of microcysts.
- The tumor cells are usually devoid of significant nuclear pleomorphism or mitotic activity.
- Very rare examples show anaplastic histomorphology in the form of increased mitotic activity, spontaneous tumor necrosis, and microvascular proliferation [3].

4.2.3 Ependymoma and Anaplastic Ependymoma

Classic (WHO grade 2) and anaplastic (WHO grade 3) ependymomas account for the majority of ependymal tumors, with reported proportions of grade 2: grade 3 ependymomas varying widely from 17:1 to 1:7.

- Classic ependymomas are composed of uniform small tumor cells in a fibrillary matrix forming perivascular pseudo-rosettes and, less frequently, true ependymal rosettes with a central lumen.
- They show pushing borders and are well demarcated from surrounding brain parenchyma.
- Posterior fossa ependymomas sometimes show a biphasic pattern with nodules of increased cellularity and mitotic activity [4].
- Supratentorial ependymomas frequently show clear cell change and a rich branching vascular pattern [5].
- Classic ependymomas (WHO grade 2) generally show bland cytology with low mitotic activity, with only focal palisading necrosis or microvascular proliferation, if present.
- Anaplastic ependymomas (WHO grade 3), in addition, show increased cellularity, frequent mitotic activity, and widespread palisading necrosis and microvascular proliferation.
- Morphological subtypes of ependymomas based on dominant histological patterns include papillary, clear cell, and tanycytic: papillary ependymomas are usually intraventricular, clear cell ependymomas are frequently supratentorial, and tanycytic ependymomas occur in the spinal cord; these subtypes are not of clinicopathological significance and may be WHO grade 2 or grade 3.
- Ependymomas show perivascular staining of the tumor cell processes for glial fibrillary acidic protein and show dot-like and ring-like staining pattern for epithelial membrane antigen representing intracytoplasmic lumen formation, a characteristic ultrastructural feature of normal ependymal cells [6].

4.2.4 Ependymoma, RELA Fusion Positive

- Genetically defined entity introduced in the WHO 2016 update defined by the presence of fusions of the *RELA* gene (v-rel avian reticuloendotheliosis viral oncogene homolog A) most commonly with *C11orf95* (ZFTA [7]) resulting in constitutive upregulation of NF-kB pathway [2, 5, 8].
- It occurs primarily in the supratentorial compartment and accounts for approximately 70% of supratentorial ependymomas.
- They occur in both children and adults, more common in children.
- These tumors frequently show a clear cell morphology with rich vascular network.
- They are more commonly of grade 3 histology and are associated with poor outcomes [5, 9].
- It is to be noted that this entity is the only ependymoma subtype that is genetically defined in the 2016 update of the WHO classification.
- The use of immunohistochemistry for p65/RelA and L1CAM, a downstream target of NF-KB signaling, is used as surrogate markers for the identification of fusions of the *RELA* gene (Fig. 4.2).
- The gold standard method of detection of *RELA* fusions is molecular analysis, by RT-PCR, FISH or DNA methylation profiling.

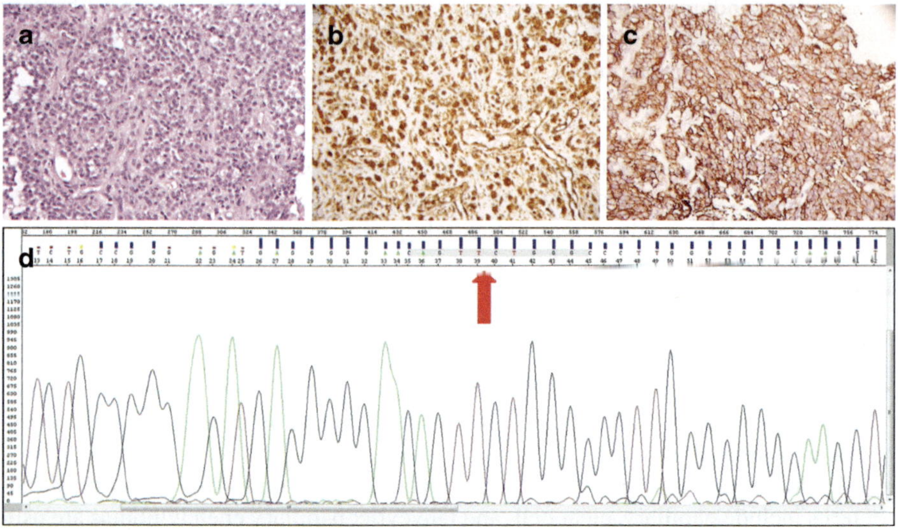

Fig. 4.2 Pathological features of supratentorial ependymoma, RELA fusion-positive. Hematoxylin and eosin stained sections from a supratentorial ependymoma show the presence of clear tumor cells with fine branching vasculature (**a**); Tumor nuclei are diffusely positive for p65/Rela protein (**b**); Tumor cells also show diffuse membranous immunoreactivity for L1CAM (**c**); Sanger sequencing of reverse transcriptase polymerase chain reaction products using primers for C11orf95 (ZFTA) and RELA gene reveals *C11orf95 (ZFTA)-RELA* fusion transcripts (**d**)

4.3 Molecular Pathology

High-throughput gene expression and DNA methylation profiling studies have identified site-specific molecular subsets of ependymomas with distinct demographic and clinicopathological associations [8–10]. Pajtler et al. performed methylation profiling of over 500 ependymal tumors and identified three distinct molecular subgroups among subependymomas based on their site of origin (supratentorial, infratentorial, and spinal). All myxopapillary ependymomas segregated to a distinct methylation profile. Other than subependymomas and myxopapillary ependymomas where methylation patterns correlated with tumor histomorphology, all grade 2 and grade 3 ependymal tumors despite histomorphological similarities delineated into at least six distinct methylation-based subgroups that were based on site of origin with distinct demographic and outcome variables [9, 11] as shown in Table 4.1.

- The two most common molecular subgroups are the PFA subgroup in the posterior fossa and the *RELA* fusion-positive subgroup in the supratentorium (previously described in Sect. 4.2.4).
- Multiple studies have confirmed the poor prognostic outcomes of PFA [9, 12–16] and *RELA* fusion-positive ependymomas, and it is important to identify these subgroups in routine diagnostic practice.

Table 4.1 Molecular subgroups of grade 2 and 3 ependymal tumors identified on DNA methylation profiling

Primary location	Subgroup	Age group	Frequency	Defining (epi) genetic alteration	Method of diagnosis	Prognosis
Posterior fossa	PF-A (pediatric-type)	2.5–5 years	~75% of all posterior fossa ependymomas	CIMP-phenotype, global DNA hypomethylation, EZHIP overexpression, H3 K27M mutation, 1q gain	Loss of H3K27me3 protein expression by immunohistochemistry or by DNA methylation profiling	Poor
	PF-B (adult-type)	Median age ~20 years	~25% of all posterior fossa ependymomas	Multiple chromosome losses and gains	Retained H3K27me3 expression on immunohistochemistry	Good
Supratentorial	Ependymoma, *RELA* fusion-positive	Children and adults; median age ~8 years	~70% of supratentorial ependymomas	Fusions of the *RELA* gene with *C11orf95 (ZFTA)* gene	Demonstration of fusion transcript by RT-PCR/FISH or immunohistochemistry for L1CAM	Poor
	Ependymomas, *YAP*1 fusion-positive	Infants	Rare	Fusions of the *YAP1* gene with *MAMLD1* or *FAM118B*	Demonstration of fusion transcript by RT-PCR or FISH/ DNA methylation profiling	Limited data
Spinal cord	Spinal ependymoma	Adults	Majority of spinal cord ependymomas	22q monosomy, *NF2* mutations	Histological diagnosis	Excellent
	Spinal ependymomas, *NMYC* amplified	Adults	Rare; intradural extramedullary	Amplification of *NMYC* (2p24)	FISH or DNA methylation profiling	Poor

4.3.1 PFA Molecular Subgroup (Pediatric Type)

- The PFA ependymomas harbor balanced genomes and are driven by epigenetic modifications.
- They show CpG island methylator phenotype, global DNA methylation, and loss of trimethylation at lysine residue at position 27 of histone H3.
- While rare PFA ependymomas harbor *H3K27M* mutations contributing to the H3K27me3 loss as seen in H3K27M diffuse gliomas, more commonly, PFA ependymomas overexpress EZHIP (EZH2 inhibitory protein) that inhibits EZH2-mediated H3K27 trimethylation leading to loss of H3K27me3 and the resultant epigenetic dysregulation [14, 17, 18].
- Chromosome 1q gain is independent prognostic marker for overall survival and progression-free survival in posterior fossa ependymomas [9, 19, 20].
- Multiple authors have found that H3K27me3 immunohistochemistry is a robust marker for the identification of PFA ependymoma subgroup [14], and this marker is now recommended by the C-IMPACT-NOW [21] and is likely to be incorporated in the upcoming edition of the WHO classification (Fig. 4.3).

4.3.2 Spinal Ependymoma, NMYC Amplified

- More recently identified molecular subgroup associated with aggressive behavior [11, 22].
- Rare spinal cord ependymomas of frequently anaplastic histology showing amplification of *NMYC* gene.
- Associated with frequent cranial and extracranial metastases.
- The other molecular subgroups, viz. spinal ependymoma and PFB subgroups, are associated with excellent outcomes [9].

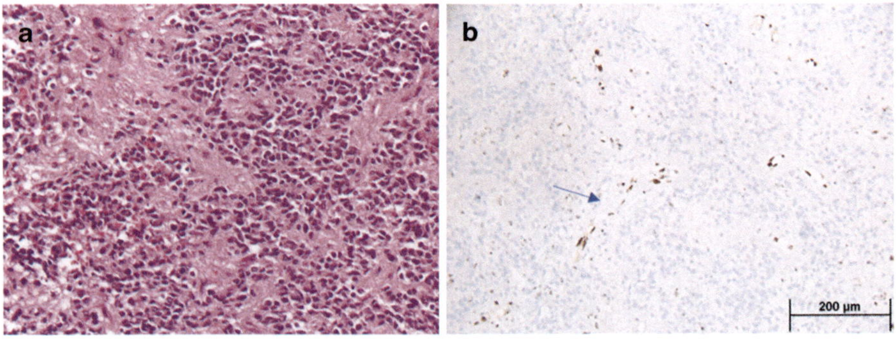

Fig. 4.3 H3K27me3 immunohistochemistry for subgrouping of posterior fossa ependymomas. Hematoxylin and eosin stained sections from a posterior fossa ependymoma (**a**) that showed loss of expression of trimethylated form of H3 histone protein at position K27 classifying as PFA molecular subgroup (**b**). Note retained expression in endothelial and inflammatory cells (arrow)

- A very rare subset of supratentorial ependymomas harbor fusions of the *YAP1* gene; these tumors are more common in infants, and there is limited data on their outcome [5, 9, 12].
- It is important to note that a small proportion of supratentorial ependymomas that lack *RELA* and *YAP1* gene fusions may actually represent high-grade CNS neuroepithelial tumors, in particular, CNS embryonal tumors harboring *BCOR* internal tandem duplications or *MN1* gene fusions, the latter comprising tumors that share ependymoma-like morphology [16, 23]. Recently, genes other than RELA have been discovered as fusion partners with C11orf95 (*ZFTA*) in supratentorial ependymomas [7].

4.4 Future Perspectives

The last decade has witnessed an improved understanding of the molecular underpinnings of the biological heterogeneity of ependymal tumors that extends beyond histological grade. The lack of objective grading criteria, marked interobserver variability in grading, and the morphological heterogeneity of these tumors limits the usefulness of WHO grade in accurate prognostication. Treatment guidelines based on molecular subgroups are evolving, and grade-based treatment decisions are likely to become obsolete in the near future. A global ependymoma consensus conference in 2015 recommended that in patients >12 months of age, maximal safe resection and focal radiotherapy should be standard of care in PF-A posterior fossa ependymomas; for the PF-B posterior fossa ependymomas, radiotherapy may be withheld in view of their good prognosis [24], heralding the era of molecular subgroup-based treatment decisions in ependymoma management. There have been contradictory reports as well [20, 25], and the prognostic impact of molecular subgrouping needs to be validated in independent cohorts.

The C-IMPACT-NOW update 7 (Consortium to Inform Molecular and Practical Approaches to CNS Tumor Taxonomy—Not Official WHO update 7) [21] has recommended updating the classification of ependymal tumors as follows:

1. Subependymoma, WHO grade 1
2. Myxopapillary ependymoma, WHO grade 2
3. Supratentorial ependymoma, *C11orf95 (ZFTA)* fusion-positive
4. Supratentorial ependymoma, *YAP1* fusion-positive
5. Supratentorial ependymoma, WHO grade 2/3
6. Posterior fossa ependymoma, PFA subgroup
7. Posterior fossa ependymoma, PFB subgroup
8. Posterior fossa ependymoma, WHO grade 2/3
9. Spinal ependymoma, *MYCN* amplified
10. Spinal ependymoma, WHO grade 2/3

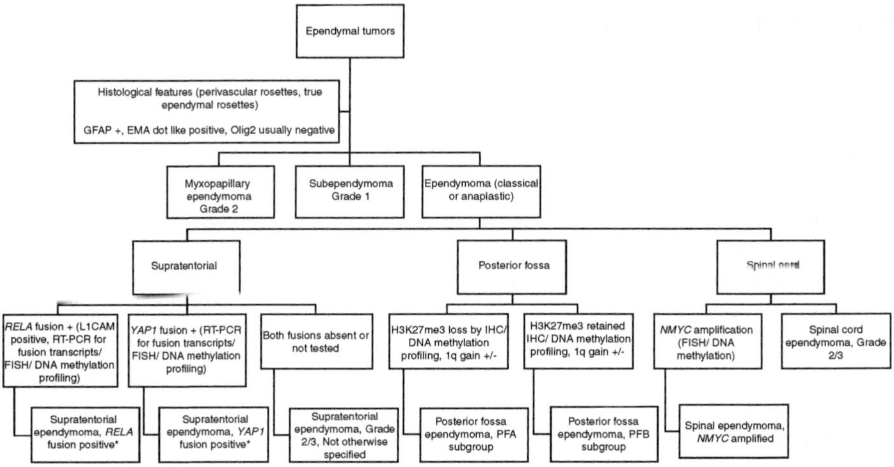

Fig. 4.4 Approach to diagnosis of ependymal tumors. Abbreviations: *GFAP* glial fibrillary acidic protein, *EMA* epithelial membrane antigen, *RELA* v-rel avian reticuloendotheliosis gene, *L1CAM* L1-cell adhesion molecule, *YAP1* yes-associated protein 1. *Only molecular entity currently recognized by the WHO classification of CNS tumors (2016 update) [2] and renamed as 'supratentorial ependymoma, *C11orf95 (ZFTA)* fusion-positive in the C-IMPACT-NOW-7 [21]

It is to be noted that 7 of the 10 proposed subtypes are defined (epi)genetically and do not have an assigned grade. The upcoming WHO classification of CNS tumors is likely to incorporate these recommendations. Routine identification of these molecular subtypes will enable inclusion in clinical trials, confirmation of the prognostic significance, and development of better therapies (Fig. 4.4).

References

1. Ostrom QT, Gittleman H, Fulop J, Liu M, Blanda R, Kromer C, Wolinsky Y, Kruchko C, Barnholtz-Sloan JS. CBTRUS statistical report: primary brain and central nervous system tumors diagnosed in the United States in 2008-2012. Neuro-Oncology. 2015;17(Suppl 4):iv1–iv62.
2. Ellison DW, McLendon R, Wiestler OD, et al. Ependymoma. In: Louis DN, Ohgaki H, Wiestler OD, Cavenee WK, editors. WHO classification of tumours of the central nervous system. 4th ed. Lyon: IARC; 2016. p. 106–12.
3. Lee JC, Sharifai N, Dahiya S, et al. Clinicopathologic features of anaplastic myxopapillary ependymomas. Brain Pathol. 2019;29(1):75–84.
4. Ellison DW, Kocak M, Figarella-Branger D, et al. Histopathological grading of pediatric ependymoma: reproducibility and clinical relevance in European trial cohorts. J Negat Results Biomed. 2011;10:7.
5. Malgulwar PB, Nambirajan A, Pathak P, et al. C11orf95-RELA fusions and upregulated NF-KB signalling characterise a subset of aggressive supratentorial ependymomas that express L1CAM and nestin. J Neuro-Oncol. 2018;138:29–39.
6. Nambirajan A, Sharma MC, Rajeshwari M, Kakkar A, Suri V, Sarkar C. A comparative immunohistochemical study of epithelial membrane antigen and NHERF1/EBP50 in the diagnosis of ependymomas. Appl Immunohistochem Mol Morphol. 2018;26(1):71–8.

7. Zheng T, Ghasemi DR, Okonechnikov K, Korshunov A, Sill M, Maass KK, Benites Goncalves da Silva P, Ryzhova M, Gojo J, Stichel D, Arabzade A, Kupp R, Benzel J, Taya S, Adachi T, Shiraishi R, Gerber NU, Sturm D, Ecker J, Sievers P, Selt F, Chapman R, Haberler C, Figarella-Branger D, Reifenberger G, Fleischhack G, Rutkowski S, Donson AM, Ramaswamy V, Capper D, Ellison DW, Herold-Mende CC, Schuller U, Brandner S, Hernaiz Driever P, Kros JM, Snuderl M, Milde T, Grundy RG, Hoshino M, Mack SC, Gilbertson RJ, Jones DTW, Kool M, von Deimling A, Pfister SM, Sahm F, Kawauchi D, Pajtler KW. Cross-species genomics reveals oncogenic dependencies in ZFTA/C11orf95 fusion-positive supratentorial ependymomas. Cancer Discov. 2021:candisc.0963.2020. https://doi.org/10.1158/2159-8290.
8. Parker M, Mohankumar KM, Punchihewa C, et al. C11orf95-RELA fusions drive oncogenic NF-κB signalling in ependymoma. Nature. 2014;506:451–5.
9. Pajtler KW, Witt H, Sill M, Jones DT, Hovestadt V, Kratochwil F, et al. Molecular classification of ependymal tumors across all CNS compartments, histopathological grades, and age groups. Cancer Cell. 2015;27:728–43.
10. Witt H, Mack SC, Ryzhova M, et al. Delineation of two clinically and molecularly distinct subgroups of posterior fossa ependymoma. Cancer Cell. 2011;20:143–57.
11. Ghasemi DR, Sill M, Okonechnikov K, et al. MYCN amplification drives an aggressive form of spinal ependymoma. Acta Neuropathol. 2019;138(6):1075–89.
12. Fukuoka K, Kanemura Y, Shofuda T, Fukushima S, Yamashita S, Narushima D, Kato M, Honda-Kitahara M, Ichikawa H, Kohno T, Sasaki A, Hirato J, Hirose T, Komori T, Satomi K, Yoshida A, Yamasaki K, Nakano Y, Takada A, Nakamura T, Takami H, Matsushita Y, Suzuki T, Nakamura H, Makino K, Sonoda Y, Saito R, Tominaga T, Matsusaka Y, Kobayashi K, Nagane M, Furuta T, Nakada M, Narita Y, Hirose Y, Ohba S, Wada A, Shimizu K, Kurozumi K, Date I, Fukai J, Miyairi Y, Kagawa N, Kawamura A, Yoshida M, Nishida N, Wataya T, Yamaoka M, Tsuyuguchi N, Uda T, Takahashi M, Nakano Y, Akai T, Izumoto S, Nonaka M, Yoshifuji K, Kodama Y, Mano M, Ozawa T, Ramaswamy V, Taylor MD, Ushijima T, Shibui S, Yamasaki M, Arai H, Sakamoto H, Nishikawa R, Ichimura K, Japan Pediatric Molecular Neuro-Oncology Group (JPMNG). Significance of molecular classification of ependymomas: C11orf95-RELA fusion-negative supratentorial ependymomas are a heterogeneous group of tumors. Acta Neuropathol Commun. 2018;6(1):134.
13. Xi S, Sai K, Hu W, Wang F, Chen Y, Wang J, Zeng J, Chen Z. Clinical significance of the histological and molecular characteristics of ependymal tumors: a single institution case series from China. BMC Cancer. 2019;19(1):717.
14. Nambirajan A, Sharma A, Rajeshwari M, Boorgula MT, Doddamani R, Garg A, Suri V, Sarkar C, Sharma M. EZH2 inhibitory protein (EZHIP/Cxorf67) expression correlates strongly with H3K27me3 loss in posterior fossa ependymomas and is mutually exclusive with H3K27M mutations. Brain Tumor Pathol. 2020; https://doi.org/10.1007/s10014-020-00385-9.
15. Panwalkar P, Clark J, Ramaswamy V, Hawes D, Yang F, Dunham C, et al. Immunohistochemical analysis of H3K27me3 demonstrates global reduction in group-a childhood posterior fossa ependymoma and is a powerful predictor of outcome. Acta Neuropathol. 2017;134:705–14.
16. Nambirajan A, Gurung N, Suri V, Sarkar C, Kumar A, Singh M, Sharma MC. C19MC amplification and expression of Lin28A and Olig2 in the classification of embryonal tumors of the central nervous system: a 14-year retrospective study from a tertiary care center. Childs Nerv Syst. 2020; https://doi.org/10.1007/s00381-020-04973-0.
17. Hübner JM, Müller T, Papageorgiou DN, et al. EZHIP/CXorf67 mimics K27M mutated oncohistones and functions as an intrinsic inhibitor of PRC2 function in aggressive posterior fossa ependymoma. Neuro-Oncology. 2019;21:878–89.
18. Jain SU, Do TJ, Lund PJ, et al. PFA ependymoma-associated protein EZHIP inhibits PRC2 activity through a H3 K27M-like mechanism. Nat Commun. 2019;10:2146.
19. Kilday JP, Rahman R, Dyer S, Ridley L, Lowe J, Coyle B, Grundy R. Pediatric ependymoma: biological perspectives. Mol Cancer Res. 2009;7(6):765–86.
20. Merchant TE, Bendel AE, Sabin ND, Burger PC, Shaw DW, Chang E, Wu S, Zhou T, Eisenstat DD, Foreman NK, Fuller CE, Anderson ET, Hukin J, Lau CC, Pollack IF, Laningham FH, Lustig RH, Armstrong FD, Handler MH, Williams-Hughes C, Kessel S, Kocak M, Ellison DW,

Ramaswamy V. Conformal radiation therapy for pediatric ependymoma, chemotherapy for incompletely resected ependymoma, and observation for completely resected, supratentorial ependymoma. J Clin Oncol. 2019;37(12):974–83.
21. Ellison DW, Aldape KD, Capper D, et al. cIMPACT-NOW update 7: advancing the molecular classification of ependymal tumors [published online ahead of print, 2020 Jun 5]. Brain Pathol. 2020; https://doi.org/10.1111/bpa.12866.
22. Swanson AA, Raghunathan A, Jenkins RB, Messing-Jünger M, Pietsch T, Clarke MJ, Kaufmann TJ, Giannini C. Spinal cord ependymomas with MYCN amplification show aggressive clinical behavior. J Neuropathol Exp Neurol. 2019;78(9):791–7.
23. Sturm D, Orr BA, Toprak UH, Hovestadt V, Jones DTW, Capper D, Sill M, Buchhalter I, Northcott PA, Leis I, Ryzhova M, Koelsche C, Pfaff E, Allen SJ, Balasubramanian G, Worst BC, Pajtler KW, Brabetz S, Johann PD, Sahm F, Reimand J, Mackay A, Carvalho DM, Remke M, Phillips JJ, Perry A, Cowdrey C, Drissi R, Fouladi M, Giangaspero F, Łastowska M, Grajkowska W, Scheurlen W, Pietsch T, Hagel C, Gojo J, Lötsch D, Berger W, Slavc I, Haberler C, Jouvet A, Holm S, Hofer S, Prinz M, Keohane C, Fried I, Mawrin C, Scheie D, Mobley BC, Schniederjan MJ, Santi M, Buccoliero AM, Dahiya S, Kramm CM, von Bueren AO, von Hoff K, Rutkowski S, Herold-Mende C, Frühwald MC, Milde T, Hasselblatt M, Wesseling P, Rößler J, Schüller U, Ebinger M, Schittenhelm J, Frank S, Grobholz R, Vajtai I, Hans V, Schneppenheim R, Zitterbart K, Collins VP, Aronica E, Varlet P, Puget S, Dufour C, Grill J, Figarella-Branger D, Wolter M, Schuhmann MU, Shalaby T, Grotzer M, van Meter T, Monoranu CM, Felsberg J, Reifenberger G, Snuderl M, Forrester LA, Koster J, Versteeg R, Volckmann R, van Sluis P, Wolf S, Mikkelsen T, Gajjar A, Aldape K, Moore AS, Taylor MD, Jones C, Jabado N, Karajannis MA, Eils R, Schlesner M, Lichter P, von Deimling A, Pfister SM, Ellison DW, Korshunov A, Kool M. New brain tumor entities emerge from molecular classification of CNS-PNETs. Cell. 2016;164(5):1060–72.
24. Pajtler KW, Mack SC, Ramaswamy V, Smith CA, Witt H, Smith A, Hansford JR, von Hoff K, Wright KD, Hwang E, Frappaz D, Kanemura Y, Massimino M, Faure-Conter C, Modena P, Tabori U, Warren KE, Holland EC, Ichimura K, Giangaspero F, Castel D, von Deimling A, Kool M, Dirks PB, Grundy RG, Foreman NK, Gajjar A, Korshunov A, Finlay J, Gilbertson RJ, Ellison DW, Aldape KD, Merchant TE, Bouffet E, Pfister SM, Taylor MD. The current consensus on the clinical management of intracranial ependymoma and its distinct molecular variants. Acta Neuropathol. 2017;133(1):5–12.
25. Tanrıkulu B, Danyeli AE, Özek MM. Is H3K27me3 status really a strong prognostic indicator for pediatric posterior fossa ependymomas? A single surgeon, single center experience. Childs Nerv Syst. 2020;36:941–9..

Pathology and Molecular Pathology of Meningioma

5

Aruna Nambirajan, Jyostna Singh, and Vaishali Suri

5.1 Introduction

Meningiomas are a group of dural-based neoplasms that are postulated to arise from the meningothelial cells of the arachnoid layer. They are the most common primary brain tumors and account for ~36% of all brain tumors. The median age at occurrence is 65 years and are more common in females than males. They occur as intradural masses in intracranial, intraspinal, or intraorbital locations, and rarely in epidural and intraventricular locations. Intracranial meningiomas occur most commonly in the cerebral convexities, particularly in parasagittal locations, followed by olfactory groove, sphenoid ridges, tuberculum sellae, optic nerve sheath, petrous ridge, tentorium, and posterior fossa. Intraspinal meningiomas most commonly involve the thoracic spine [1].

5.2 Pathology

5.2.1 Gross Pathology

- Firm, lobulated masses with a broad area of dural attachment.
- Bone invasion is common and usually leads to reactive hyperostosis of the involved bone.
- Extracranial extension into scalp and brain parenchymal involvement may also be seen.
- En plaque pattern of spread, i.e., carpet-like growth pattern is sometimes seen in sphenoid wing meningiomas.

A. Nambirajan (✉) · J. Singh · V. Suri
Department of Pathology, AIIMS, New Delhi, India

S. Mallick et al. (eds.), *Evidence based practice in Neuro-oncology*,
https://doi.org/10.1007/978-981-16-2659-3_5

- Atypical and anaplastic meningiomas may show frank necrosis and are usually larger in size.

5.2.2 Histopathology and WHO Grade

- According to the WHO classification of central nervous system tumors, revised fourth edition [1], meningiomas correspond to WHO histological grades 1–3.
- Majority of meningiomas are WHO grade 1, being typically benign and slow growing.
- Approximately 10–25% of meningiomas belong to WHO grade 2 or 3 histology and are associated with atypical clinical outcomes in the form of frequent recurrences and rarely, distant extracranial metastases (most commonly to lung, pleura, bone, and liver).
- WHO classification recognizes 15 histopathological subtypes of meningiomas (Fig. 5.1; Table 5.1).
- Any histopathological variant exhibiting unequivocal brain invasion or increased mitotic activity >4 per 10 high-power fields or any three of the following five features, i.e., increased cellularity, sheet-like growth pattern, prominent nucleoli, spontaneous necrosis, and small cell change are assigned WHO grade 2.
- Any histopathological subtype with mitotic activity >20 per 10 high-power fields or with overt resemblance to a carcinoma, sarcoma, or melanoma are assigned as WHO grade 3.

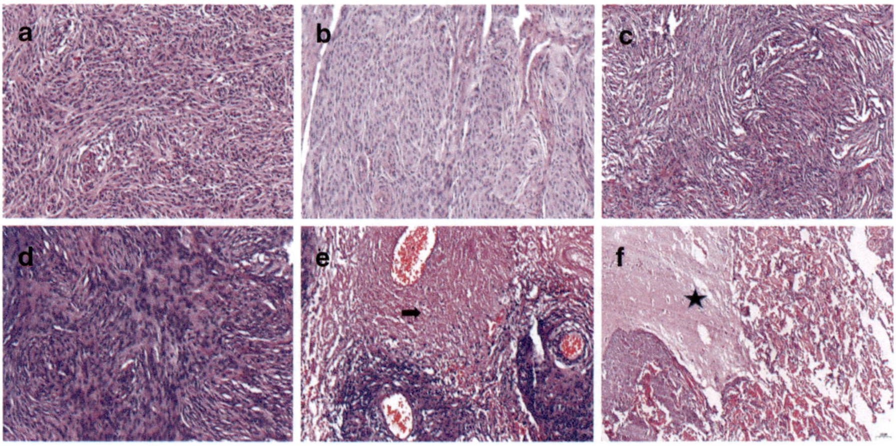

Fig. 5.1 Histopathological subtypes of meningiomas. Hematoxylin and eosin stained sections from a transitional meningioma (**a**), meningothelial meningioma (**b**), and fibroblastic meningioma (**c**). Atypical meningioma shows increased cellularity and increased mitoses (**d**). Anaplastic meningioma showing large areas of tumor necrosis (arrow) along with frequent mitoses and anaplasia (**e**). Brain invasion (star) in a meningioma (**f**)

Table 5.1 WHO classification of meningiomas [1]

Subtype	Grade	Histopathological characteristics	Clinicoradiological characteristics, if any
Meningothelial meningioma	1	Lobules of epithelioid meningothelial cells arranged in syncytial patterns; whorls and psammoma bodies infrequent	Common variant Frequent in anterior skull base
Fibrous meningioma	1	Spindled cells in fascicles in a collagenized matrix	Common variant Frequent in convexity
Transitional meningioma	1	Combination of meningothelial and fibrous patterns with frequent whorls and psammoma bodies	Common variant Frequent in convexity
Psammomatous meningioma	1	Predominance of psammoma bodies forming confluent masses frequently obscuring tumor cells	Uncommon Characteristically seen in the thoracic spine of middle aged to elderly females
Angiomatous meningioma	1	Prominent small- to medium-sized blood vessels with variable thickening and hyalinization of vessel wall obscuring tumor cells	Uncommon Peri-tumoral cerebral edema frequent
Microcystic meningioma	1	Thin elongated tumor cells with microcystic change and cobweb-like tumor appearance	Uncommon Peri-tumoral cerebral edema frequent
Secretory meningioma	1	Intracytoplasmic lumina formation in tumor cells with periodic acid Schiff-positive eosinophilic secretions and cytokeratin immunoreactivity in surrounding tumor cells	Uncommon Peri-tumoral cerebral edema frequent May be associated with elevated serum CEA levels
Lymphoplasmacyte-rich meningioma	1	Prominent lymphoplasmacytic infiltrate obscuring tumor cells	Rare Peri-tumoral cerebral edema prominent with leptomeningeal thickening May be associated with hypergammaglobulinemia or anemia
Metaplastic meningioma	1	Focal or widespread lipomatous, osseous, cartilaginous, or other mesenchymal differentiation	Uncommon
Chordoid meningioma	2	Predominated by cords or trabeculae of tumor cells in a myxoid/mucoid matrix	Rare May be associated with hematological manifestations such as Castleman disease

(continued)

Table 5.1 (continued)

Subtype	Grade	Histopathological characteristics	Clinicoradiological characteristics, if any
Clear cell meningioma	2	Predominated by clear glycogen rich round to polygonal tumor cells in sheets with prominent collagen deposition in perivascular and interstitial zones	Rare Children and young adults affected Propensity to involve the cerebellopontine angle and cauda equina
Atypical meningioma	2	Any meningioma with (1) brain invasion or (2) mitoses >4 per 10 HPF or (3) harboring three of the following five features: increased cellularity, sheet-like growth pattern, prominent nucleoli, spontaneous necrosis, and small cell change	Roughly 20–25% of meningiomas are grade 2 Higher rates in males Involve cerebral convexities and non-skull base locations Metastases in ~2% cases [2]
Papillary meningioma	3	Predominated by a perivascular pseudo-papillary pattern with a perivascular nucleus-free zone; frequently admixed with rhabdoid areas	Rare Young children High propensity for brain invasion, recurrence, lung metastases High risk of death
Rhabdoid meningioma	3	Predominated by the presence of rhabdoid phenotype cells with eccentric nucleus, prominent nucleoli, and paranuclear cytoplasmic inclusions; frequently admixed with papillary areas	Uncommon
Anaplastic meningioma	3	Any meningioma with mitotic activity >20 per 10 high-power fields or with overt resemblance to a carcinoma, sarcoma, or melanoma	Rare; 1–6% of all meningiomas Higher rates in males Involve cerebral convexities and non-skull base locations Metastases in ~9% cases [2]

Adapted from WHO classification of CNS tumors [1]

- Certain histopathological subtypes are assigned higher grades: clear cell and chordoid meningiomas are assigned grade 2 while papillary and rhabdoid meningiomas are assigned grade 3.
- Most common subtypes are the meningothelial, fibrous, and transitional meningiomas with the former predominating at skull base and the latter two predominating in cerebral convexities.

5.2.3 Immunophenotype

- Immunopositive for epithelial membrane antigen in a membranous pattern and vimentin, albeit with poor diagnostic specificity and sensitivity (EMA is frequently focal or absent in atypical or anaplastic meningiomas).
- Staining for somatostatin receptor 2A is highly sensitive diagnostic marker for meningiomas, positive in atypical and anaplastic examples as well, although is not specific being positive in neuroendocrine tumors as well [5].
- Progesterone receptor is frequently expressed in WHO grade 1 meningiomas, with increasing loss in atypical and anaplastic meningiomas [1].

5.2.4 Molecular Pathology

5.2.4.1 *NF2* Meningiomas

- Majority (>50%) of adult and pediatric meningiomas harbor monosomy 22q12.2 locus, harboring the *NF2* gene and inactivating *NF2* mutations, leading to loss of heterozygosity (LOH) of the tumor suppressor gene *NF2* that drives tumorigenesis.
- *NF2* encodes for merlin that functions as a membrane-cytoskeleton scaffolding protein involved in the regulation of multiple pathways, including the mammalian hippo, PI3K/mTORC1/Akt, and receptor dependent mitogenic signaling pathways.
- NF2 inactivation leads to dysregulation of the above-mentioned pathways resulting in increased proliferation.
- Increased expression of *FAK* (focal adhesion kinase) involved in invasion and migration has been noted in NF2 meningiomas and is a potential therapeutic target (NCT02523014).
- *NF2* LOH is observed in all grades of meningiomas with additional cytogenetic abnormalities such as loss of 1p, 6, 10q, 14q, 18, and 9p (harboring *CDKN2A/B* loci) seen mainly in higher grade meningiomas, indicating that while *NF2* LOH is a founder mutation/early event, progressive accumulation of additional cytogenetic events, in particular, *CDKN2A/B* (cyclin-dependent kinase inhibitor 2A/2B) homozygous deletion, contributes to tumor progression (Fig. 5.2a–f).
- Co-deletion of 1p and 14q has been associated with increased risk of recurrence in an otherwise grade 1 meningioma [3].
- *NF2*-meningiomas predominate in cerebral convexities and the most frequent histopathological subtypes are transitional, fibrous, and psammomatous meningiomas.
- NF2 inactivation is overrepresented in atypical and anaplastic meningiomas, and these tumors typically show the additional cytogenetic abnormalities described above.

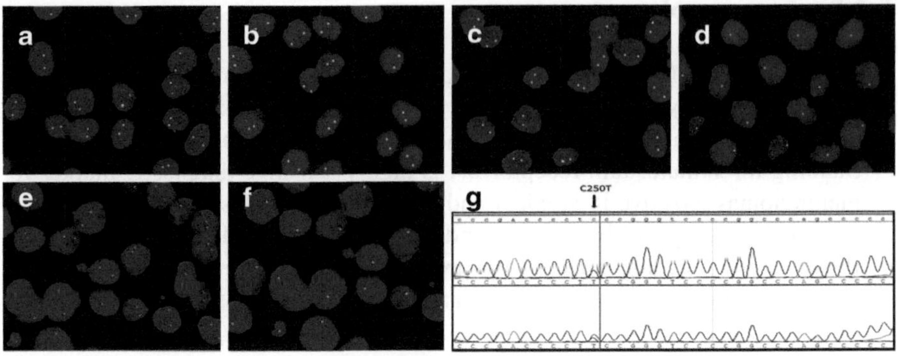

Fig. 5.2 Cytogenetic and molecular alterations of pathogenetic significance in meningiomas. Photomicrograph of florescence in situ hybridization analysis for 1p (**a**, **b**), 14q (**c**, **d**), and 22q (**e**, **f**). (**a**) Control showing normal 1p status: nuclei show two red (test) and two green (control) signals (1000×). (**b**) Case showing 1p deletion: Majority of nuclei show one red (test) and two green (control) signals (1000×). (**c**) Control showing normal 14p status: nuclei show two green (test) and two red (control) signals (1000×). (**d**) Case showing 14q deletion: Majority of nuclei show one green (test) and two red (control) signals (1000×); (**e**) Control showing normal 22q status: nuclei show two red (test) and two green (control) merged signals (1000×). (**f**) Case showing 22q deletion status: nuclei show one red (test) and one green (control) signal. (**g**) Electropherogram of TERT promoter mutation

5.2.4.2 Non-NF2 Meningiomas

- A number of recent studies have identified numerous genetic alterations in meningiomas lacking *NF2* alterations [11–14].
- *TRAF7* (tumor necrosis factor [TNF] receptor associated factor 7) mutations identified in 20% of meningiomas, are mutually exclusive with *NF2* alterations, often with concomitant mutations in either *AKT1* (v-Akt murine thymoma viral oncogene homolog1) or *KLF4* (Krüppel-like factor 4).
- *TRAF7*, an E3 ubiquitin ligase that interacts with MAPK pathway, is involved in modulation of NF-kB pathway and activation of apoptosis, and *TRAF7* mutations in meningiomas result in TRAF7 inactivation.
- AKT1 is a central component of the PI3K signaling pathway, and activating mutations in meningiomas (c.49G>A, $AKT1^{E17K}$) lead to constitutive PI3K signaling and cellular proliferation and are observed in 7–12% of all meningiomas with concomitant *TRAF7* mutations in half of them.
- *KLF4* mutations (c.1225A>C, $KLF4^{K409Q}$) are also activating in nature, leading to increased cellular proliferation, detected in 9–12% of meningiomas usually in association with *TRAF7* mutations or *PIK3CA* mutations and are mutually exclusive with *AKT1* mutations.
- *PIK3CA* (phosphatidylinositol-4,5-bisphosphate 3-kinase catalytic subunit alpha) activating mutations ($\mathit{PIK3CA}^{H1047R}$ and $\mathit{PIK3CA}^{E545K}$) result in AKT1 activation and are observed in 4–7% of meningiomas, mutually exclusive with *AKT1* mutations and commonly co-occurring with *KLF4* or *TRAF7* mutations;

they are overrepresented in meningiomas arising in patients receiving long-term progestins and high-dose androgen inhibitor therapies [4].

- *TRAF7, AKT1, KLF4,* and *PIK3CA* mutations are seen mainly in anterior and middle fossae skull base meningiomas and are almost always of grade 1 histology, and specific mutations profiles are characteristically of some histopathological subtypes: *TRAF7* + *KLF4* mutations in nearly 100% of secretory meningiomas and *AKT1* mutations in meningothelial and transitional meningiomas.
- Mutations in two related genes: *SMO* (smoothened frizzled class receptor) and *SUFU* (suppressor of fused homolog), regulators of the hedgehog pathway, are seen in 3–5% and 1% of all meningiomas, mutually exclusive with all other mutations and enriched in anterior fossa skull base grade 1 meningiomas; SMO also represents a potential therapeutic target under clinical evaluation (NCT02523014).
- Mutations in *SMO, AKT, KLF4,* and *TRAF7* (exon 17) have not been detected in pediatric meningiomas [18].
- Mutations in *POLR2A* (RNA polymerase II) observed in 6% of meningiomas and associates with grade 1 histology, skull base tuberculum sellae location, and meningothelial subtype.
- Loss of function germline or somatic mutations in *SMARCE1* gene, a core subunit of the mammalian SWI/SNF complex, is the genetic hallmark of clear cell meningiomas; inactivating mutations/deletions in another SWI/SNF subunit, PBRM1, has been observed in a high proportion of aggressive papillary meningiomas [17]; patients with germline mutations in *SMARCB2,* another core subunit of SWI/SNF complex, harbor multiple meningiomas and schwannomas; somatic *SMARCB2* mutations are rare and seen in some Grade 1 and 2 *NF2*-meningiomas.
- Loss of function mutations in *BAP1* (BRCA-associated protein 1), both germline and somatic, in association with chromosome 3p deletion leading to loss of BAP1 protein is exclusively seen in the aggressive rhabdoid meningiomas.
- Radiation-induced meningiomas show complex karyotypes, harbor lower frequencies of *NF2* mutations, and typically lack the non-*NF2* mutations seen in sporadic meningiomas; recent high-throughput studies have however discovered high incidence of structural aberrations (intronic gene rearrangements and fusions) involving the *NF2* locus [16].

5.2.4.3 Familial Syndromes

- Germline mutations in *NF2* (neurofibromatosis type 2), *SMARCE1* (familial syndrome with increased incidence of clear cell meningiomas), *SMARCB2* (rhabdoid tumor predisposition syndrome, Coffin–Siris syndrome, schwannomatosis), *BAP1* (BAP1 tumor predisposition syndrome), *PTCH1* (patched), and *SUFU* (Gorlin syndrome/nevoid basal cell carcinoma syndrome) predispose to increased incidence of meningiomas [14].

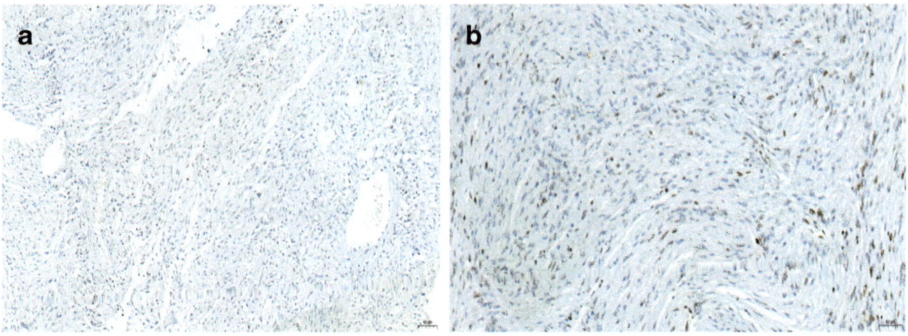

Fig. 5.3 H3K27me3 staining in meningiomas. Loss of trimethylation at position K27 of histone H3 in meningiomas (**a**, **b**) as manifested by loss of staining in tumor cells and retained staining of normal inflammatory cells and endothelial cells

5.2.4.4 TERT Promotor Mutations

- *TERT* promoter mutations (hotspot regions C228T and C250T) are seen in 6.5–11% of meningiomas (Fig. 5.2g), occur in both *NF2* and non-*NF2* meningiomas, and correlate with higher rates of progression and poor outcomes [6].

5.2.4.5 DMD Mutations

- Somatic intragenic deletions of the dystrophin-encoding and muscular dystrophy-associated DMD gene is enriched in progressed meningioma and correlates with poor outcomes [15].

5.2.4.6 Epigenetics

- Two major subgroups and six minor subgroups have been identified on DNA methylation profiling with all non-*NF2* meningiomas grouping into a single methylation subgroup with a benign outcome [9].
- Loss of trimethylation of histone H3 at position K27 (Fig. 5.3) has been observed in meningiomas with poor outcome [10].

5.3 Prognosis and Outcome

- Histological grade and extent of resection are the most important prognostic variables.
- WHO grade 1 meningiomas show a 5-year recurrence rate of 5–10%; WHO grade 2 meningiomas show recurrence rates of 50% at 5 years, while WHO grade 3 meningiomas invariably recur [1, 13].
- Simpson grades I–III resection (gross total) associates with 10-year recurrence rates of 9–29% as opposed to 44–100% recurrence rates of meningiomas undergoing excisions of Simpson grades IV or V (subtotal) [1, 13].
- Skull base meningiomas, despite commonly being of WHO grade 1 histology, are prone to recurrence due to higher rates of subtotal excisions.

- Ki-67 staining identifies hot spots for mitotic counts; higher proliferation indices associate with higher grades although no specific cut-offs are validated.
- Microscopic brain invasion is associated increased risk of recurrence and is a criterion for WHO grade 2 meningioma diagnosis [7].
- Bone invasion has been associated with worsened outcome in grade 2 meningiomas [8].
- Cytogenetic alterations such as 1p and 14q deletions, *TERT* promotor mutations, homozygous deletions of *CDKN2A/B*, and *DMD* mutations are enriched in recurrent and progressed meningiomas.
- *PBRM1* and *BAP1* mutations are enriched in papillary and rhabdoid meningiomas and correlate with aggressive biological behavior.

5.4 Future Perspectives

Although histological grade and extent of resection still remain the main prognostic variables determining treatment and outcome, recent years have seen a tremendous increase in our understanding of the molecular pathogenesis of meningiomas. In particular, clinical trials are underway evaluating inhibitors to FAK and SMO in *NF2* and non-*NF2* meningiomas, respectively. The routine diagnosis of meningiomas is still based on histopathology. In cases with diagnostic confusion with other tumors, the diagnosis can be supported by DNA methylation, mutation analysis for the classic drivers: *NF2*, *TRAF7*, *AKT1*, *KLF4*, *SMO*, *PIK3CA*, *SMARCE1*, or *BAP1* (SMARCE1 and BAP1 protein loss in tumor cells can be surrogate immunohistochemical markers for *SMARCE1* and *BAP1* inactivation), or cytogenetic abnormalities such as 22q monosomy, 1p and 14q deletions. *TERT* promoter mutations and *CDKN2A* homozygous deletions identify meningiomas with increased rates of progression to higher grade and supplement histological grade in accurate prognostication. Routine identification of these established molecular and cytogenetic markers will enable inclusion in clinical trials, confirmation of the prognostic significance, and development of better therapies.

References

1. Perry A, Louis DN, Budka H, von Deimling A, Sahm F, Rushing EJ, Mawrin C, Claus EB, Loeffler J, Sadetzki S. Meningioma. In: Louis DN, Ohgaki H, Wiestler OD, Cavenee WK, editors. WHO classification of tumours of the central nervous system, revised. 4th ed. Lyon: IARC; 2016. p. 232–7.
2. Dalle Ore CL, Magill ST, Yen AJ, Shahin MN, Lee DS, Lucas CG, Chen WC, Viner JA, Aghi MK, Theodosopoulos PV, Raleigh DR, Villanueva-Meyer JE, McDermott MW. Meningioma metastases: incidence and proposed screening paradigm. J Neurosurg. 2019;5:1–9. https://doi.org/10.3171/2019.1.JNS181771.
3. Kakkar A, Kumar A, Das A, Pathak P, Sharma MC, Singh M, Suri A, Sarkar C, Suri V. 1p/14q co-deletion: a determinant of recurrence in histologically benign meningiomas. Indian J Pathol Microbiol. 2015;58(4):433–8.

4. Peyre M, Gaillard S, de Marcellus C, et al. Progestin-associated shift of meningioma mutational landscape. Ann Oncol. 2018;29(3):681–6.
5. Menke JR, Raleigh DR, Gown AM, Thomas S, Perry A, Tihan T. Somatostatin receptor 2a is a more sensitive diagnostic marker of meningioma than epithelial membrane antigen. Acta Neuropathol. 2015;130(3):441–3.
6. Gautagny S, Nault JC, Mallet M, et al. High incidence of activating TERT promoter mutations in meningiomas undergoing malignant progression. Brain Pathol. 2014;24(2):184–9.
7. Perry A, Scheithauer BW, Stafford SL, et al. "Malignancy" in meningioms: a clinicopathologic study of 116 patients, with grading implications. Cancer. 1999;85(9):2046–56.
8. Gabeau-Lacet D, Aghi M, Betensky RA, Barker FG, Loeffler JS, Louis DN. Bone involvement predicts poor outcome in atypical meningioma. J Neurosurg. 2009;111(3):464–71.
9. Sahm F, Schrimpf D, Stichel D, Jones DTW, Hielscher T, Schefzyk S, Okonechnikov K, Koelsche C, Reuss DE, Capper D, Sturm D, Wirsching HG, Berghoff AS, Baumgarten P, Kratz A, Huang K, Wefers AK, Hovestadt V, Sill M, Ellis HP, Kurian KM, Okuducu AF, Jungk C, Drueschler K, Schick M, Bewerunge-Hudler M, Mawrin C, Seiz-Rosenhagen M, Ketter R, Simon M, Westphal M, Lamszus K, Becker A, Koch A, Schittenhelm J, Rushing EJ, Collins VP, Brehmer S, Chavez L, Platten M, Hänggi D, Unterberg A, Paulus W, Wick W, Pfister SM, Mittelbronn M, Preusser M, Herold-Mende C, Weller M, von Deimling A. DNA methylation-based classification and grading system for meningioma: a multicentre, retrospective analysis. Lancet Oncol. 2017;18(5):682–94.
10. Katz LM, Hielscher T, Liechty B, Silverman J, Zagzag D, Sen R, Wu P, Golfinos JG, Reuss D, Neidert MC, Wirsching HG, Baumgarten P, Herold-Mende C, Wick W, Harter PN, Weller M, von Deimling A, Snuderl M, Sen C, Sahm F. Loss of histone H3K27me3 identifies a subset of meningiomas with increased risk of recurrence. Acta Neuropathol. 2018;135(6):955–63.
11. Clark VE, Erson-Omay EZ, Serin A, Yin J, Cotney J, Ozduman K, Avşar T, Li J, Murray PB, Henegariu O, Yilmaz S, Günel JM, Carrión-Grant G, Yilmaz B, Grady C, Tanrikulu B, Bakircioğlu M, Kaymakçalan H, Caglayan AO, Sencar L, Ceyhun E, Atik AF, Bayri Y, Bai H, Kolb LE, Hebert RM, Omay SB, Mishra-Gorur K, Choi M, Overton JD, Holland EC, Mane S, State MW, Bilgüvar K, Baehring JM, Gutin PH, Piepmeier JM, Vortmeyer A, Brennan CW, Pamir MN, Kiliç T, Lifton RP, Noonan JP, Yasuno K, Günel M. Genomic analysis of non-NF2 meningiomas reveals mutations in TRAF7, KLF4, AKT1, and SMO. Science. 2013;339(6123):1077–80.
12. Brastianos PK, Horowitz PM, Santagata S, Jones RT, McKenna A, Getz G, Ligon KL, Palescandolo E, Van Hummelen P, Ducar MD, Raza A, Sunkavalli A, Macconaill LE, Stemmer-Rachamimov AO, Louis DN, Hahn WC, Dunn IF, Beroukhim R. Genomic sequencing of meningiomas identifies oncogenic SMO and AKT1 mutations. Nat Genet. 2013;45(3):285–9.
13. Suppiah S, Nassiri F, Bi WL, Dunn IF, Hanemann CO, Horbinski CM, Hashizume R, James CD, Mawrin C, Noushmehr H, Perry A, Sahm F, Sloan A, Von Deimling A, Wen PY, Aldape K, Zadeh G. International consortium on meningiomas. Molecular and translational advances in meningiomas. Neuro-Oncology. 2019;21(Suppl 1):i4–i17. https://doi.org/10.1093/neuonc/noy178.
14. Lee YS, Lee YS. Molecular characteristics of meningiomas. J Pathol Transl Med. 2020;54(1):45–63.
15. Juratli TA, McCabe D, Nayyar N, Williams EA, Silverman IM, Tummala SS, Fink AL, Baig A, Martinez-Lage M, Selig MK, Bihun IV, Shankar GM, Penson T, Lastrapes M, Daubner D, Meinhardt M, Hennig S, Kaplan AB, Fujio S, Kuter BM, Bertalan MS, Miller JJ, Batten JM, Ely HA, Christiansen J, Baretton GB, Stemmer-Rachamimov AO, Santagata S, Rivera MN, Barker FG 2nd, Schackert G, Wakimoto H, Iafrate AJ, Carter SL, Cahill DP, Brastianos PK. DMD genomic deletions characterize a subset of progressive/higher-grade meningiomas with poor outcome. Acta Neuropathol. 2018;136(5):779–92.
16. Agnihotri S, Suppiah S, Tonge PD, Jalali S, Danesh A, Bruce JP, Mamatjan Y, Klironomos G, Gonen L, Au K, Mansouri S, Karimi S, Sahm F, von Deimling A, Taylor MD, Laperriere NJ, Pugh TJ, Aldape KD, Zadeh G. Therapeutic radiation for childhood cancer drives structural aberrations of NF2 in meningiomas. Nat Commun. 2017;8(1):186.

17. Williams EA, Wakimoto H, Shankar GM, Barker FG 2nd, Brastianos PK, Santagata S, Sokol ES, Pavlick DC, Shah N, Reddy A, Venstrom JM, Alexander BM, Ross JS, Cahill DP, Ramkissoon SH, Juratli TA. Frequent inactivating mutations of the PBAF complex gene PBRM1 in meningioma with papillary features. Acta Neuropathol. 2020;140(1):89–93.
18. Battu S, Kumar A, Pathak P, Purkait S, Dhawan L, Sharma MC, Suri A, Singh M, Sarkar C, Suri V. Clinicopathological and molecular characteristics of pediatric meningiomas. Neuropathology. 2018;38(1):22–33.

Pathology and Molecular Biology of Medulloblastoma

6

Kavneet Kaur and Chitra Sarkar

6.1 Introduction

Medulloblastoma (MB) is the most common embryonal tumor of central nervous system which corresponds to WHO grade IV. It is the most common malignant brain tumor in pediatric age group accounting for approximately 25% of all childhood intracranial neoplasms. Almost one fourth of all MBs occur in adults; however, they form <1% of all intracranial neoplasms in adults. MBs arise from cerebellum and less commonly from dorsal brain stem [1].

Medulloblastomas show substantial heterogeneity at histological as well as molecular levels. According to the current consensus and WHO classification of CNS tumors, MBs should be classified according to both histological and molecular features. Molecular classification is based on clustering of MBs by transcriptome profiling and more recently by DNA methylation profiling [2, 3].

Medulloblastomas can develop in the setting of various inherited cancer syndromes with germline mutations of PTCH1 and SUFU (nevoid basal cell carcinoma syndrome/Gorlin syndrome), APC (Familial adenomatous polyposis), TP53 (Li–Fraumeni syndrome), mismatch repair cancer syndrome, Rubinstein–Taybi syndrome, and so on [4].

6.2 Clinical Profile

- Age and sex distribution: Peak incidence at 3 and 7 years of age with median age of 9 years with a male preponderance (Male:Female = 1.7:1).
- Location: Most MBs arise from cerebellum or rarely from dorsal brain stem and grow in the fourth ventricle while some arise laterally in the cerebellar hemispheres.

K. Kaur (✉) · C. Sarkar
Department of Pathology, AIIMS, New Delhi, India

S. Mallick et al. (eds.), *Evidence based practice in Neuro-oncology*,
https://doi.org/10.1007/978-981-16-2659-3_6

- Clinical features: The tumors which grow in the fourth ventricle block cerebrospinal fluid pathways and cause an increase in intracranial pressure (ICP). Hence, patients usually present with symptoms of raised ICP including severe headaches and sudden bouts of vomiting. Tumors arising in the cerebellum cause cerebellar ataxia which is also a common symptom. Rarely, patients can present with symptoms of cranial nerve compression.
- Spread: MBs have a tendency to spread through cerebrospinal fluid and metastasize along the neuraxis. Rarely, metastasis outside CNS has also been reported especially to bones and lymphatic system.

6.3 Integrated Diagnosis

The updated classification provides optimal prognostic and predictive information, hence, an integrated approach is recommended whenever feasible. An integrated diagnosis is multi-layered which includes histological typing, grading according to WHO classification, molecular subgrouping, and further genetic information which has relevant clinical utility [5].

6.4 Medulloblastomas, Histologically Defined

MB is an embryonal tumor comprising of small, round, blue, poorly differentiated tumor cells with a high nuclear to cytoplasmic ratio and scant cytoplasm with high proliferative indices including frequent mitosis, apoptosis, and karyorrhexis. Differential diagnosis include other embryonal tumors including embryonal tumor with multi-layered rosettes which show characteristic LIN28A immunopositivity, atypical teratoid/rhabdoid tumors which are identified by loss of INI1 immunoexpression, small cell glioblastoma (glial fibrillary acidic protein IHC positive), etc.

The four histological variants of medulloblastoma include classic, desmoplastic/nodular, medulloblastoma with extensive nodularity and large cell/anaplastic [6].

6.4.1 Classic Medulloblastoma (Fig. 6.1a)

- Classic morphology is the most common and accounts for approximately 70% of all MBs.
- Can occur in any age group from infancy to adulthood.
- Midline location in cerebellum or dorsal brain stem.
- Archetypal small round blue cell, highly cellular densely packed tumor.
- Lacks significant desmoplasia.
- Homer Wright rosettes seen in ~40% cases.
- Can belong to any molecular group.
- Prognosis: variable depending on the molecular group.

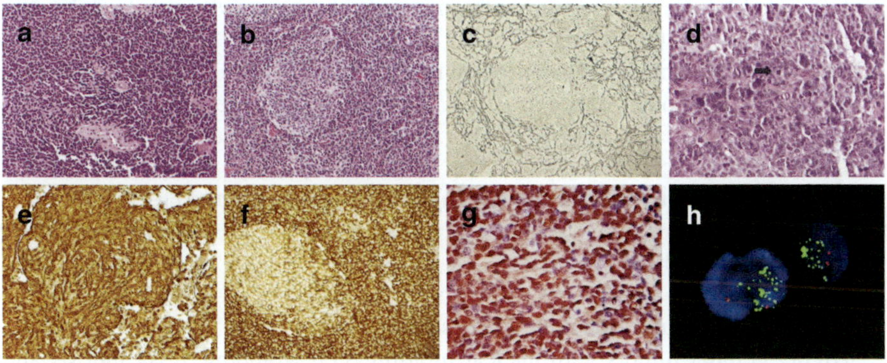

Fig. 6.1 (**a**–**d**) Histopathological types of medulloblastoma: Classic MB with characteristic small round blue carrot shaped cells in sheets (**a**); Desmoplastic/nodular medulloblastoma with a pale nodule with neurocytic differentiation which is reticulin poor (**c**) surrounded by cellular areas with tumor cells having high nuclear:cytoplasmic ratio and scant cytoplasm and are reticulin rich (**c**) (**h** and **e**, **b**; Reticulin, **c**); Large cell/anaplastic medulloblastoma comprising of tumor cells with marked nuclear pleomorphism and brisk mitosis (arrow). (**e**–**f**) Molecular subgrouping using immunohistochemistry and fluorescence in-situ hybridization: A WNT-activated tumor showing nuclear positivity for beta-catenin (**e**); Strong and diffuse cytoplasmic positivity of GAB1 in SHH subgroup tumor (**f**); Nuclear immunopositivity for YAP1 in SHH-activated MB (**g**); MYC amplification by FISH showing numerous green signals (MYC) in contrast to two orange signals (CEP 8) in a Group 3 MB

6.4.2 Desmoplastic/Nodular (D/N) Medulloblastoma (Fig. 6.1b, c)

- Constitutes ~20% of all MBs.
- Most common in infants (<3 years) and adults.
- Lateral location in the cerebellar hemispheres.
- D/N is a biphasic tumor with uniform pale nodules surrounded by poorly differentiated cells.
  - Nodules show neurocytic differentiation and lower proliferative indices, and are reticulin poor.
  - Internodular zones are densely packed, with high mitotic count and Ki67 proliferative index; dense desmoplasia as demonstrated by reticulin-positive collagen fibers is seen.
- Almost all D/N align to SHH group.
- Favorable clinical outcome in infants; intermediate in other age groups.

6.4.3 Medulloblastoma with Extensive Nodularity (MBEN)

- Accounts for ~3–4% of all MBs.
- Almost exclusively seen in infants (<3 years).
- Midline location with characteristic grape-like morphology on imaging.

- Expanded lobular architecture with large irregular neurocytic pale nodules and narrow internodular regions.
- Best prognosis among all histological types.
- Molecular subgroup: SHH.

6.4.4 Large Cell/Anaplastic (LC/a) Medulloblastoma (Fig. 6.1d)

- Accounts for ~10% of all MBs.
- Can occur in any age group.
- Midline location in cerebellum or dorsal brain stem.
- Highly cellular tumors with nuclear molding, cell wrapping, brisk mitosis and nuclear anaplasia and enlargement.
- Worst prognosis among all histological variants.
- Molecular group: Group 3 and SHH activated, TP53 mutant tumors.

6.4.5 Patterns

Any of the histological types can rarely show melanotic differentiation or myogenic differentiation and then the terms "Melanocytic medulloblastoma" or "Medullomyoblastoma" are used [7, 8].

6.5 Medulloblastomas, Molecularly Defined

The WHO 2016 classification divides medulloblastoma into three major molecular groups: WNT, SHH, and Non-WNT/non-SHH (Fig. 6.2) [6, 9].

6.5.1 Medulloblastoma, WNT-Activated

Embryonal tumors which arise from cerebellum or dorsal brain stem and show Wingless signaling pathway activation. Around 10% of all MBs are WNT-activated. More than 85% of cases show mutations in exon 3 of CTNNB1 or monosomy of chromosome 6.

- Peak incidence in children between 7 and 14 years; rare in infants and adults.
- Location: Midline, or with extension to cerebellar peduncle/cerebellopontine angle.
- Imaging: Many of the tumors reveal close contact to brain stem [10].
- Histological type: Mostly classic; rarely LC/A.
- Diagnosis:
  - Immunohistochemistry (IHC): Nuclear positivity for beta-catenin (Fig. 6.1e); focal for YAP1 while immunonegative for GAB1.
  - Sequencing for CTNNB1 gene for mutations in exon 3.

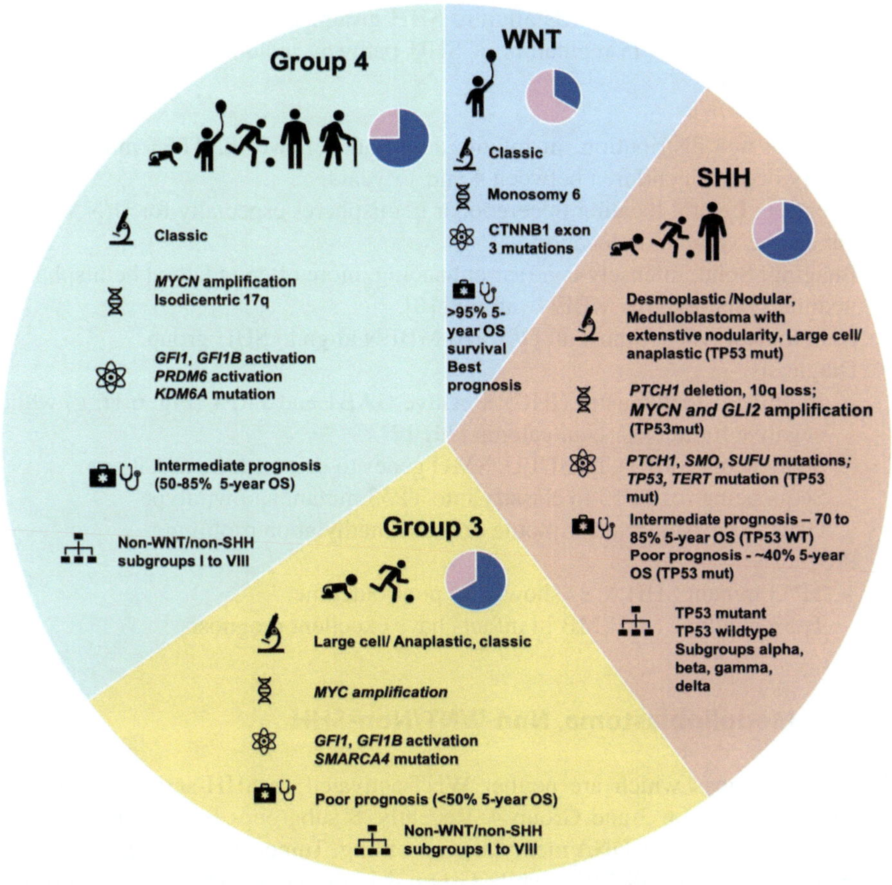

Fig. 6.2 Clinical and molecular profile of different molecular groups of medulloblastoma; Most common histopathological type; Frequent copy number alterations; Frequent genetic alterations; Clinical outcome; Subgroups within molecular groups

  - Demonstration of monosomy 6 by fluorescence in-situ hybridization (FISH) or copy number profiling.
  - Gene expression/transcriptome or DNA methylation profiling.
- Prognosis: Excellent prognosis with overall survival close to 100%; however, the prognosis of WNT tumors in adults is not as favorable [11, 12].

6.5.2 Medulloblastoma, SHH-Activated

Medulloblastomas which arise from cerebellum and show evidence of SHH signaling pathway activation. SHH MBs are further divided into TP53 mutant and TP53 wildtype depending on presence or absence of germline or somatic TP53 mutations.

Approximately 30% of all MBs align to SHH group. TP53 wildtype characteristically show alterations in regulators of SHH pathway including PTCH, SUFU, or SMO genes [13].

- Bimodal age distribution, in infants and adults; however, TP53 mutant commonly occurs in children between 4 and 17 years.
- Location: Lateral location in cerebellar hemispheres especially for D/N variant, but can be vermian in location.
- Imaging: Solid, intensely contrast enhancing, more often in lateral hemispheres, nodular grape-like for MBEN variant [10].
- Histological type: Almost all D/N and MBEN align to SHH group.
- Diagnosis:
  - Immunohistochemistry (IHC): Positive GAB1 and YAP1 (Fig. 6.1f, g) while negative for nuclear beta-catenin [11, 12].
  - Sequencing for PTCH, SUFU, SMO genes to demonstrate mutations.
  - Sequencing for TP53 to classify into TP53 mutant and wildtype.
  - Gene expression/transcriptome or DNA methylation profiling.
- Prognosis:
  - TP53 mutant SHH MBs show very poor outcome.
  - Tp53 wildtype SHH MB in infants have excellent prognosis.

6.5.3 Medulloblastoma, Non-WNT/Non-SHH

Medulloblastomas which are neither WNT-activated or SHH-activated, and are classified into Group 3 and Group 4. Recently, 8 subgroups have been identified which are delineated by DNA methylation profiling. Tumors which align to group 3 account for ~25% of all MBs, while Group 4 is the largest molecular group and accounts for approximately 40% of all MBs. Group 3 tumors have MYC amplification in ~20% of cases. Group 4 tumors have recurrent alterations in KDM6A, SNCAIP, etc.

- A significant number of Group 3 MBs are seen in infants and are extremely rare in adults; Group 4 occurs in all age groups with peak incidence between 5 and 15 years [9, 14].
- Location: Both groups occur in midline in inferior portion of cerebellum.
- Metastatic disease is present at the time of diagnosis in ~40% of Group 3 MBs.
- Recurrence of non-WNT/non-SHH usually occurs as leptomeningeal metastasis [15].
- Histological type: Mostly classic; LC/A predominantly in Group 3 [6].
- Diagnosis:
  - Immunohistochemistry (IHC): Negative for all immunohistochemical markers including nuclear beta-catenin, GAB1, and YAP 1 [11, 12].

- MYC by fluorescence in-situ hybridization (FISH) for MYC amplification.
- Gene expression/transcriptome or DNA methylation profiling.
- Prognosis:
  - Group 3 with MYC amplification: Very poor prognosis.
  - Group 4 tumors: Intermediate prognosis.
  - Group 4 tumors with loss of chr 11: Better prognosis [16].

6.6 Risk Stratification

6.6.1 Clinical Risk Stratification: Chang Staging System

- Classify patients into high and average risk groups based on age, extent of surgical resection, and metastatic status.
- High risk is assigned if the age of the patients is <3 years (infants), or there is leptomeningeal dissemination at the time of diagnosis or if there is residual tumor ($\geq$1.5 cm^2) post-surgical resection.

6.6.2 Refined Risk Stratification [17]

Combination of both histopathological types and molecular groups.

6.6.2.1 According to WHO 2016 Classification [1]

Combination of clinical, histological, and molecular profiles has been used to refine risk stratification which divides risk groups according to survival:

- Low risk group (>90% survival)
  - Non-metastatic WNT tumors
  - Non-metastatic average risk Group 4 tumors with loss of chr 11 or whole gain of chr 17
- Standard risk group (75–90% survival)
  - Non-metastatic SHH, TP53 WT, non MYCN amplified tumors
  - Non-metastatic Group 3, non MYC amplified tumors
  - Non-metastatic Group 4 tumors without loss of chr 11
- High risk group (50–75% survival)
  - Metastatic or non-metastatic SHH, MYCN amplified
  - Metastatic SHH, TP53 WT
- Very high risk group (<50% survival)
  - Metastatic or non-metastatic SHH, TP53 mutant
  - Metastatic or non-metastatic Group 3, MYC amplified
  - Metastatic Group 3 tumors

6.6.3 Benefits of Risk Stratification

- Low risk tumors: De-escalation of radiotherapy/chemotherapy or less aggressive surgery
- High and very high risk tumors: More aggressive therapy with addition of chemotherapeutic drugs
- Targeted therapy, for example, small molecule inhibitors against smoothened (SMO) homologue which is SHH pathway co-receptor like vismodegib

6.7 Management

Some of the ongoing Phase I and II clinical trials include PNET5 (NCT02066220), SJMB12 ((NCT01878617), and ACNS1422 (NCT02724579), which are offering de-escalation of radiotherapy (craniospinal irradiation) for classical WNT cases, addition of carboplatin in non-WNT/non-SHH high risk cases (in PNET5 trial), and addition of alkylating chemotherapy in very high risk cases. In SJMB12, they are investigating role of pemetrexed and gemcitabine in high risk Group 3 or group 4 MBs. In addition, targeted therapy against SHH pathway inhibitors is also being tried [18, 19].

6.8 Approach to Diagnosis

Immunohistochemistry: A panel of three immunohistochemical markers can be used to classify MBs into WNT, SHH, and non-WNT/non-SHH. Nuclear positivity for beta-catenin characterizes WNT tumors, they also show focal expression of YAP1. SHH tumors show cytoplasmic expression of GAB1 and nuclear positivity for YAP1 while non-WnT/non-SHH tumors are immunonegative for all three markers [6, 11, 12].

Fluorescence in-situ *hybridization:* Many genetic and cytogenetic events which have diagnostic and prognostic value within the main molecular groups can be detected by FISH which can supplement the molecular subgrouping by immunohistochemistry [16].

- WNT: monosomy 6 for confirmation of WNT activation.
- SHH: NMYC and GLI2 amplification confers worse prognosis.
- Non-WNT/non-SHH: MYC amplification in Group 3 associated with a worse prognosis.

Gene Sequencing by Sanger sequencing: Single hotspot mutation testing by Sanger sequencing can be used to supplement immunohistochemistry.

- WNT: CTNNB1 exon 3 mutations for confirmation of WNT signature.
- SHH: PTCH/SMO/ SUFU for confirmation of SHH pathway activation; TP53 mutations for segregation of SHH group into TP53 mutant and wildtype.

Nanostring Assay: A high throughput method that uses RNA derived from formalin-fixed paraffin-embedded tissue (FFPE) to analyze 22 subgroup-specific genes and classifies MB into WNT, SHH, Group 3, and Group 4 [13, 20].

Gene expression profiling: Transcriptome profiling is the technique which first identified the molecular groups and is the gold standard; however, it requires fresh-frozen tissue and is quite expensive [13].

DNA methylation profiling by 850k methylation array is the gold standard and the best high throughput method which precisely identifies all molecular groups and subgroups and can be used on DNA derived from fresh-frozen tissue or FFPE tissue [2, 20].

The choice of method is left to the discretion of the neuropathologist depending on availability of resources, economic viability, and technical expertise.

6.9 Future Perspectives

Subgroups within molecular groups have been delineated by DNA methylation or transcriptome profiling which have distinct clinical profile including outcome [21]. Within WNT group, two preliminary subgroups have been identified, viz. WNT-α and WNT-β. Tumors which belong to WNT-α are the more common, arise in children and have monosomy 6, while the WNT-β tumors are seen in older children and young adults and do not show monosomy 6. SHH tumors axiomatically segregate into 4 subgroups which have been tentatively named SHH-alpha, beta, gamma, and delta. Beta and gamma subgroups occur predominantly in infants, desmoplastic/nodular histology dominates in these subgroups with MBEN seen in SHH-gamma tumors and presence of diagnostic PTCH/SUFU/SMO mutations/deletions. SHH-alpha occurs in older children and adolescents, harbors a preponderance of TP53 mutations along with MYCN and/or GLI2 amplification, shows evidence of chromothripsis, and LC/A histology is most commonly seen in this group. It is associated with worse prognosis. SHH-delta is identified more commonly in adults, and show D/N more than classic variant histology, and harbour PTCH or SMO alterations. These patients being adults with SHH activated tumors are excellent candidates for SMO inhibitors as SHH signaling is essential during development and hence, not recommended in skeletally immature patients.

Non-WNT/non-SHH group can be delineated into 8 subgroups which have been provisionally termed G3/4-I to VIII in a large meta-analysis. Subgroups II/III have the worst outcome, MYC amplification, and more LC/A histology [22].

References

1. Ellison DW, Eberhart CG, Pietsch T. Embryonal tumors. In: Louis DN, Ohgaki H, Wiestler OD, Cavenee WK, editors. WHO classification of tumors of the central nervous system, Revised. 4th ed. Lyon: IARC; 2016. p. 184–200.
2. Capper D, Jones DTW, Sill M, et al. DNA methylation-based classification of central nervous system tumours. Nature. 2018;555:469–74.
3. Taylor MD, Northcott PA, Korshunov A, et al. Molecular subgroups of medulloblastoma: the current consensus. Acta Neuropathol. 2012;123:465–72.

4. Waszak SM, Northcott PA, Buchhalter I, et al. Spectrum and prevalence of genetic predisposition in medulloblastoma: a retrospective genetic study and prospective validation in a clinical trial cohort. Lancet Oncol. 2018;19:785–98.
5. Louis DN, Perry A, Burger P, et al. International society of neuropathology-Haarlem consensus guidelines for nervous system tumor classification and grading. Brain Pathol. 2014;24:429–35.
6. Ellison DW, Dalton J, Kocak M, et al. Medulloblastoma: clinicopathological correlates of SHH, WNT, and non-SHH/WNT molecular subgroups. Acta Neuropathol. 2011;121:381–96.
7. Phillips CL, Miles L, Jones BV, et al. Medulloblastoma with melanotic differentiation: case report and review of the literature. J Neuro-Oncol. 2011;103:759–64.
8. Rajeshwari M, Kakkar A, Nalwa A, et al. WNT-activated medulloblastoma with melanotic and myogenic differentiation: report of a rare case. Neuropathology. 2016;36:372–5.
9. Kool M, Korshunov A, Remke M, et al. Molecular subgroups of medulloblastoma: an international meta-analysis of transcriptome, genetic aberrations, and clinical data of WNT, SHH, Group 3, and Group 4 medulloblastomas. Acta Neuropathol. 2012;123:473–84.
10. Perreault S, Ramaswamy V, Achrol AS, et al. MRI surrogates for molecular subgroups of medulloblastoma. AJNR Am J Neuroradiol. 2014;35:1263–9.
11. Kaur K, Kakkar A, Kumar A, et al. Integrating molecular subclassification of medulloblastomas into routine clinical practice: a simplified approach. Brain Pathol. 2016;26:334–43.
12. Kaur K, Kakkar A, Kumar A, et al. Clinicopathological characteristics, molecular subgrouping, and expression of miR-379/miR-656 cluster (C14MC) in adult medulloblastomas. J Neuro-Oncol. 2016;130:423–30.
13. Northcott PA, Shih DJ, Peacock J, et al. Subgroup-specific structural variation across 1000 medulloblastoma genomes. Nature. 2012;488:49–56.
14. Remke M, Hielscher T, Northcott PA, et al. Adult medulloblastoma comprises three major molecular variants. J Clin Oncol. 2011;29:2717–23.
15. Ramaswamy V, Remke M, Bouffet E, et al. Recurrence patterns across medulloblastoma subgroups: an integrated clinical and molecular analysis. Lancet Oncol. 2013;14:1200–7.
16. Shih DJ, Northcott PA, Remke M, et al. Cytogenetic prognostication within medulloblastoma subgroups. J Clin Oncol. 2014;32:886–96.
17. Ramaswamy V, Remke M, Bouffet E, et al. Risk stratification of childhood medulloblastoma in the molecular era: the current consensus. Acta Neuropathol. 2016;131:821–31.
18. Bautista F, Fioravantti V, de Rojas T, et al. Medulloblastoma in children and adolescents: a systematic review of contemporary phase I and II clinical trials and biology update. Cancer Med. 2017;6:2606–24.
19. Robinson GW, Orr BA, Wu G, et al. Vismodegib exerts targeted efficacy against recurrent sonic hedgehog-subgroup medulloblastoma: results from phase ii pediatric brain tumor consortium studies PBTC-025B and PBTC-032. J Clin Oncol. 2015;33:2646–54.
20. Kaur K, Jha P, Pathak P, et al. Approach to molecular subgrouping of medulloblastomas: comparison of NanoString nCounter assay versus combination of immunohistochemistry and fluorescence in-situ hybridization in resource constrained centres. J Neuro-Oncol. 2019;143:393–403.
21. Cavalli FMG, Remke M, Rampasek L, et al. Intertumoral heterogeneity within medulloblastoma subgroups. Cancer Cell. 2017;31:737–54.e6.
22. Sharma T, Schwalbe EC, Williamson D, et al. Second-generation molecular subgrouping of medulloblastoma: an international meta-analysis of Group 3 and Group 4 subtypes. Acta Neuropathol. 2019;138:309–26.

Part II

Clinical Examination, Neuro-Imaging and Basics of Radiotherapy in Neuro-Oncology

Radiology in Modern Neuro-Oncology Practice

7

Ashima Chander and S. Leve Joseph Devarajan

Part I: Overview of Diagnostic Modalities and Basic Neuroanatomy

Knowledge of anatomy and familiarity with the principles of image formation in various modalities are two basic requirements, among other things, for understanding images in radiology and deciphering clinically useful information out of them. A description of physical principles of various imaging modalities is beyond the scope of this chapter. Yet we attempt to provide a brief overview on them with stress on the clinical role of each imaging modality in the modern neuro-oncology practice. We have also presented the sectional anatomy of the brain as a pictorial essay.

7.1 Plain Radiography

Contrast in plain-film radiography is based on the differential attenuation of the x-ray beam by various tissues, with denser tissues causing greater attenuation of an x-ray beam and appearing lighter on the x-ray film image.

Plain radiography has a very limited role in neuroradiology in the current era. Skull and spine radiography is often used as a part of skeletal survey in patients with multiple myeloma or with other known malignancy to look for skeletal metastases. Majority of skeletal metastases are lytic; however, sclerotic metastases are seen in certain malignancies like prostate, breast, lymphoma, etc. Presence of a lytic or sclerotic lesion on skull or spine radiography can occasionally be the first indicator of a malignant neoplasm in a patient.

A. Chander · S. L. J. Devarajan (✉)
Department of Neuro-Radiology, AIIMS, New Delhi, USA

S. Mallick et al. (eds.), *Evidence based practice in Neuro-oncology*,
https://doi.org/10.1007/978-981-16-2659-3_7

7.2 Computed Tomography (CT)

7.2.1 Physical Principles

CT like plain X-rays also works on the principle of differential X-ray beam attenuation and uses a thin beam of X-rays from an X-ray tube, that pass through the patient. These x-ray photons are collected by an array of detectors placed on the other side of the patient, which show a differential rate of intensity on a grey scale depending on the degree of absorption along the x-ray beam, that further depends on the density of the object the beam is passing through. Rotation of this tube-detector assembly permits multiple projections in a single slab. Complex mathematical algorithms are used by the computer to obtain an absorption value for each unit area (pixel) within a CT slice and reconstruct a cross-sectional image [1].

The CT absorption values are called Hounsfield Units and represent a relative quantitative measure of radiodensity, with water having a value of zero Hounsfield unit, tissues denser than water having positive values, and tissues less dense than water having negative values (Fig. 7.1). Low density tissues are assigned darker color and high density tissues are assigned brighter colors on the grey scale (Figs. 7.2, 7.3, and 7.4).

7.3 Basic Cross-Sectional Neuroanatomy

Here we present a short pictorial review of the normal cross-sectional imaging anatomy of the brain (Figs. 7.1, 7.2, 7.3, and 7.4). We recommend a systematic approach while interpreting scan images. The following points are pertinent to be noted.

1. CT scan of the head is acquired in the axial plane with the slices parallel to the orbitomeatal line (Fig. 7.1). The thickness of the slices may vary between 2 mm

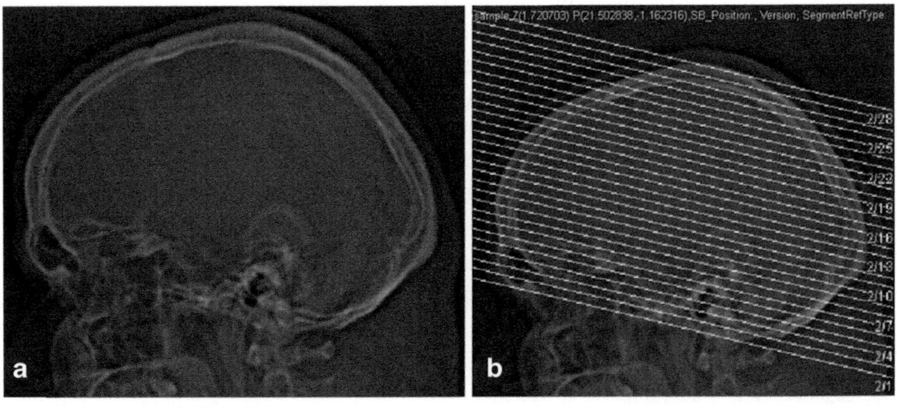

Fig. 7.1 (**a**) Shows scanogram of the head on which the axial sections are planned (**b**). Note the section lines are parallel to the line connecting superior rim of the orbit to the external meatus of the ear

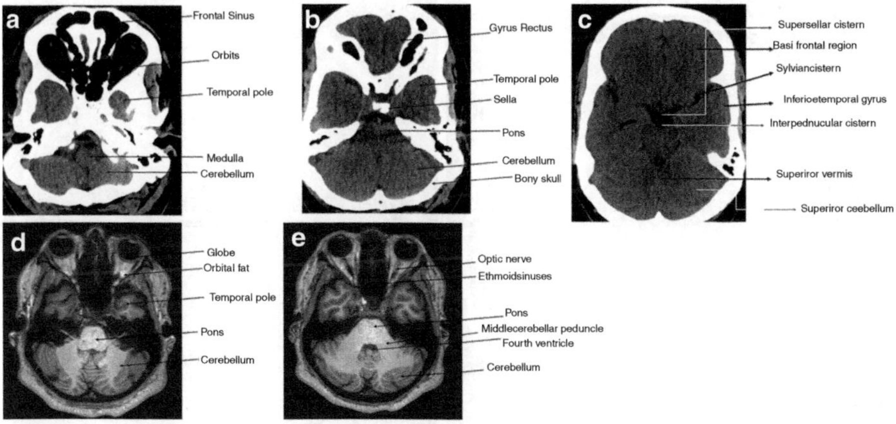

Fig. 7.2 Axial CT and T1W MR images at the level of posterior fossa and brainstem demonstrating normal anatomy. (**a**–**c**) The bony structures appear markedly hyperdense on CT with very high HU values. The CSF appears very hypodense with HU values consistent with fluid density (0–20). The orbital fat appears very hypodense. (**d**, **e**) T1WI MR images throughout the level of pons. The grey matter appears grey and white matter appears white, and fluid appears hypointense on T1W images. Fat appears hyperintense on T1WI, as demonstrated in the orbital fat

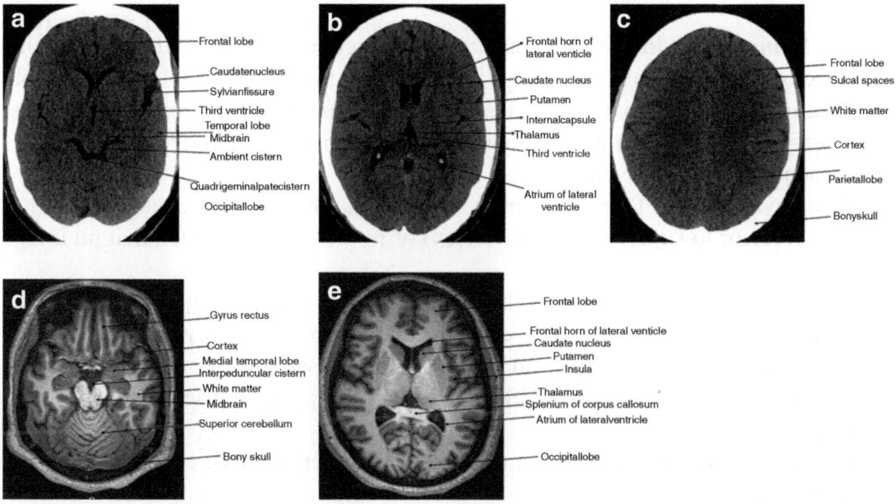

Fig. 7.3 Axial CT and T1W MR images at the level of midbrain (**a**, **d**), basal ganglia (**b**, **e**) and at the supraganglionic levels (**c**). (**a**–**c**) The grey matter on CT appears brighter than white matter which appears relatively hypodense. (**d**, **e**) On T1W MRI, the deep grey nuclei (grey matter) appear grey and the corona radiata, corpus callosum and internal capsules (white matter) appear white

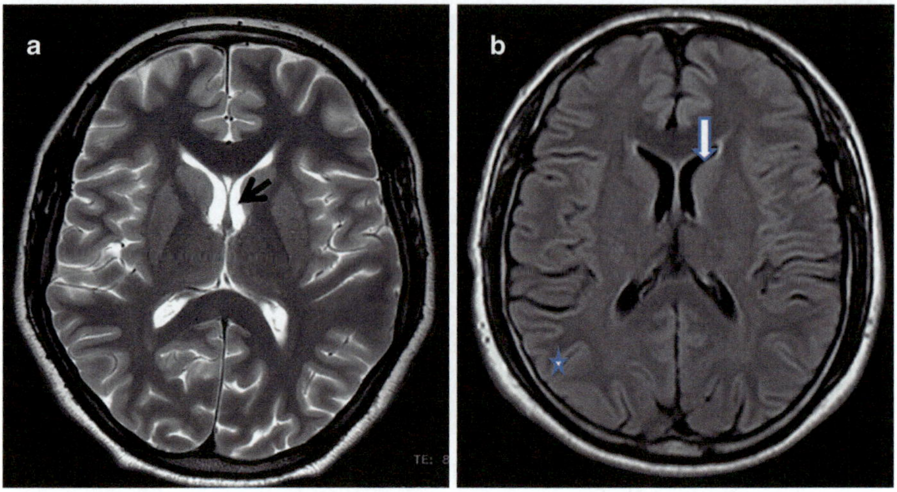

Fig. 7.4 T2W and FLAIR axial images at the level of basal ganglia. (**a**) T2WI showing hyperintense (bright) appearing CSF in the lateral ventricles (black arrow) and the sulcal spaces. Grey matter in the cortex and the deep grey nuclei appear grey and white matter appears dark on T2WI. (**b**) FLAIR image showing dark CSF in the ventricles (thick white arrow) and sulci, with grey matter (white star) appearing brighter than the white matter

and 5 mm depending on the institutional practice. With the advent of multidetector CT scans, much thinner slices (e.g., 0.6 mm) are possible at reasonable radiation cost, such that these images can be reconstructed in coronal or sagittal or any plane without much distortion; CT angiograms and image guidance CTs are acquired in this fashion.

2. While evaluating CT scans of brain, images may be studied sequentially either from inferior to superior (base to vertex) or in the other way such that no image is missed.
3. In each slice every structure visible should be seen carefully either from outside to inwards or the reverse. For example the scalp, bony calvarium, meninges including subarachnoid space and the contained vessels, the cortex, white matter and deep grey nuclei, and the CSF spaces including the ventricles and basal cisterns, etc.
4. MRI offers excellent soft tissue resolution with better tissue contrast and grey-white differentiation as compared to CT apart from being multiplanar.

7.4 Role of CT in Neuro-Oncologic Imaging

- CT is often the first imaging modality used to image patients who present to the emergency with acute-subacute symptoms, and in such patients CT may detect/rule out intracranial SOLs. In the acute setting, CT helps in detection of intracranial herniations or hydrocephalus secondary to mass lesions, which may require urgent neurosurgical intervention (Figs. 7.5 and 7.6).

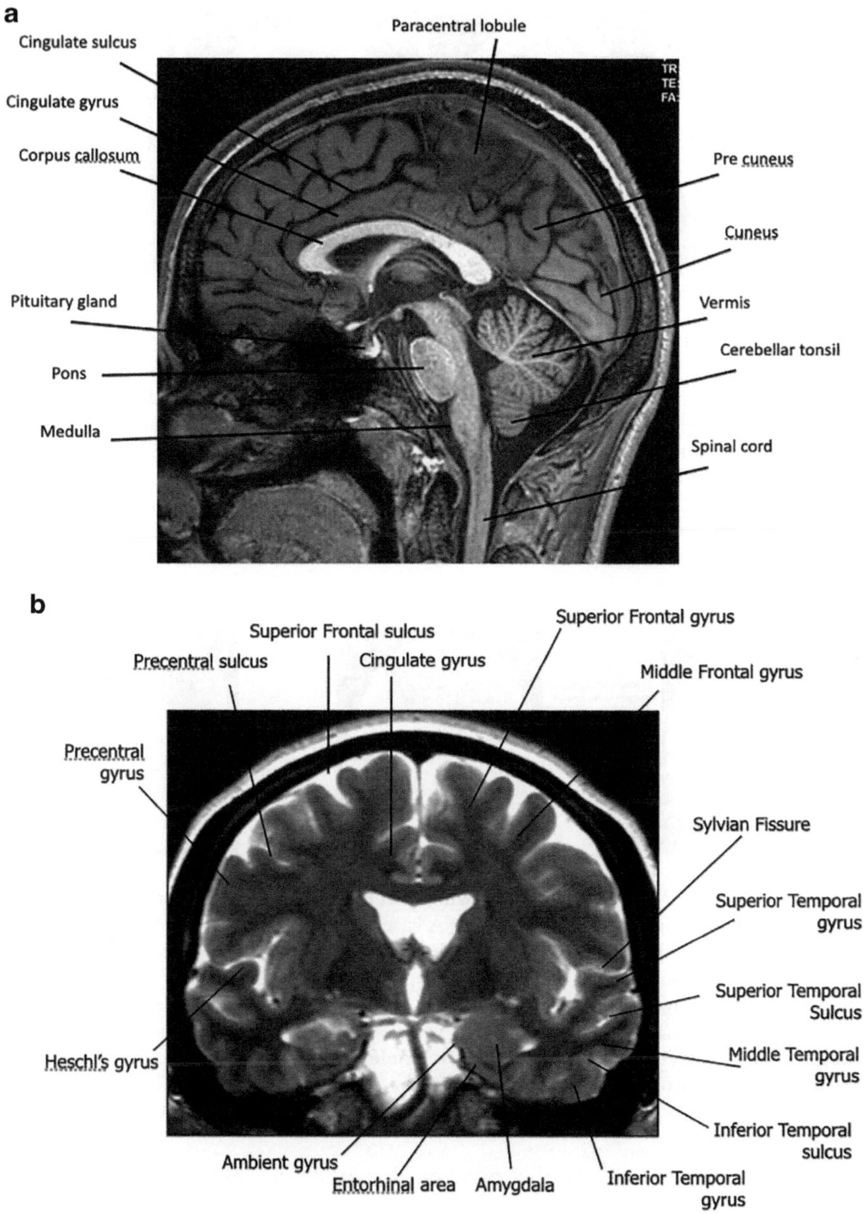

Fig. 7.5 (**a**) Sagittal T1W image showing the midline structures; (**b**) Coronal T2W image

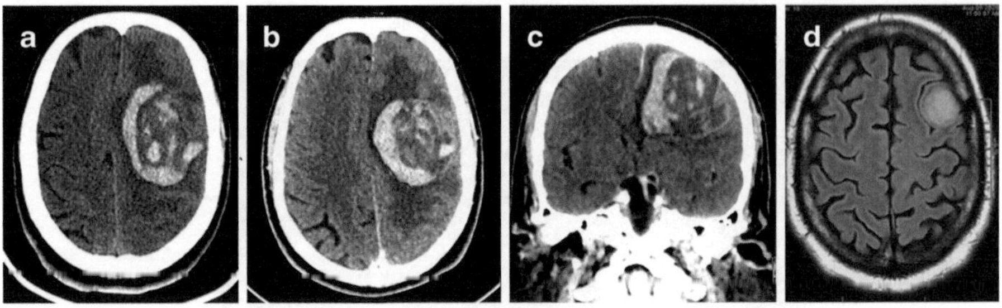

Fig. 7.6 A 56 year old male patient presented to the emergency department with acute onset headache and vomiting. NCCT and CECT was done in the emergency setting. (**a**) Axial NCCT image shows a large heterogeneous appearing lesion, with multiple hyperdense areas within consistent with acute intratumoral haemorrhage. (**b, c**) Contrast enhanced axial and coronal reformatted images showing hemorrhagic tumor with few peripheral areas of enhancement within the lesion as well as non-enhancing necrotic areas within the lesion. (**d**) FLAIR axial image done 4 months before this CT, shows a relatively defined homogeneously hyperintense lesion in the left middle frontal gyrus. The patient was diagnosed as a case of left frontal low grade glioma. There is increase in size of the lesion with increase in mass effect with a midline shift to the right side in the CT images, that was not seen in previous scan

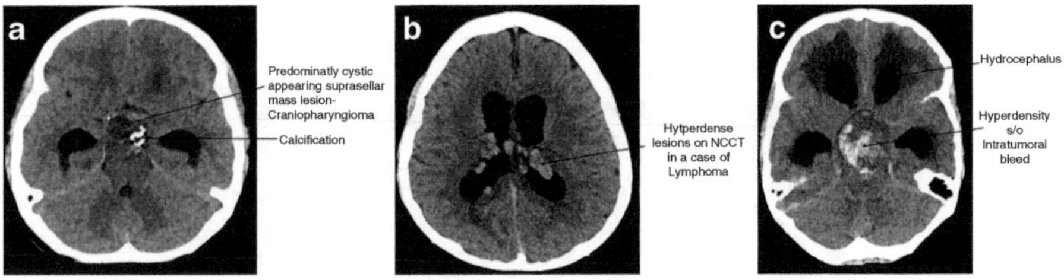

Fig. 7.7 (**a**) A case of craniopharyngioma with axial CT image showing a predominantly cystic lesion with foci of peripheral calcification. (**b**) A case of multifocal lymphoma, with non contrast axial CT images showing the lesion appearing hyper dense. Lymphomas appear dense on CT due to dense cellularity. (**c**) A young adult male who presented to the emergency with headache and vomiting, NCCT shows a heterogeneous suprasellar lesion with hyperdense areas within s/o tumoral bleed. The mass effect on the third ventricle caused hydrocephalus which is seen as enlarged frontal and temporal horns of lateral ventricle with periventricular ooze. Histopathology revealed Pilocytic astrocytoma

- Provides information on the number, location and associated oedema and mass effect.
- Density of the lesion may indicate its cellularity and hence sometimes the histological type—e.g., hyper-density in lymphomas (Fig. 7.7b) and meningiomas, whereas gliomas appear hypodense on NCCT.
- Can detect acute hemorrhage as well as necrosis within the lesions (Figs. 7.6 and 7.7c).

- Has highest sensitivity for calcification—increases diagnostic specificity for some CNS tumors (e.g., craniopharyngioma (Fig. 7.7b), oligodendroglioma, retinoblastoma, pineoblastoma).
- CT is critical in evaluation of chondroid and osseous lesions of the calvarium and skull base, and can well depict chondroid calcification and ossification in these lesions. Hyperostosis associated with meningiomas is also well depicted with CT.
- The anatomy of paranasal sinuses is best detected with CT.
- Contrast-enhanced CT requires use of an intravenous iodinated contrast medium, to detect areas of abnormal blood–brain barrier breakdown where contrast will seep into the parenchyma. Pattern of contrast enhancement, whether solid homogeneous enhancement, heterogeneous, or ring enhancement aids in narrowing the differential diagnosis (e.g., solid near uniform enhancement seen in lymphomas, thick irregular rim enhancement in glioblastomas, absence of enhancement in low grade gliomas) [1].

CT scan involves substantial radiation exposure and hence CT scans should be used judiciously in clinical practice.

7.5 Magnetic Resonance Imaging

MRI is the mainstay of imaging workup in neuro-oncology. Exquisite soft tissue contrast, high spatial resolution and capability to provide functional information make MRI an indispensable tool. Understanding the principles of image formation in MRI involves some complicated higher physics and may not be needed for a clinician. However, familiarity with various MR sequences and the appearance of normal brain and spine structures and their common pathologies is very helpful for a practicing neuro-oncologist.

Contrast-enhanced MRI is the gold standard imaging modality and provides highly sensitive anatomical information about a neoplasm in the nervous system. Advanced imaging modalities provide complementary information about brain tumor metabolism, blood flow, and ultrastructure, and are being increasingly incorporated into routine clinical sequences [2].

7.6 Basic MRI Sequences

7.6.1 T1 Weighted Image

Grey matter appears grey and white matter appears relatively white, ventricles and CSF spaces are dark on T1 (Figs. 7.2 and 7.3). T1 images are better suited for studying anatomy. Majority of pathologies appear hypointense or dark on T1 images. Few things that appear bright on T1 are fat (Fig. 7.2), hemorrhage, soft calcification, melanin, and concentrated protein.

7.6.2 T2 Weighted Image

Grey matter appears grey and white matter appears relatively dark, ventricles and CSF spaces are bright on T2 (Fig. 7.4a). T2 images are better suited for studying pathology in a given case. Majority of pathologies appear hyperintense or bright on T2 images. Few things that appear dark on T2 are certain stages of hemorrhage, soft calcification, melanin, and protein.

7.6.3 FLAIR

Fluid attenuated inversion recovery sequence. It is basically a T2 weighted image in which CSF (water) signal has been selectively suppressed. Therefore ventricles and CSF spaces are dark on FLAIR images, whereas pathological fluid appears bright (Fig. 7.4b).

7.6.4 SWI

Susceptibility weighted imaging is utilized in clinical practice for detection of calcification and bleed. Both show blooming (increased dark signal) on SWI images (Fig. 7.11).

7.6.5 DWI

Diffusion weighted imaging studies the random Brownian motion of water molecules in the body. Water molecules outside the body exhibit free diffusion with constant random Brownian motion, whereas the movement of water molecules in biologic tissues is restricted because of the interactions with cell membranes and macromolecules. Diffusion contrast is based on difference in the motion of water molecules in various tissues, with intracellular diffusion being more restricted than extracellular diffusion (because of cell membrane).

Restriction of diffusion is seen as a high signal on DWI and a correspondingly low signal on ADC images. ADC (apparent diffusion coefficient) maps also provide a quantitative measure of diffusion [3].

7.6.6 STIR

Short tau inversion recovery sequence suppresses bright T1 signal from fat. It is used to suppress fat signal and is useful in spine imaging.

7.6.7 Post Contrast-Enhanced MRI

T1 weighted post gadolinium images are used to characterize lesions based on their enhancement pattern.

7.7 Conventional/Structural MRI Evaluation in Brain Tumors

The excellent soft tissue contrast offered by MRI makes it indispensable in the non-invasive evaluation of brain and spine tumors.

MRI helps in differentiating between intra- and extra-axial mass lesions, which is usually the first step in arriving at a reasonable differential.

The number of lesions and mass effect is evaluated, with multiple lesions representing metastatic disease, or more aggressive disease as in patients with multifocal/multicentric gliomas, or reflecting presence of non-neoplastic processes such as infection and demyelination.

Further characterization of brain tumors is dependent on the various imaging features they present. Masses can appear solid, cystic, mixed solid-cystic, or cystic with a solid nodule. Presence or absence of necrosis and hemorrhage helps in grading of gliomas, indicating the organ of origin in metastatic tumors (e.g., cystic/necrotic metastases seen in squamous histology).

Signal intensity on T1, T2, and FLAIR can provide relevant information.

- Most tumors are hypointense or isointense on T1WI. T1 hyperintensity may be seen in lesions with hemorrhage or early calcification or in melanoma metastases. Fat containing lesions like dermoids are also T1 hyperintense, as are lesions with high protein content like epidermoids and colloid cysts.
- Most tumors appear hyperintense on T2/FLAIR. T2 hypointensity may be seen in densely cellular tumors like lymphomas (Fig. 7.8). T2 hypointensity may also be seen secondary to hemorrhage within tumors. Increased T2 flow-voids representing enlarged or increased vessels are seen in vascular tumors like hemangioblastomas, meningiomas, and hemangiopericytomas.
- T2-FLAIR mismatch sign is highly useful in identifying the IDH mutation status in diffuse astrocytomas (IDH mutant, 1p19q non-codeleted).
- Increasing FLAIR hyperintensity on follow-up imaging in posttreatment gliomas is considered an indicator of disease progression.
- DWI:
  - DWI provides useful information about the cellularity. Hypercellular tumors with densely packed cells and with high nucleus/cytoplasmic ratios, e.g., lymphomas, medulloblastomas, show restricted diffusion (appear bright on DWI and dark on ADC maps) (Fig. 7.8) [2].
  - DWI is also useful in differentiating high grade necrotic tumors showing peripheral ring enhancement (e.g., GBM) from intracranial abscesses (Figs. 7.9 and 7.10).

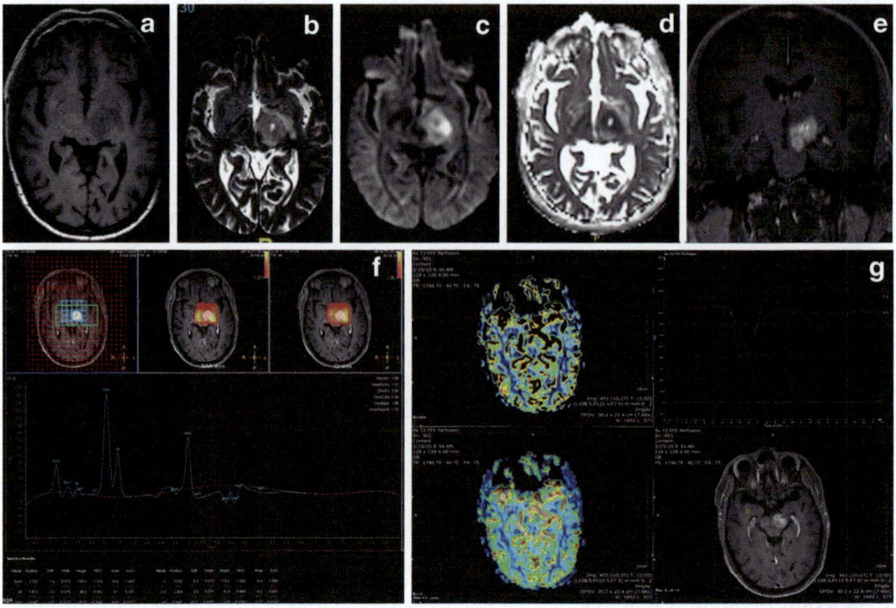

Fig. 7.8 (**a–g**) Multiparametric imaging in the evaluation of brain tumor. A71 year old male, histopathologically proven c/o Lymphoma. (**a–d**) Axial T1W, T2W, DWI and ADC images. The lesion appears relatively hypointense on T2W, and shows restriction of diffusion, both of these imaging features favouring a diagnosis of lymphoma which shows diffusion restriction because of its dense cellularity. (**e**). Post contrast coronal T1W image showing solid, near uniform enhancement within the lesion, again a feature supporting the diagnosis of lymphoma. (**f**) MR Spectroscopy showing raised Choline within the lesion, with a Cho/NAA ratio of 1.96. (**g**). DSC perfusion shows mildly elevated perfusion within the lesion, with the percentage signal recovery curve overshooting the baseline, a classical feature of lymphomas

  - DWI allows for the assessment of peritumoral oedema, regions of tumor hypoxia, integrity of white matter tracts (as in DTI), and in postoperative injury.
  - In postoperative follow-up in early and subacute postoperative period, diffusion plays an important role in distinguishing between the enhancing post-surgical ischemic injury and the enhancing residual/recurrent tumor [4].
- SWI: SWI detects presence of hemorrhage and calcification within lesions, although CT is more specific for detection of calcification. Presence of hemorrhage is a feature of high grade gliomas and hence helps in differentiation of low grade from high grade gliomas (Fig. 7.11).
- Contrast enhancement: Most tumors show some form of contrast enhancement. Contrast enhancement characteristics of a few common tumors are listed below.
  - Low grade diffuse astrocytomas do not show enhancement.
  - High grade glioma shows heterogeneous post contrast enhancement.

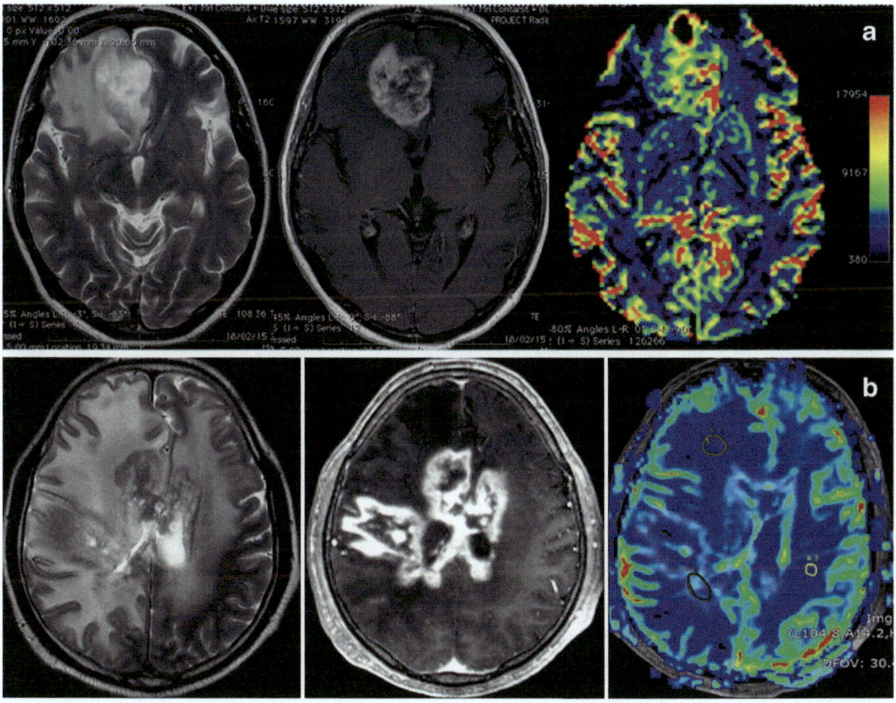

Fig. 7.9 (**a**, **b**) Role of MR perfusion for differentiating between tumor recurrence and radiation necrosis. (**a**) Recurrent tumor in a post op, post RT case of high grade glioma. Axial T2W image shows a heterogeneously hyperintense mass lesion with perilesional edema and axial T1 post contrast image showing heterogeneous enhancement with few central non enhancing necrotic areas. DSC Perfusion shows raised perfusion within the lesion, consistent with recurrence. (**b**) Heterogeneous intensity lesion seen on T2WI in a post op, post RT case of high grade glioma. Post contrast images show peripheral rim enhancing lesions with central necrotic areas. DSC Perfusion does not show significantly raised perfusion, consistent with radiation necrosis

- IDH wild-type glioblastomas show thick irregular ring enhancement, with a central non-enhancing necrotic core (Fig. 7.11).
- Lymphomas show solid homogeneous enhancement, except for posttreatment lymphomas which may appear necrotic with heterogeneous or ring enhancement (Fig. 7.8).
- Dural tail in post contrast images is specific for meningiomas.

Post contrast images are essential to detect leptomeningeal and ependymal spread of tumors. Post gadolinium screening of spine is hence recommended for tumors that are known to show leptomeningeal drop metastases like glioblastomas, medulloblastomas, ependymomas, germinomas, and pineoblastomas.

In patients with brain metastases, use of contrast-enhanced imaging also helps in detecting additional small metastases that may otherwise be missed on routine non-contrast sequences.

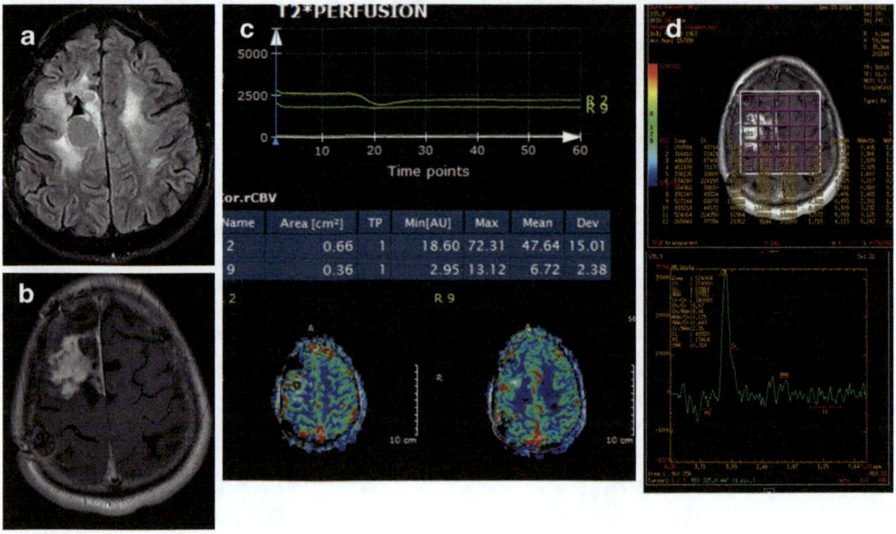

Fig. 7.10 (**a–d**) 42 year male operated for Rt. frontal glioma post RT. MRI reveals heterogeneously enhancing mass with perilesional edema. The lesion shows raised perfusion on DSC perfusion study, a finding consistent with recurrence. Multivoxel MR Spectroscopy through the lesion shows raised Choline s/o recurrence. There is absence of a lipid-lactate peak, which is seen more commonly in radiation necrosis

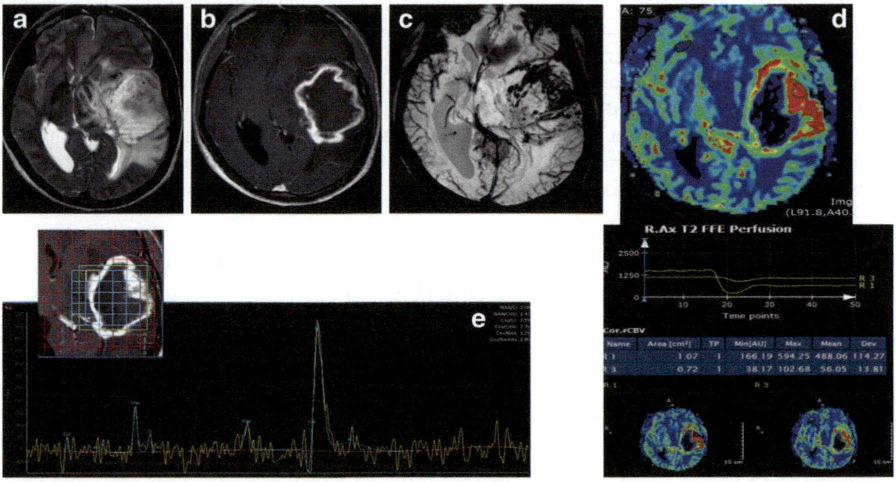

Fig.7.11 (**a–e**) A 57 year male with seizures, speech difficulty and right sided weakness X 4 months . (**a–c**) Conventional MRI reveals necrotic lesion with irregular thick peripheral enhancing rim. SWI shows multiple areas of blooming within the lesion s/o haemorrhage within the lesion - a feature of higher grade gliomas. (**d**) DSC perfusion study showing markedly enhanced perfusion within the solid peripheral enhancing component of the lesion. (**e**). MRSI shows elevated choline, reduced NAA and a lipid-lactate peak - features supporting the diagnosis of a glioblastoma

7.8 Advanced Cross-Sectional Imaging Techniques in Neuro-Oncology

7.8.1 CT Perfusion

While conventional cross-sectional imaging techniques provide anatomical information, in vivo perfusion imaging provides additional information regarding the tumor vascularity and hemodynamics, which in turn helps in better characterization of the tumor, helps in assessing the grade and hence the prognosis [5]. Perfusion imaging is also an important tool in assessing treatment response on imaging and differentiating treatment related changes from tumor progression. Perfusion imaging in brain tumors is usually performed with MR imaging and can be done as part of the multi-parametric baseline MR imaging. However, CT perfusion has its own advantages as described below and is poised to perform much more significant role in the future.

- CT perfusion allows assessment of blood volume and permeability in the same study, with a single injection of contrast, whereas studying blood volume and permeability on MR requires two different perfusion techniques with usage of contrast separately for each technique [6].
- Perfusion CT is especially useful in posttreatment follow-up of brain tumors since the susceptibility artifacts seen secondary to hemorrhage and mineralization can compromise the evaluation of these patients by MR perfusion. Radiation necrosis shows lower CBV and lower PS (permeability surface area product) than recurrent tumors, and hence is useful in differentiating between them [5].
- For similar reason, perfusion CT may be helpful in hemorrhagic tumors, and hence may complement MR perfusion as a problem solving tool.

7.8.1.1 Technique

With dynamic perfusion imaging, serial images at very short intervals are acquired for a region of interest after rapid intravenous administration of a contrast agent. Imaging biomarkers are then obtained using pharmacokinetic models.

- Cerebral blood volume (CBV): Total volume of blood in a given volume of brain, measured in ml/100 g brain tissue
- Cerebral blood flow (CBF): Volume of blood moving through a given volume of brain per unit time, ml/100 g brain tissue/min
- Mean transit time: Average transit time through a given brain region and represents time between inflow and outflow of contrast through a given brain region
- Time to peak (TTP): Time taken by contrast to achieve maximum enhancement
- Time to drain (TTD): Time taken for washout of contrast
- Permeability surface area product (PS)

- Flow extraction product (FEP): Represents the volume transfer constant (k-trans), which reflects the passage of contrast between blood and the extravascular-extracellular space [7].

Regional tumor blood volume is an assessment of the tumor vasculature and is considered a surrogate marker for microvascular density. Tumor vascular permeability is also used as a marker for assessing the tumor grade, since increased permeability is seen with immature blood vessels seen in neoangiogenesis [5].

7.8.2 Advanced MRI Techniques

In the present era, the role of MRI in brain tumors is no longer limited to providing anatomical images, but has progressed to providing information on tumor physiology as well.

Advanced techniques including perfusion, spectroscopy, tractography, and functional MRI are used for non-invasive characterization of brain tumors and pretreatment evaluation, as well as for post-therapeutic evaluation to differentiate treatment response and effects from tumor progression.

7.8.3 MRS

Magnetic resonance spectroscopy enables us to determine in vivo the biochemical nature of either lesions or brain parenchyma.

While conventional MRI provides an anatomical image, MR spectroscopy provides a spectrum rather than an image, the MR spectrum consisting of metabolite peaks (normal and abnormal) that represent signal intensities/concentration (y-axis) as a function of frequency expressed as parts per million (x-axis). Spectra are obtained either from one selected brain region in the case of single-voxel spectroscopy (SVS) or from multiple brain regions in the case of MR spectroscopic imaging (MRSI).

Proton MR spectroscopy is used in clinical practice and shows three predominant metabolite peaks representing the compounds creatine and phosphocreatine (Cr) associated with cellular energy metabolism, choline containing compounds (Cho) as markers of cell membrane density and integrity, and N-acetyl aspartate (NAA) which is considered to be a marker of neuronal integrity [8].

Other metabolites such as lipids, lactate, myoinositol, glutamate/glutamine, alanine, etc. may also be observed depending upon the pathology.

7.8.3.1 Clinical Applications in Brain Tumor Imaging

- Brain tumor evaluation: Proton MR spectroscopy in brain tumors generally shows elevated choline as a result of increased cell density and cell membrane turnover and reduced NAA due to neuronal destruction [8].

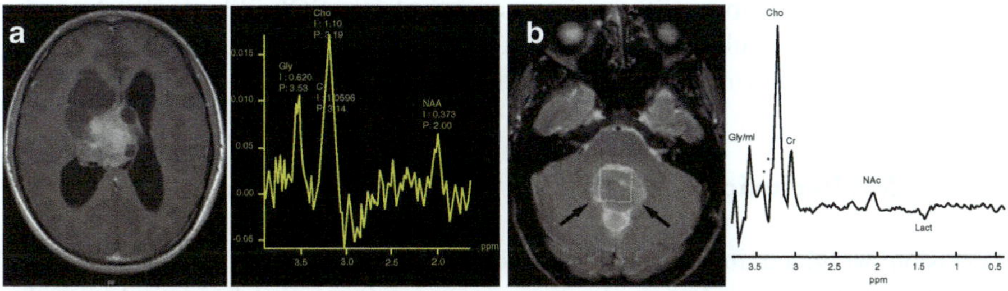

Fig. 7.12 Specific MRS peaks in different tumors. (**a**) Single voxel spectroscopy (SVS) showing Glycine peak at 3.55 ppm in central neurocytoma, besides elevated choline. (**b**). SVS showing a Taurine peak in medulloblastoma at 3.4 ppm besides raised choline at 3.2 ppm

- Specific metabolite peaks can be seen in certain tumors, like alanine peak in meningiomas, taurine peak in medulloblastoma, and glycine peak in high grade pediatric tumors and central neurocytomas (Fig. 7.12) [9].
- Grading of glial tumors: Presence of lactate and lipid peaks indicate a higher grade of malignancy, the exception to the rule being pilocytic astrocytomas, that despite being low grade tumors can demonstrate lipid-lactate peaks.
- Differentiation between IDH mutant and wild-type astrocytomas: The 2016 WHO classification of brain tumors classifies astrocytomas into IDH mutant and wild types, with the IDH mutant astrocytomas showing a 2-hydroxyglutarate (2-HG) peak resonating at 2.25 ppm (not seen with wild-type tumors) [9, 10].
- Tumor extent in gliomas: Brain tumors show perilesional hyperintensity that generally reflects perilesional vasogenic oedema [9]. The detection of an increased Cho/NAA ratio in the non-enhancing peritumoral T2/FLAIR hyperintensity reflects tumor invasiveness in infiltrative lesions like gliomas, as in vasogenic oedema no abnormal metabolic ratios are seen.
- Differentiation of gliomas from metastases: Intralesional spectroscopy findings in metastases of increased choline with reduced NAA are similar to gliomas. MRS can aid in differentiating gliomas which are infiltrative and can extend beyond enhancing margins and show abnormal peritumoral spectra on MRS, from brain metastases that are well-defined and exhibit a near-normal spectrum in the peritumoral region.
- Differentiation of recurrent tumor from posttreatment changes and radiation necrosis: The role of MRS has been widely studied in differentiating posttreatment changes and radiation necrosis from recurrent gliomas. Elevated Cho/NAA and Cho/Cr have been shown to reliably differentiate recurrent tumor from posttreatment injury, while lipid and lactate peaks can be seen in both conditions (Fig. 7.10) [4, 11].
- MRS can help guide biopsy, the Cho/NAA ratio can help differentiate areas of solid tumor with the highest cell density from oedema, and hence reflect appropriate site for biopsy and reduce sampling errors [9].

- MRS can also help to differentiate non neoplastic SOLs like brain abscesses (pyogenic, tubercular, fungal) and tumefactive demyelination from necrotic brain tumors showing peripheral rim enhancement.

7.8.4 MR Perfusion

Perfusion weighted imaging brings information on cerebral physiology at the microvasculature level. It provides non-invasive estimates of functional hemodynamic parameters of normal and pathological brain tissue. It is a measure of the volume of blood passing through the microvasculature of a volume of tissue in a given time.

Three main techniques are currently employed for perfusion imaging, namely dynamic susceptibility contrast (DSC), dynamic contrast-enhanced (DCE) T1-weighted perfusion, and arterial spin-labelled (ASL) imaging, each having its own particular advantages and disadvantages. Both DSC and DCE perfusion require intravenous administration of gadolinium-based contrast agents (GBCAs) and provide good signal to noise ratio.

7.8.4.1 DSC Perfusion

DSC perfusion is the most commonly used MR perfusion technique, is rapid and exploits the T2* effects of gadolinium. It is based on the first pass of bolus of intravascular contrast, and after the injection of contrast, rapid sequential imaging is used to measure the change in MR signal. The susceptibility effect of gadolinium contrast agent leads to a fall in signal intensity. A time versus intensity curve is plotted based on the change in signal, from which various perfusion parameters are calculated. Two commonly applied DSC parameters are relative cerebral blood volume (rCBV), used to estimate perfusion normalized to contralateral white matter, and percentage signal recovery (PSR), used to estimate leakiness by measuring the recovery of the signal intensity curve to baseline [12].

7.8.4.2 DCE Perfusion

DCE perfusion also requires dynamic imaging after the intravenous administration of gadolinium contrast. DCE perfusion exploits the leakiness of pathologic capillaries to provide useful information about microvascular structure. Abnormal tumor vasculature permits leakage of contrast from the intravascular space to the extravascular interstitial space, which is measured dynamically before, during, and after the administration of gadolinium on T1 weighted images. Leakage of contrast (gadolinium) from the intravascular to the extravascular-extracellular space (EES) leads to an increase in signal intensity on T1. The volume transfer constant or k-trans, a measure of capillary permeability, is the primary metric derived from DCE perfusion. K-trans can be used to grade tumors, particularly gliomas, as gliomas with increased capillary permeability are more likely to be higher grade than lower grade [11].

7.8.4.3 ASL Perfusion

ASL perfusion imaging does not require the intravenous administration of gadolinium contrast, and hence provides an advantage where there is a concern for renal toxicity and also where repeated imaging is required amidst rising concerns of gadolinium deposition in the brain. ASL however is limited by its long acquisition time and decreased spatial resolution.

7.8.4.4 Technique

Both DSC and DCE perfusion require administration of a fast bolus of gadolinium-based contrast agent, hence contrast is injected through a power injector at the rate of 3–5 ml/sec, followed by a saline flush and requires an 18G or 20G cannula. Both perfusion studies can be performed in a single MRI protocol, with the DCE sequence performed before the DSC sequence, with an interval of 5–8 min between the studies. Higher contrast dose is required with contrast administered twice, once before each perfusion study [13].

7.8.4.5 Clinical Applications

- Brain tumor imaging: Perfusion parameters can be used to differentiate neoplastic lesions with high grade/vascular tumors showing raised perfusion from non-neoplastic space occupying lesions like tumefactive demyelination and brain abscesses that will show decreased perfusion.
- Grading of gliomas: Perfusion imaging helps in grading of gliomas, the general dictum being that higher rCBV values represent increased and more aggressive neoangiogenesis and hence correlate with higher grade of tumor (Fig. 7.10). Similarly, higher k-trans values on DCE are seen with higher-grade tumors. This however is not true for oligodendrogliomas which show higher rCBV values irrespective of grading, and also gives spurious results in pilocytic astrocytomas or extra-axial tumors like meningioma that may show raised perfusion. Also, the perfusion parameters show overlap of rCBV values between different glioma grades, and hence may not always be accurate [11].
- Differentiation of glioblastomas from metastases: Perfusion may help in distinguishing between glioblastoma and metastases. While both will show raised rCBV, perfusion abnormalities in the T2-hyperintense peritumoral region immediately surrounding the tumor will go in favor of primary brain tumor which is usually more infiltrative, whereas the T2 hyperintensity surrounding metastases represents bland oedema [11].
- Differentiation of high grade glioma from lymphoma: PWI can also help distinguish between high grade glioma and lymphoma. Higher rCBV is seen in high grade gliomas, with increased leakiness and less percentage signal recovery (PSR), whereas lymphomas show lower rCBV values, and high PSR, with return of signal intensity curve to above baseline (Figs. 7.8 and 7.11) [11].
- For guiding biopsy: Perfusion imaging is useful to identify region(s) of higher-grade pathology (areas of higher perfusion abnormality) within an individual heterogeneous tumor. This is helpful in guiding biopsy, directing tissue sampling at the most malignant portion of the lesion.

- Evaluating treated brain tumors: PWI can help differentiate between true tumor progression, pseudoprogression, and radiation injury, with raised perfusion seen in contrast enhancing lesions representing true tumor progression vis-à-vis treatment related changes (Fig. 7.9) [4, 14].

7.8.5 Functional MRI (fMRI)

Functional MRI is an advanced MRI tool that helps in the localization of various functional areas in the brain such as motor or speech area in relation to a given pathology.

fMRI is an indirect measure of neuronal activity and uses the blood oxygen level dependent (BOLD) signal as contrast, and relies on the T2*effect of deoxyhemoglobin for contrast. The local blood flow in a particular region of the brain is dependent upon the neuronal activity. In response to activation by a task, there is increased oxygen and energy consumption leading to formation of deoxyhemoglobin (paramagnetic and appears dark on T2* images). However, there is also concomitant increase in blood flow to the functioning tissue, and this oxygen delivery is more than the oxygen requirement, with washout of the deoxyhemoglobin and increase in oxyhemoglobin leading to increase in signal intensity. These signal changes although small, can be picked up by MRI [15].

7.8.5.1 Technique

In task-based fMRI, the patient alternates between a passive resting state and task performance, while relative changes in BOLD signal are measured and areas of signal change that correlate with the stimulus are identified and subsequently compared with high resolution T1W anatomical images [4]. Post-processing techniques provide colored maps of eloquent areas that can be superimposed on anatomical images.

7.8.5.2 Clinical Applications

- In clinical practice, attempt is made to localize the motor strip, speech areas, and memory centers using fMRI in patients who are undergoing resections of adjacent tumors. The information has important implications for presurgical planning, intraoperative navigation to maximize tumor resection while preserving important brain function and helps in preoperative counselling of patients (Fig. 7.13) [16].

7.8.6 DTI

Diffusion tensor imaging is a further advanced application of diffusion imaging, where dependency of diffusion on the direction of white matter (anisotropy) can be mapped spatially, from which fiber track maps can be generated. DTI based

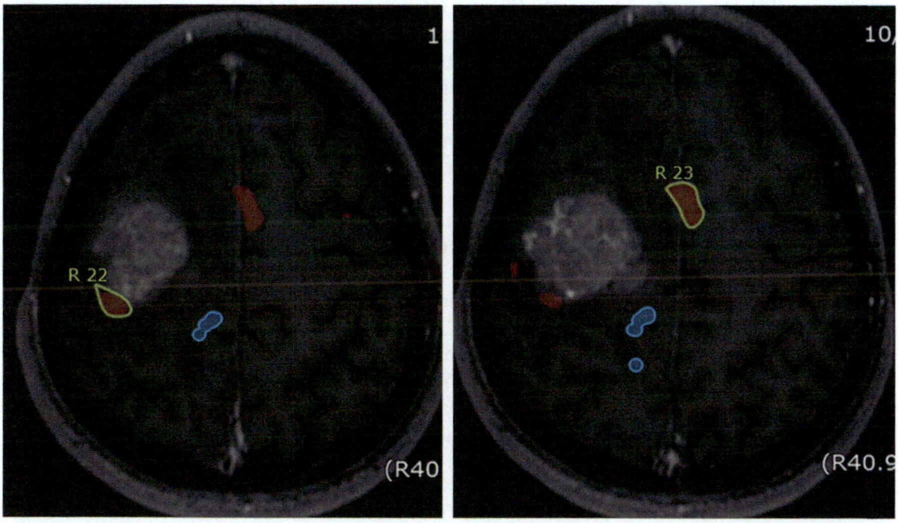

Fig. 7.13 fMRI in a case of tumor located near the motor strip, shows the hand motor area (R22) abutting the lesion along its posterolateral aspect. Note R23 representing activity in the normal leg areawhich is well away from the tumor margins

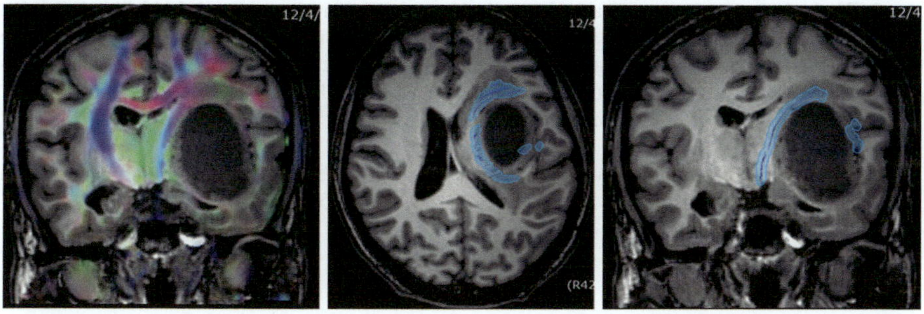

Fig. 7.14 DTI in a case of Glioblastoma shows displacement of the left corticospinal tracts

tractography can be used in presurgical planning, by analyzing the integrity of white matter tracts to determine whether there is tumor invasion or tumor displacement of the adjacent white matter tracts (Fig. 7.14).

References

1. International Atomic Energy Agency. Diagnostic radiology physics. Vienna: IAEA; 2014.
2. Villanueva-Meyer JE, Mabray MC, Cha S. Current clinical brain tumor imaging. Clin Neurosurg. 2017;81(3):397–415.
3. Koh DM, Collins DJ. Diffusion-weighted MRI in the body: applications and challenges in oncology. Am J Roentgenol. 2007;188(6):1622–35.

4. Strauss SB, Meng A, Ebani EJ, Chiang GC. Imaging glioblastoma posttreatment: progression, pseudoprogression, pseudoresponse, radiation necrosis. Radiol Clin N Am. 2019;57(6):1199–216. https://doi.org/10.1016/j.rcl.2019.07.003.
5. Jain R. Perfusion CT imaging of brain tumors: an overview. Am J Neuroradiol. 2011;32(9):1570–7.
6. Shankar JJS, Woulfe J, Da Silva V, Nguyen TB. Evaluation of perfusion CT in grading and prognostication of high-grade gliomas at diagnosis: a pilot study. Am J Roentgenol. 2013;200(5):504–9.
7. Hakim, Arsany & Zaharchuk, Greg & Wiest, Roland. (2019). A Practical Guide to CT Brain Perfusion. https://doi.org/10.26044/ecr2019/C-0936.
8. Soares DP, Law M. Magnetic resonance spectroscopy of the brain: review of metabolites and clinical applications. Clin Radiol. 2009;64(1):12–21. https://doi.org/10.1016/j.crad.2008.07.002.
9. Bradley WG. MR spectroscopy of the brain for radiologists. Biomed Imaging Interv J. 2007;270(3):658–79. https://doi.org/10.1148/radiol.13130531.
10. Laino ME, Young R, Beal K, Haque S, Mazaheri Y, Corrias G, et al. Magnetic resonance spectroscopic imaging in gliomas: clinical diagnosis and radiotherapy planning. BJR Open. 2020;2(1):20190026.
11. Zhang H, Ma L, Wang Q, Zheng X, Wu C, Xu BN. Role of magnetic resonance spectroscopy for the differentiation of recurrent glioma from radiation necrosis: a systematic review and meta-analysis. Eur J Radiol. 2014;83(12):2181–9.
12. Zhang J, Liu H, Tong H, Wang S, Yang Y, Liu G, et al. Clinical applications of contrast-enhanced perfusion MRI techniques in gliomas: recent advances and current challenges. Contrast Media Mol Imaging. 2017;2017:7064120.
13. Essig M, Shiroishi MS, Nguyen TB, Saake M, Provenzale JM, Enterline D, et al. Perfusion MRI: the five most frequently asked technical questions. Am J Roentgenol. 2013;200(1):24–34.
14. Fatterpekar GM, Galheigo D, Narayana A, Johnson G, Knopp E. Treatment-related change versus tumor recurrence in high-grade gliomas: a diagnostic conundrum - use of dynamic susceptibility contrast-enhanced (DSC) perfusion MRI. Am J Roentgenol. 2012;198(1):19–26.
15. Glover GH. Overview of functional magnetic resonance imaging. Neurosurg Clin N Am. 2011;22(2):133–9.
16. Bogomolny DL, Petrovich NM, Hou BL, Peck KK, Kim MJJ, Holodny AI. Functional MRI in the brain tumor patient. Top Magn Reson Imaging. 2004;15(5):325–35.

PET in Brain Tumors

8

Sneha Prakash, Meghana Prabhu, and Nishikant Avinash Damle

8.1 Introduction

Positron emission tomography (PET) is a nuclear medicine imaging method with increasing relevance for the diagnosis, prognostication, and monitoring of gliomas, especially glioblastomas. PET (a molecular/functional imaging technique) provides additional insight beyond MRI (a morphological/anatomical imaging technique) into the biology of gliomas, which can be used for non-invasive grading, differential diagnosis, delineation of tumor extent, planning of surgery, and radiotherapy as well as post-treatment monitoring. Owing to the progress in PET imaging using radiolabelled amino acids, amino acid PET has been recommended as an additional tool in the diagnostic assessment of brain tumors by the Response Assessment in Neuro-Oncology (RANO) working group that develops new standardized response criteria for clinical trials in brain tumors [1, 2]. Table 8.1 lists the primary goals of PET imaging in brain tumors.

8.2 Significance of the Blood–Brain Barrier

Brain tumors pose specific issues that are related to the unique characteristics of the organ they sit in. Most of the brain is separated from the blood by the blood–brain barrier (BBB) that exerts a much more restrictive control over substances that are allowed to pass and thus, many tracers that easily reach tumors in the body would reach brain tumors only when:

S. Prakash · N. A. Damle (✉)
Department of Nuclear Medicine, AIIMS, New Delhi, India

M. Prabhu
Department of Nuclear Medicine, Institute of Liver and Biliary Sciences, New Delhi, India

S. Mallick et al. (eds.), *Evidence based practice in Neuro-oncology*,
https://doi.org/10.1007/978-981-16-2659-3_8

Table 8.1 Primary goals of PET imaging

Integrity of blood–brain barrier(BBB)
Lesion detection, localization, delineation of extent, therapy planning, and assessing response to therapy and recurrence
Low-grade tumors—Monitor anaplastic transformation into high-grade tumors and evaluate recurrent disease
High-grade/metastatic tumors—To differentiate between recurrent tumor and post-therapy changes

(a) the tumor caused a disruption of the BBB, like glioblastomas
(b) the tumor developed from intracranial tissues that do not have a BBB, like meningioma

Thus, the break of the BBB is regarded as the main diagnostic indicator for malignant gliomas, meningiomas, and brain metastases, which can easily be detected on contrast-enhanced magnetic resonance imaging (MRI) and computed tomography (CT). There has been more focus in the development and evaluation of brain tumor radiotracers that do not depend on BBB damage, such as labeled amino acids, as they are being transported by large-capacity specific transporters across an undamaged BBB [3].

There are brain tumors with an intact BBB such as most low-grade gliomas, which may be missed on a CECT scan but are easily detected by increased signal on T_2-weighted MRI scans without contrast enhancement. Nuclear imaging's role is prominent in the arena of biological characterization of brain tumors, especially for therapy planning.

One of the important roles of imaging is grading the malignancy. MRI and CT rely on BBB damage (frequent in grade 3 and 4 but absent in grade 2) and morphologic appearance (e.g., presence of necrosis, vascularity) for grading and although this is mostly sufficient in untreated gliomas, the damage and necrosis to BBB resulting from formation of reactive tissue post therapy renders it unreliable in treated tumors. Here, imaging methods that distinguish tumor from reactive non-neoplastic tissue will contribute significantly to clinical decision making. Contrast enhancement also cannot provide proper grading in brain tumors without BBB, such as meningiomas and lymphomas. Generally, functional measures related to tumor proliferation deliver more reliable information on prognosis than anatomical imaging methods.

8.3 Advantages of PET in Brain Tumor Imaging

Brain tumors frequently exhibit different degrees of anaplasia in different tumor parts and surgical biopsies, especially when taken stereotactically, may miss the most malignant tumor part underestimating the tumor grade. To overcome this, targeting of biopsies toward the most malignant tumor part is an important clinical utility of PET imaging.

Another big problem with gliomas is their potential to infiltrate normal brain that causes tumor recurrences even after complete tumor removal and local therapies. The degree and extent of infiltration are important information for tailoring therapy (e.g., by conformal or stereotactic radiotherapy) individually and even though MR fluid-attenuated inversion recovery (FLAIR) sequences and diffusion tensor imaging have good sensitivity, they lack specificity. Molecular imaging techniques that provide sensitivity and specificity would be a major step forward.

There are large variations in the response of tumors to therapy within tumor types and often even in different areas of the same tumor. With morphologic imaging, it is difficult to determine whether a tumor is responding and only much later after completion of therapy, the outcome becomes evident. Especially with chemotherapy, monitoring of therapeutic efficacy is a major goal to modify inefficient therapy before the patient's condition deteriorates so much that it reduces any more therapeutic options. Molecular imaging techniques provide improved outcome parameters for therapy monitoring, which also is relevant for conducting efficient clinical trials of new therapeutics.

Differentiating tumor recurrence from radiation necrosis is particularly problematic in morphological imaging as radiation necrosis also produces disruption in the BBB by vascular and astrocytic damage, and thus contrast enhancement, edema, and cortical dysfunction that are indistinguishable from recurrent tumor on conventional CT or MRI. Radiotracer uptake seen in tumor lesions can be utilized as a major advantage in diagnosing recurrence by PET [4].

8.4 Role of PET

8.4.1 Tumor Detection, Delineation of Tumor, and Differential Diagnosis of Brain Tumors

During initial workup, MRI is the first choice of investigation and is irreplaceable. Functional imaging must be used for tumor characterization when MRI results are equivocal.

^{18}F-FDG PET/CT has a moderate sensitivity and specificity for differentiating brain tumors from non-tumor lesions, owing to its high physiological uptake [5]. ^{18}F-FDG is nonspecific in that its uptake is increased in inflammatory/infectious lesions as well as in certain benign lesions (brain abscess, acute inflammatory demyelination, etc.). Two steps critical for improving the accuracy of interpretation include MR imaging co-registration and delayed imaging at 3 to 8 h after injection whereby the tumor to normal brain uptake ratio increases with time following injection [6, 7]. In human immunodeficiency virus-infected patients, ^{18}F-FDG PET/CT can help to discriminate CNS lymphoma from toxoplasmosis with a high degree of accuracy [8].

Zhao et al. in their meta-analysis demonstrated a pooled sensitivity and specificity of 95% and 83%, respectively in detecting brain tumors with ^{11}C-methionine PET/CT and also a higher diagnostic accuracy values compared to ^{18}F-FDG PET/

CT, likely due to the higher ^{11}C-methionine uptake in brain tumors and lower accumulation in normal brain tissue [9]. Recent study demonstrated that radiolabelled amino acid PET/CT may ameliorate the delineation of high-grade gliomas compared to standard MRI [10]. ^{18}F-FDOPA PET/CT shows a pooled sensitivity and specificity of 71% and 86%, respectively in detecting newly diagnosed gliomas [11]. ^{18}F-FET PET/CT demonstrated a pooled sensitivity and specificity of 82% and 76%, respectively for diagnosis of a brain tumor. ^{68}Ga-DOTANOC PET/CT shows high meningioma to background ratios and it has provided valuable additional information on the extent of meningiomas beneath osseous structures, especially at the skull base [12].

Evaluation of reliability of PET/CT tracers in primary brain tumors has been done more as compared to brain metastases.

8.4.2 Tumor Grading

As compared to ^{11}C-methionine and ^{18}F-FET PET/CT, a meta-analysis concluded that ^{18}F-FDG PET/CT has lower sensitivity but higher specificity in distinguishing between high-grade and low-grade gliomas [13]. Low-grade gliomas and high-grade gliomas generally show different degrees of FDG uptake.

A prospective study compared ^{18}F-FMISO and ^{18}F-FDG PET/CT in 14 grade IV and 9 grade II or III glioma patients and concluded an improved ability of ^{18}F-FMISO to distinguish glioblastoma from lower grades as compared with ^{18}F-FDG [14].

8.4.3 Disease Prognosis

A recent meta-analysis demonstrated that poor overall survival could be indicated by an increased TBR (tumor to background ratio) of ^{18}F-FDG PET, ^{11}C-methionine PET, and ^{18}F-FET PET [15]. TBR and metabolic tumor volume are also significant prognostic parameters for patients with gliomas. Patients with an increased TBR have a higher mortality risk and patients with an increased metabolic tumor volume have a higher adverse events/mortality risk [16]. Anaplastic transformation is reliably indicated by a high FDG uptake in a previously known low-grade tumor [17]. Di Chiro et al. studied 45 patients proven high-grade brain tumors after surgery, radiation, and chemotherapy and concluded that poorly differentiated tumors showed significantly higher glucose metabolism than more differentiated ones [18]. Tumor-to-blood ratio above 1.2 identifies ^{18}F-FMISO uptake within hypoxic tissue above background normal tissue, hypoxic volume and tumor to-blood maximum value correlate with worsened prognosis in glioblastoma independently of other prognostic factors [19].

8.4.4 Diagnosis of Recurrent Brain Tumors and Differentiating between Post-therapy Radiation Necrosis from Residual/Recurrent Brain Tumor

Differentiating recurrent brain tumors from non-tumoral lesions after radiation therapy and/or chemotherapy is crucial for further management. Although characteristic appearances have been described on gadolinium enhanced MR imaging, it cannot reliably distinguish active tumor from post-radiation changes [20]. A meta-analysis showed a pooled sensitivity and specificity of 78% and 77%, respectively for ^{18}F-FDG PET/CT in detecting brain tumor recurrence [21]. There is no clear guideline with regard to optimal time for performing ^{18}F-FDG PET/CT after radiation; however, the general recommendation is that it should not be performed before 6 weeks after the completion of radiation treatment for this purpose [22]. One important point of caution is in patients receiving corticosteroids, as it results in generalized reduction in cerebral metabolic rate for glucose [23].

^{11}C-methionine PET/CT revealed a pooled sensitivity and specificity of 92% and 87%, respectively in detecting brain tumor recurrence and with higher values compared to ^{18}F-FDG PET/CT. Furuse et al. reported increased diagnostic performance of ^{18}F-FET PET/CT in this setting compared to ^{18}F-FDG and ^{11}C-methionine PET/CT [9]. ^{11}C-choline PET/CT has been shown to have a good diagnostic accuracy for differentiating glioma recurrence from radiation induced changes in a recent meta-analysis, with a pooled sensitivity and specificity of 87% and 82%, respectively [24]. Quantitative ^{18}F-FLT PET with kinetic modeling may distinguish tumor recurrence from radionecrosis [21].

8.4.5 Planning Stereotactic Biopsy

^{18}F-FDG PET/CT guided stereotactic brain biopsy has proven to be useful for improved delineation of anaplastic regions and better identification of neoplastic residues, thereby increasing the diagnostic yield of brain biopsy [25, 26].

Table 8.2 lists the role of PET in primary brain tumors and post-treatment.

Table 8.2 Lists the role of PET in primary brain tumors

Primary tumors	Post-treatment
To assess the grade of malignancy	To assess the grade of malignancy
Guides the site of biopsy	To identify residual high-grade malignancy
Planning for stereotactic surgery	Monitoring of progression
Gives prognostic information Effects of tumor on surrounding or remote brain areas information	Differentiating radiation induced necrosis from recurrence

8.5 PET Tracers

8.5.1 Glucose Metabolism

- ^{18}F-Fluorodeoxyglucose(^{18}F-FDG)

8.5.2 Amino Acid Transport

- ^{11}C-Methionine (^{11}C -MET)
- ^{18}F-Fluoroethyl tyrosine (^{18}F-FET)
- ^{18}F-dihydroxy phenylalanine (^{18}F-F DOPA)
- ^{18}F-Fluciclovone (^{18}F-FACBC)

8.5.3 Cellular Proliferation

- ^{18}F-Fluorothymidine (^{18}F-FLT)

8.5.4 Hypoxia Imaging

- ^{18}F-Fluoromisonidazole (^{18}F-FMISO)

8.5.5 Theragnostic Agent

- ^{68}Ga-Prostate Specific Membrane Antigen (^{68}Ga-PSMA) [27].

8.5.6 ^{18}F-FDG

- Useful properties of ^{18}F-FDG for brain tumor imaging:
  - GLUT1 is active at the normal BBB, which allows the radiotracer to reach the entire tumor volume.
  - Tumor cells uptake
    High glycolytic rates with high levels of enzymes that control glycolysis (Warburg effect)
    Low levels of glucose 6-phosphatase and changes in glucose transport rate
    Significant increase in functional GLUT transporters at the cell surface (especially GLUT 3) in aggressive tumors [27–29]
  - The uptake is flow independent and is irreversible.
  - The amount of radiotracer uptake by gliomas correlates with tumor grade and prognosis.

  - Has greater uptake in recurrent high-grade gliomas than in radiation necrosis and thus helps in distinguishing recurrent tumor from post-radiation necrosis [22].
- Limitations of ^{18}F-FDG in brain tumor imaging:
  - High normal brain tissue uptake that results in reduced tumor to background ratio.
  - Tumors with only a modest increase in glucose metabolism, i.e., low-grade tumors, may be difficult to identify.
  - In low-grade gliomas, uptake of FDG may be similar to that of normal white matter. Uptake in high-grade gliomas may be lower than or similar to that of normal gray matter. Because the similarities in glucose metabolic activity between tumors and normal gray and white matter cause difficulties in the analysis of FDG PET images, several studies have shown that a delayed scan acquisition about 3 h post FDG injection significantly increases the contrast between malignant brain tumors and normal brain tissue [30].
  - Heterogeneity of ^{18}F-FDG uptake within a single tumor further decreases the sensitivity.
  - Various non-malignant intracerebral lesions also have varying levels of increased FDG uptake (e.g., with inflammatory or infectious causes) which makes differentiating between malignant and non-malignant causes of increased FDG uptake difficult.
- Applications of ^{18}F-FDG in brain tumor imaging:
  - Prognostication and grading of tumor—high degree of FDG uptake is usually seen in high-grade tumors and correlates with a poorer prognosis. Glioblastoma can be excluded by a negative FDG PET.
  - Delineation of point of highest uptake for stereotactic biopsy.
  - Assessment of tumor infiltration.
  - Differentiating recurrent high-grade glioma or brain metastases (presenting with hypermetabolism) from radiation necrosis (that have areas of hypometabolism).
  - Assessment of early recurrence.
  - Post-therapy response assessment. Contrast-enhanced MR imaging is unreliable in the post-therapy assessment.
  - Co-registration of FDG PET with MR imaging—allows better spatial resolution and anatomic correlation and improves the diagnostic performance.
  - In immunocompromised patients—differentiating tumor recurrence from non-malignant or benign lesions. For example, lymphoma versus toxoplasmosis (infectious/non-lymphomatous lesions—central hypometabolism in the setting of an enhancing lesion. Ideally lymphoma is more intense than toxoplasmosis but variability is seen leading to lower diagnostic accuracy in clinical practice).
  - Anaplastic transformation is reliably indicated by a high FDG uptake in a previously known low-grade tumor and is strong prognostic marker. Not sensitive for detecting recurrent low-grade tumors without anaplastic transformation [31].

  - Determination of residual malignant tissue after surgery.
- Evaluation of lesion on FDG PET:
  - Lesion activity should not be evaluated by absolute uptake value or ratio to untreated normal brain tissue.
  - Lesion activity should be evaluated by whether it is above the expected background activity based on referencing to the uptake in the adjacent brain region and to have the MRI structural information available for correlation [32].
  - Suspicious lesion is any area of FDG uptake greater than the uptake of expected background activity using adjacent brain as reference or any FDG uptake in a region showing contrast enhancement on the co-registered MR images.
  - Pattern of FDG uptake
    Grade IV brain tumor (GBM)—heterogeneous high uptake
    Pilocytic astrocytoma, ganglioglioma—benign tumors with high ^{18}F-FDG uptake
    Meningiomas—uptake is variable
    Low-grade tumors—uptake usually similar to that of normal white matter
    High-grade tumors—uptake less than or similar to that of normal gray matter
    Other malignant tumors with high ^{18}F-FDG uptake—PNET, medulloblastomas, malignant lymphomas, and brain metastases from systemic cancers

8.6 Amino Acid Tracers

- Advantages and applications of amino acid tracers:
  - Amino acid uptake, especially in gliomas, is related to increased facilitated transport mediated by type L amino acid carriers (LAT). Tumors increase transporter expression in their vasculature. Increased uptake in most low-grade gliomas in the absence of BBB damage is a substantial advantage over CT, conventional MRI, and ^{18}F-FDG PET [30].
  - Radiolabelled AAs are considered superior to ^{18}F-FDG PET for almost all neuro-oncology applications. The most significant advantage of these tracers is related to their markedly lower background activity in normal brain tissue compared with FDG enabling the detection of smaller tumors (primary or recurrent).
  - Because amino acid tracer uptake in brain tumors does not depend on the BBB, delineation of tumor extent is reportedly more accurate, and areas with increased uptake on PET are often larger than areas with contrast-enhanced lesions on MRI Therefore, amino acid PET can be useful for treatment planning and better outcomes for patients have been reported with radiotherapy planned on the basis of tumor extent as defined using amino acid PET [28].
  - ^{11}C-MET and ^{18}F-FET provide more accurate estimates of gross tumor volume and margins and guidance for tumor biopsies than contrast-enhanced MR imaging and CT.

- For the diagnosis and management of patients, especially the ones with low-grade tumors, methionine PET is significantly superior to FDG PET. Methionine PET can also detect recurrent tumors with higher sensitivity and specificity than FDG PET, allowing differentiation between tumor recurrence and radio-necrosis.
- Because of the short half-life of ^{11}C (20 min), ^{18}F-labeled (half-life—110 min) amino acid tracers were developed, such as FDOPA and FET. ^{18}F-FET can distinguish high-grade from low-grade gliomas as this tracer exhibits differences in the time activity curves of tracer uptake depending on tumor grade. FET has high specificity and sensitivity in primary brain tumors. Using ^{18}F-FET-PET recurrent tumors can be differentiated from non-neoplastic post-therapeutic changes, with a sensitivity and specificity of approximately 90% [22].
- FDOPA, like the other amino acid tracers, is also more sensitive and specific than FDG and is particularly valuable for diagnosing recurrent low-grade gliomas. It is also useful for differentiating radiation necrosis from recurrent tumors. Figure 8.1 shows the ^{18}F-FDG PET/CT and ^{18}F-FDOPA PET/CT images of a 56-year-old female with a history of gliomatosis cerebri who

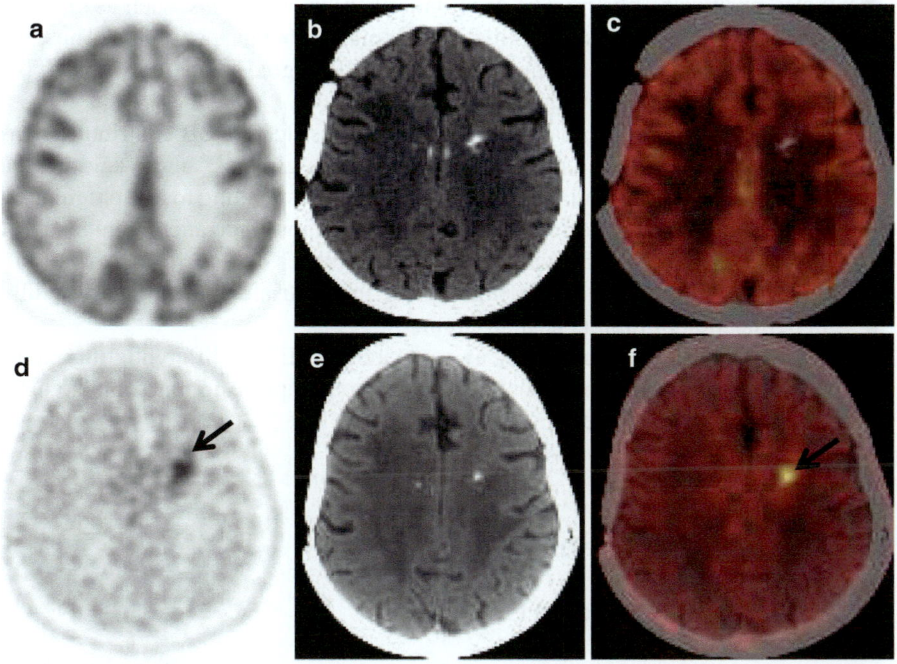

Fig. 8.1 (**a**) Maximum intensity projection (MIP) image of 18F FDG PET-CT showing physiologic tracer uptake in the brain no abnormal tracer uptake. (**b**) Axial CT brain showing few irregular calcific densities in left frontal region showing no significant FDG uptake on fused PET-CT image (**c**). (**d**) Maximum intensity projection (MIP) image of 18F FDOPA PET-CT showing focal increased tracer uptake in left frontal region corresponding to increased DOPA uptake in the calcified lesion seen in the left frontal region on axial CT (**e**) and fused PET-CT (**f**) (black arrows)

underwent right temporo-parietal craniotomy, chemotherapy, and radiotherapy for the same and after resolution, presented with new onset seizures. Recurrence was suspected but MRI was reported as negative for any recurrent lesions. FDG PET scan was done that showed no hypermetabolic lesions. An FDOPA PET/CT done for correlation showed a recurrent hypermetabolic lesion in the left frontal cortex for which patient underwent another surgery and was confirmed as recurrence of histopathology. Therefore, in this case the MRI and FDG PET were false negative while FDOPA PET was true positive.
  - The diagnostic value of MRI and CT concerning changes in tumor size or contrast enhancement in response to therapy is limited because the known reactive transient BBB alterations with consecutive contrast enhancement may mimic tumor progression which is called pseudo-progression that might lead to over-treatment. Amino acid tracers are valuable tools for treatment monitoring because they help in assessing treatment response early (response suggested by deactivation of amino acid transport) and evaluating patient prognosis after therapy [29].
- Limitations of amino acid tracers [22, 30–32]:
  - False-positive results may be seen under benign conditions, such as cases of demyelination, leukoencephalitis, hematomas, ischemia, or abscess. Thus, the predictive value is limited, and a biopsy is still the gold standard in most of the unknown space-occupying lesions in the brain.
  - In high-grade gliomas, tracer leakage from a disrupted BBB contributes considerably to amino acid uptake but in low-grade gliomas, amino acid uptake occurs without substantial BBB breakdown, corresponding to an upregulation of LAT1. Hence, the correlation between the tumor grade and the intensity of amino acid radiotracer uptake remains doubtful.
  - ^{18}F-FDOPA has an inherently high uptake in the striatum leading to difficulty in delineation of tumors near striatum.
  - ^{11}C-MET has lower specificity as compared to FET in differentiating recurrence from radiation necrosis as it shows some uptake in macrophage also. The sensitivity however is similar.
  - ^{11}C-MET has a short half-life.

8.7 Other Radiopharmaceuticals

8.7.1 ^{18}F-Fluorothymidine (18F-FLT)

It relies on the upregulation of DNA synthetic pathways, due to increased cellular proliferation in tumor cell compared with the normal brain cells. There is upregulation of TK1 expression inside tumor or other rapidly proliferating cells. It serves primarily as a marker of the fraction of tumor cells in the S-phase of cell cycle at the time of imaging (proliferation rate) [30–32].

Limitation: The cellular uptake is not specific to tumor cells, and any proliferative process may show increased ^{18}F-FLT uptake.

Uses:

- Tumor grading: high correlation between ^{18}F-FLT uptake and the high-grade gliomas but significant variability in uptake in low-grade gliomas.
- Recurrence: more sensitive compared to FDG.
- Response assessment: positive correlation between uptake (SUVmax) and Ki-67 (marker of cell proliferation) suggesting that 18F-FLT PET can serve as an early predictor for response to therapy in high-grade gliomas. >25% reduction in tumor FLT uptake (as measured by SUV) was found to be a predictive metabolic response.

8.7.2 ^{18}F-FMISO

Has been developed as a PET tracer for hypoxia imaging.

Hypoxia leads to resistance to radiotherapy and tumor progression.

Enters tumor cells by passive diffusion, where it is reduced by nitroreductase enzymes to become trapped exclusively in hypoxic cells.

Uptake is independent of perfusion and BBB disruption and is found in high-grade gliomas but not in low-grade gliomas [27]. FMISO only concentrates in viable cells but not dead, necrotic cells. Can be used as a prognostic agent for treatment outcomes and has a role in directing and monitoring targeted therapy.

8.7.3 ^{11}C-Choline (CHO) and ^{18}F-Fluorocholine (FCHO)

Cellular uptake of ^{11}C-CHO is mediated by choline transporters that are upregulated in many human cancers. CHO transporters are active at the BBB allowing access of ^{11}C-CHO to non-enhancing regions within gliomas. The entry of CHO into cells is followed by phosphorylation by choline kinases and subsequent use in phospholipid metabolic pathways. Thus, ^{11}C-CHO accumulation in tumor cells indicates increased demand for phospholipid synthesis, including membrane synthesis. ^{11}C-CHO has uptake in gliomas that may assist in tumor grading and for visualizing the non-enhancing regions within glioma [29]. Apart from the short half-life of C^{11}, its limitation is that it shows uptake in normal brain structures such as the choroid plexus and pituitary gland. It is a better agent than FDG for differentiating recurrence from radiation necrosis [31].

^{18}F-FCHO has poor penetration of the BBB, likely due to differences in recognition by transporters at the BBB, limiting the utility of ^{18}F-FCHO in gliomas.

8.7.4 SSTR Ligands (^{68}Ga DOTA-TOC and DOTA-NOC)

These are somatostatin analogs useful in localizing meningiomas because of high expression of the somatostatin receptor subtype 2 in meningiomas. These agents are helpful in visualizing the location and extent of meningiomas near bony structures especially at the base of the skull and dural sinuses [28, 29].

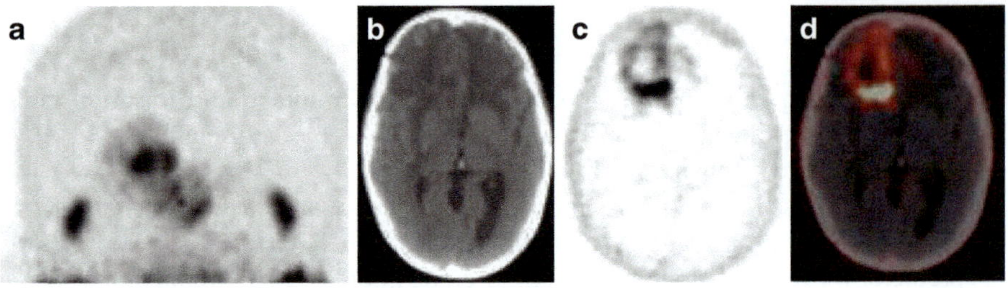

Fig. 8.2 (**a**) Maximum intensity projection (MIP) image of 68 Ga PSMA PET-CT showing increased tracer uptake in the right frontal region. (**b**) Axial CT section of brain showing ill-defined lesion in the right frontal region showing increased tracer uptake in PET only (**c**) and fused PET-CT images (**d**)

8.7.5 ^{68}Ga-PSMA

In the evaluation of gliomas, Ga-PSMA-11 PET/CT brain imaging is a potentially useful imaging tool and recent studies have shown that it can be used for diagnosing recurrent gliomas [33]. Absence of physiological uptake of this tracer in the normal brain parenchyma results in high tumor to background values and consequently better visualization of glioma lesions. Figure 8.2 shows the ^{68}Ga PSMA PET/CT images of a 25-year-old female with primary frontal glioblastoma that was treated with frontal craniotomy and radiotherapy. Patient was on follow-up when she presented with new onset seizures, generalized weakness that led to her being wheelchair bound. MRI was done and was suggestive of recurrence. ^{68}Ga PSMA PET/CT showed uptake in the right frontal region suggestive of recurrence. Both MRI and PSMA PET/CT were concordant in this case. PSMA can be potentially used as a therapeutic agent by chelating it to Lutetium^{177} but further studies are required.

8.8 PET-MRI

Potential advantages of PET/MRI include high soft-tissue contrast and functional MRI capability (additional imaging options such as spectroscopy, functional MRI, diffusion- and perfusion-weighted imaging complementary to PET) [28].

8.9 Conclusion

PET imaging with oncologic radiotracers is a very useful modality for diagnostic and therapeutic planning in patients with brain tumors. Future advancements in PET tracers and imaging technologies will broaden the applications of PET imaging in the field of brain tumors even further.

References

1. Petrirena GJ, Goldman S, Delattre J-Y. Advances in PET imaging of brain tumors: a referring physician's perspective. Curr Opin Oncol. 2011;23(6):617–23.
2. Herholz K, Langen K-J, Schiepers C, Mountz JM. Brain tumors. Semin Nucl Med. 2012;42(6):356–70.
3. Herholz K. Brain tumors: an update on clinical PET research in gliomas. Semin Nucl Med. 2017;47(1):5–17.
4. Chen W. Clinical applications of PET in brain tumors. J Nucl Med. 2007;48(9):1468–81.
5. Treglia G, Muoio B, Trevisi G, Mattoli MV, Albano D, Bertagna F, et al. Diagnostic performance and prognostic value of PET/CT with different tracers for brain tumors: a systematic review of published meta-analyses. Int J Mol Sci. 2019;20:4669.
6. Janus TJ, Kim EE, Tilbury R, Bruner JM, Yung WK. Use of [18F]fluorodeoxyglucose positron emission tomography in patients with primary malignant brain tumors. Ann Neurol. 1993;33:540–8.
7. Spence AM, Muzi M, Mankoff DA, O'Sullivan SF, Link JM, Lewellen TK, et al. 18F-FDG PET of gliomas at delayed intervals: improved distinction between tumor and normal gray matter. J Nucl Med. 2004;45:1653–9.
8. Hoffman JM, Waskin HA, Schifter T, Hanson MW, Gray L, Rosenfeld S, et al. FDG-PET in differentiating lymphoma from nonmalignant central nervous system lesions in patients with AIDS. J Nucl Med. 1993;34:567–75.
9. Zhao C, Zhang Y, Wang J. A meta-analysis on the diagnostic performance of (18) F-FDG and (11) C-methionine PET for differentiating brain tumors. AJNR Am J Neuroradiol. 2014;35:1058–65.
10. Verburg N, Hoefnagels FWA, Barkhof F, Boellaard R, Goldman S, Guo J, et al. Diagnostic accuracy of neuroimaging to delineate diffuse gliomas within the brain: a meta-analysis. AJNR Am J Neuroradiol. 2017;38:1884–91.
11. Xiao J, Jin Y, Nie J, Chen F, Ma X. Diagnostic and grading accuracy of (18) F-FDOPA PET and PET/CT in patients with gliomas: a systematic review and meta-analysis. BMC Cancer. 2019;19:767.
12. Henze M, Schuhmacher J, Hipp P, Kowalski J, Becker DW, Doll J, et al. PET imaging of SSTR using [68Ga]-DOTA-D Phe1-Tyr3- octreotide (DOTA-TOC): first results in meningioma- patients. J Nucl Med. 2001;42:1053–6.
13. Katsanos AH, Alexiou GA, Fotopoulos AD, Jabbour P, Kyritsis AP, Sioka C. Performance of 18F-FDG, 11C-methionine, and 18F-FET PET for glioma grading: a meta-analysis. Clin Nucl Med. 2019;44:864–9.
14. Hirata K, Terasaka S, Shiga T, Hattori N, Magota K, Kobayashi H, et al. 18F-fluoromisonidazole positron emission tomography may differentiate glioblastoma multiforme from less malignant gliomas. Eur J Nucl Med Mol Imaging. 2012;39:760–70.
15. Zhang Q, Gao X, Wei G, Qiu C, Qu H, Zhou X. Prognostic value of MTV, SUVmax and the T/N ratio of PET/CT in patients with glioma: a systematic review and meta-analysis. J Cancer. 2019;10:1707–16.
16. Kim YI, Kim Y, Lee JY, Jang SJ. Prognostic value of the metabolic and volumetric parameters of (11) C-methionine positron-emission tomography for gliomas: a systematic review and meta-analysis. AJNR Am J Neuroradiol. 2018;39:1629–34.
17. Borgwardt L, Højgaard L, Carstensen H, Laursen H, Nowak M, Thomsen C, et al. Increased fluorine-18 2-fluoro-2-deoxy-D-glucose (FDG) uptake in childhood CNS tumors is correlated with malignancy grade: a study with FDG positron emission tomography/magnetic resonance imaging coregistration and image fusion. J Clin Oncol. 2005;23:3030–7.
18. Di Chiro G. Positron emission tomography using [18F] fluorodeoxyglucose in brain tumors. A powerful diagnostic and prognostic tool. Investig Radiol. 1987;22:360–71.

19. Spence AM, Muzi M, Swanson KR, O'Sullivan F, Rockhill JK, Rajendran JG, et al. Regional hypoxia in glioblastoma multiforme quantified with [18F]fluoromisonidazole positron emission tomography before radiotherapy: correlation with time to progression and survival. Clin Cancer Res. 2008;14:2623–30.
20. Kumar AJ, Leeds NE, Fuller GN, Van Tassel P, Maor MH, Sawaya RE, et al. Malignant gliomas: MR imaging spectrum of radiation therapy- and chemotherapy-induced necrosis of the brain after treatment. Radiology. 2000;217:377–84.
21. Li Z, Yu Y, Zhang H, Xu G, Chen L. A meta-analysis comparing 18F-FLT PET with 18F-FDG PET for assessment of brain tumor recurrence. Nucl Med Commun. 2015;36:695–701.
22. Basu S, Alavi A. Molecular imaging (PET) of brain tumors. Neuroimaging Clin N Am. 2009;19:625–46.
23. Fulham MJ, Brunetti A, Aloj L, Raman R, Dwyer AJ, Di Chiro G. Decreased cerebral glucose metabolism in patients with brain tumors: an effect of corticosteroids. J Neurosurg. 1995;83:657–64.
24. Gao L, Xu W, Li T, Zheng J, Chen G. Accuracy of 11C-choline positron emission tomography in differentiating glioma recurrence from radiation necrosis: a systematic review and meta-analysis. Medicine (Baltimore). 2018;97:e11556.
25. Levivier M, Goldman S, Pirotte B, Brucher JM, Balériaux D, Luxen A, et al. Diagnostic yield of stereotactic brain biopsy guided by positron emission tomography with [18F]fluorodeoxyglucose. J Neurosurg. 1995;82:445–52.
26. Pirotte B, Goldman S, Brucher JM, Zomosa G, Baleriaux D, Brotchi J, et al. PET in stereotactic conditions increases the diagnostic yield of brain biopsy. Stereotact Funct Neurosurg. 1994;63:144–9.
27. Jung JH, Ahn BC. Current radiopharmaceuticals for positron emission tomography of brain tumors. Brain Tumor Res Treat. 2018;6(2):47–53.
28. Galldiks N, Lohmann P, Albert NL, Tonn JC, Langen K-J. Current status of PET imaging in neuro-oncology. Neuro-Oncol Adv. 2019;1(1):vdz010.
29. Shooli H, Dadgar H, Wáng YJ, Vafaee MS, Kashuk SR, Nemati R, et al. An update on PET-based molecular imaging in neuro-oncology: challenges and implementation for a precision medicine approach in cancer care. Quant Imaging Med Surg. 2019;9:1597–610.
30. Nuclear LI. Medicine imaging of brain tumors. In: Hodler J, von Schulthess GK, CHL Z, editors. Diseases of the brain, head and neck, spine 2012–2015. Milano: Springer Milan; 2012. p. 231–4.
31. Wong TZ, van der Westhuizen GJ, Coleman RE. Positron emission tomography imaging of brain tumors. Neuroimaging Clin N Am. 2002;12(4):615–26.
32. Suchorska B, Tonn JC, Jansen NL. PET imaging for brain tumor diagnostics. Curr Opin Neurol. 2014;27(6):683–8.
33. Sasikumar A, Kashyap R, Joy A, CharanPatro K, Bhattacharya P, Reddy Pilaka VK, et al. Utility of 68Ga-PSMA-11 PET/CT in imaging of glioma-a pilot study. Clin Nucl Med. 2018;43(9):e304–9.

Fundamentals of Radiation for Neuro-Oncology

9

V. R. Anjali, Supriya Mallick, and Rony Benson

9.1 Introduction

9.1.1 Important Events

1895: Wilhelm Conrad Roentgen discovered X-rays.
1896: Henri Becquerel discovered natural radioactivity.
1922: Claudius Regaud laid the concept of fractionation in radiotherapy.

9.1.2 Mechanism of Radiation Action

Radiotherapy utilizes ionizing radiation to kill tumour cells by causing DNA damage. Ionizing radiation possess enough kinetic energy to eject electron from the atoms or molecules, producing ions. The ionizations can be produced by direct action (protons, electron) or by indirect action. The indirectly acting photons (X-rays, gamma rays) interact with other atoms or molecules (usually water molecule) causing excitation, ionization and free radical formation. Hydroxyl ion is one of the most important free radicals generated with high potential for breaking the chemical bond and leading to DNA damage. The principal target for biological effects of radiation is DNA. In mammalian cell, almost two-third of DNA damage is caused by hydroxyl ions.

V. R. Anjali (✉)
Department of Radiation Oncology, All India Institute of Medical Sciences, New Delhi, India

S. Mallick
Department of Radiation Oncology, National Cancer Institute, AIIMS, Haryana, India

R. Benson
Mar Sleeva Medicity, Cherpunkal, Kerala, India

S. Mallick et al. (eds.), *Evidence based practice in Neuro-oncology*,
https://doi.org/10.1007/978-981-16-2659-3_9

Radiation damage can be divided into three types:

1. Lethal damage: irreversible and irreparable damage, inevitably leads to cell death.
2. Potentially lethal damage: damages modifiable by changing cellular microenvironment.
3. Sublethal damage: Usually repaired if adequate time is given.

Principle of therapeutic radiation:

- Dose of radiation producing complete and permanent regression of tumour in irradiated area is lethal dose.
- Optimum balance should be there between normal tissue toxicity and tumour cell killing.
- Ratio of normal tissue complication probability to tumour control probability is therapeutic index.
- If lethal dose is much higher than normal tissue tolerance, delivery of curative dose will be at the cost of toxicity.

Different types of radiation fractionation schedules are summarized below (Fig. 9.1).

X axis: Number of fractions per week.
Y axis: Dose (Gy) per fraction.
4Rs of radiobiology—Basis of radiation

- Repair
- Repopulation
- Redistribution
- Reoxygenation

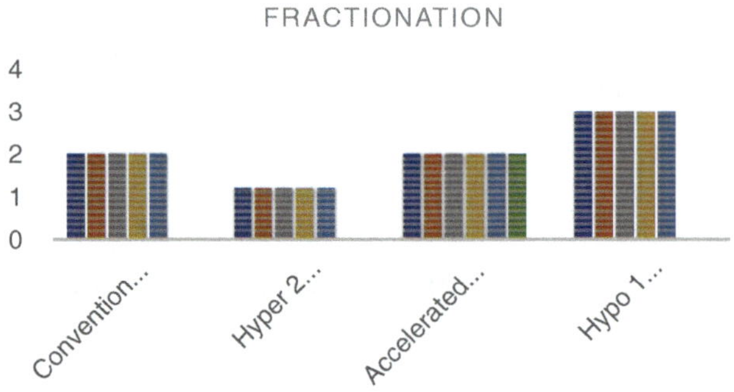

Fig. 9.1 Different radiation fractionation schedules

9.1.3 Evolution of Radiation Delivery

- Pre 1940s: Superficial and orthovoltage beams. In the kilovoltage era, treating deep seated tumour was difficult and skin toxicity was significant.
- 1950s: Megavoltage treatment by Cobalt 60(Co^{60}) machine, Gamma rays emitted by Co^{60} has an average energy of 1.25 MeV.
- 1970: LINAC produces X-rays with energy ranging from 6 to 20 MV and electron.
- Advantage of megavoltage: Better sparing of skin, subcutaneous tissue and there by lesser skin toxicity (Fig. 9.2).

9.1.4 Process of Radiation Treatment Planning

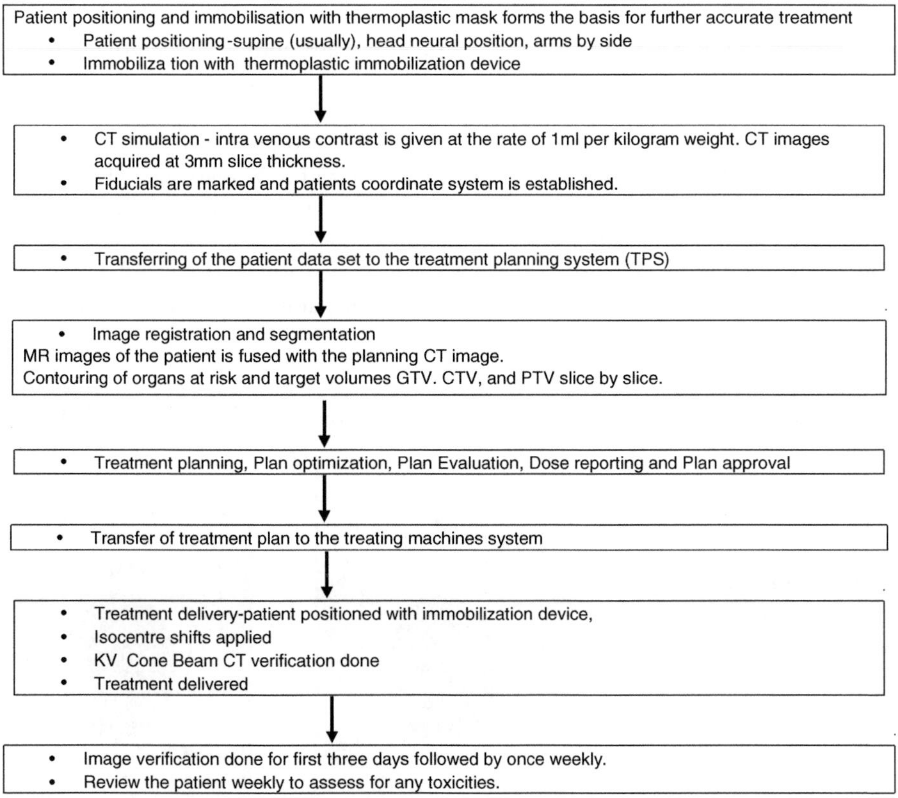

Fig. 9.2 The evolution of radiation techniques

9.2 Treatment Target Volumes and Organ at Risk

International Commission on Radiation Units and Measurements (ICRU) 50: defined GTV, CTV, PTV

ICRU 62: defined setup margin, internal margin

- GTV—Gross tumour volume is the gross macroscopic disease extent of tumour which is determined clinically as well as radiologically. It is the region of highest tumour load.
- CTV—Clinical target volume contains GTV and/or the microscopic subclinical tumour infiltrates that surround the primary tumour. This volume has to be treated adequately to achieve disease cure. CTV margins are not always given uniform expansion, it should be edited from the anatomical natural barriers where subclinical disease won't spread.
- PTV—Planning target volume incorporates the uncertainties in patient setup and internal organ motion (Fig. 9.3).

ORGAN AT RISK (critical normal structures) are normal tissue/organs which are in the vicinity of target volumes, whose radiation sensitivity may significantly influence treatment planning/prescribed dose.

In 1991, normal tissue tolerance for organ at risk was published by Emami et al. [1] TD 5/5 is the probability of 5% complication within 5 years and TD 50/5 the probability of 50% complication within 5 years of treatment. Modern treatment techniques have led to non-uniform radiation exposure to the normal tissue. Quantitative Analyses of Normal Tissue Effects in the Clinic (QUANTEC) provided normal tissue dose constrains based on dose-volume parameters (Tables 9.1 and 9.2).

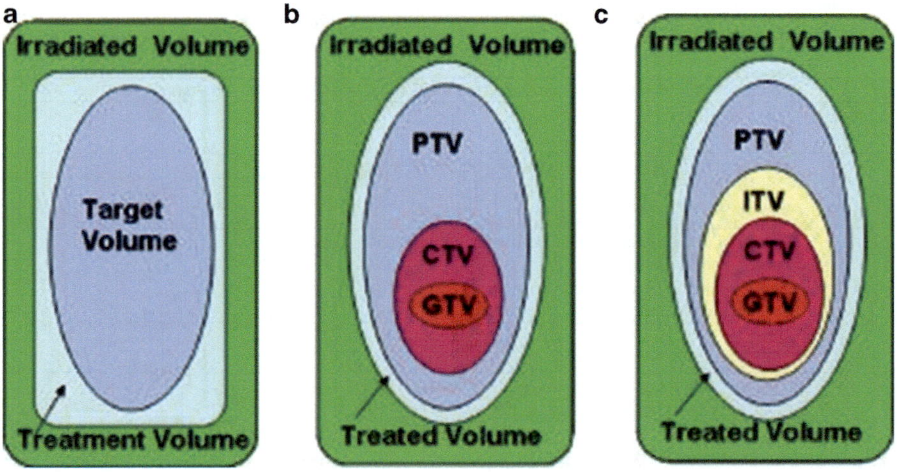

Fig. 9.3 The evolution of target volume. (**a**) ICRU 29 volumes; (**b**) ICRU 50 volumes; (**c**) ICRU 62 volumes

Table 9.1 Describes the OARs for brain irradiation and the tolerance dose limit according to Emami et al.

Emami normal tissue tolerance							
	TD 5/5 volume			TD 50/5 volume			
Organ at risk	1/3	2/3	3/3	1/3	2/3	3/3	End point
Brain	6000	5000	4500	7500	6500	6000	Necrosis infarction
Brain stem	6000	5300	5000	–	–	6500	Necrosis infarction
Optic nerve and optic chiasma	–	–	5000	–	–	6500	Blindness
Eye retina	–	–	4500	–	–	6500	Blindness
Middle ear	3000	300	3000	4000	4000	4000	Acute serous otitis
	5500	5500	5500	6500	6500	6500	Chronic serous otitis
Lens	–	–	1000	–	–	1800	Cataract requiring intervention
Parotid	–	3200	3200	–	4600	4600	Xerostomia
Spinal cord	5 cm	10 cm	20 cm	5 cm	10 cm	20 cm	Myelitis
	5000	5000	4700	7000	7000	–	Necrosis

Table 9.2 Describes the OARs for brain irradiation and the tolerance dose limit according to QUANTEC

QUANTEC: Dose-volume outcome for OARs					
Organ	Volume segmented	Technique	Dose/Volume	Rate (%)	End point
Brain	Whole organ	3DCRT	QQQ$D_{max} < 60$	<3	Symptomatic necrosis
			$D_{max} = 72$	5	
			$D_{max} = 90$	10	
Brain stem	Whole organ	3DCRT	$D_{max} < 54$	<5	Permanent cranial neuropathy or necrosis
			D10cc ≤ 59	<5	
Spinal cord	Partial organ	3DCRT	$D_{max} = 50$	0.2	Myelopathy
			$D_{max} = 60$	6	
			$D_{max} = 69$	50	
Cochlea	Whole organ	3DCRT	$D_{mean} < 45$	<30	Sensory neutral hearing loss
Optic nerve and optic chiasma	Whole organ	3DCRT	$D_{max} < 55$	<3	Optic neuritis
			D_{max} 55–60	3–7	

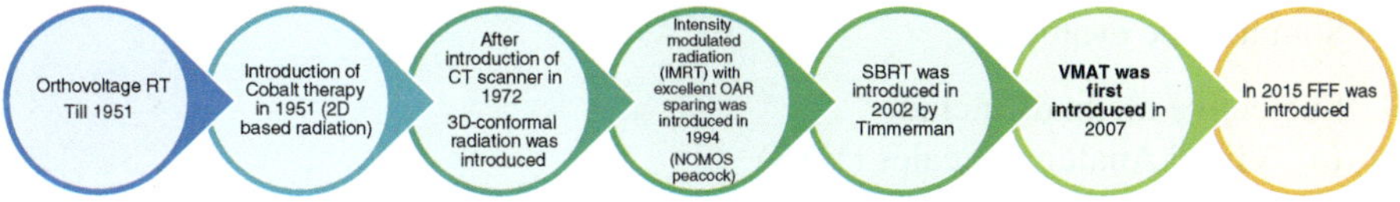

9.3 Radiation Induced Brain Injury

Radiation to the brain in therapeutic dose can cause significant injury.

9.3.1 Pathophysiology

	Hypothesis	Site	Outcome
1	Vascular hypothesis	Vessels	Ischemia and secondary white matter necrosis
2	Parenchymal hypothesis	Oligodendrocyte	Oligodendrocyte type 2 astrocyte fails to maintain myelin sheaths
3	Microglia	Microglia	Production of reactive oxygen species
4	Astrocytes	Astrocytes	Acute cellular swelling and chronic hypertrophy hyperplasia
5	*Neurons*	*Neurons*	DNA damage

Broadly classified as (a) acute effects, (b) early delayed effects, (c) late effects

(a) Acute effects: occur during or shortly after the radiation exposure which is secondary to oedema and disruption of the blood–brain barrier (BBB).

	Clinical manifestation	Management
1	Fatigue characterized by lack of improvement by rest	Methylphenidate 10 mg BD, escalating to 30 mg BD Side effect: Anxiety and insomnia
2	Dizziness	Midodrine 10 mg TDS Fludrocortisone 0.1 mg OD All these drugs cause hypertension, hypokalaemia
3	Increased intracranial pressure	Dexamethasone 8 mg TDS with PPI If not controlled mannitol infusion 1 mg/kg every 8 h Acetazolamide 250 mg BD
4	Hair loss	Regrowth can occur
5	Skin erythema	During radiation saline dressing, steroid skin cream may be applied Once radiation is completed topical emollients should be used

(b) Early delayed effects: Occurs 6–12 weeks after radiation exposure; due to transient demyelination; reversible.
   1. Generalized weakness.
   2. Somnolence.

Somnolence evaluation

1. Subjective methods include
   (a) Visual Analogue Scales (VAS)
   (b) Epworth Sleepiness Scale
2. Objectively
   (a) Physician rated scale
   (b) Littman Somnolence Syndrome (LSS) scale
   (c) Electro-encephalographic (EEG)

   These are reversible.

   Oral dexamethasone is used 4–8 mg TDS with PPI.

(c) Late effects: more than 1 year after radiation exposure leads to irreversible neurological consequences.
   1. Cognitive deficits
   2. Necrosis of brain
   3. Seizures
   4. Cranial neuropathy

Management policy for late delayed effects:

- Oral dexamethasone is prescribed most common: Initial dose of 4 mg BD for mild symptoms and 4 mg QID or 8 mg TDS for moderate to severe symptoms.
- If life threatening symptoms appear dose of dexamethasone 10–25 mg bolus followed by 4–10 mg QID.
- The patients should receive PPI along with dexamethasone therapy to reduce chances of gastritis.
- After symptomatic improvement dexamethasone should be tapered by 50% dose every third day starting 5–7 days after therapy initiation.
- For prolonged steroid therapy or if symptoms arise of oral thrush anti-fungal therapy should be used. Fluconazole 200 mg for first day subsequently 100 mg daily for 6 days should be used.
- Cognitive dysfunction occurs with total doses 20 Gy in adults and 24 Gy in children (conventional fractions).

9.4 Necrosis of Brain

1. Incidence of radiation necrosis varies widely as it is difficult to estimate the true diagnosis. The authors who used histopathology as the method of diagnosis concluded incidence 5–7%, whereas by radiological modality the incidence ranges from 20% to 25%.
2. Risk factors
   (a) Dose-volume: As the lesion size increases, deliverable dose should be reduced accordingly to limit the risk of toxicity.

Lesion size	(SRS) Safe dose (Gy)
≤20 mm	24
21–30 mm	18
31–40 mm	15

(b) Previous radiation exposure increases the chances of radiation necrosis.

Modality	Risk (%)
Previous SRS	20
WBRT	4
No WBRT	3

(c) Certain parts of brain are at higher risk of radiation necrosis. Frontal cortex has highest risk while the brain stem has least chances. Few studies have also highlighted that superficial lesions are at a lower risk.

(d) Margin for defining planning target volume should be minimized with help of image guidance.

3. Diagnosis and investigations for patients with suspected RN
   (a) While certain enhancement patterns described in the literature as "Swiss cheese", "soap bubble" or "cut green pepper" was initially thought to favour radiation necrosis, but these have only a 25% positive predictive value (Fig 9.4).
4. Management:
   - Not always an irreversible process.
   - Oral corticosteroid is the preferred first line.
   - VEGF inhibitor: bevacizumab appears to be a promising agent.
   - Oral pentoxifylline and vitamin E are also being attempted.
   - Laser interstitial thermal therapy (LITT) also appears to be a promising approach.
   - Surgical resection can be an option in selected cases.

Other late effects include:

- Endocrinopathies: Growth hormone deficiency followed by gonadotropin, TSH and ACTH deficiency
- Cataract
- Deafness
- Radiation induced second malignancy: 20-year cumulative incidence 3.3% vs. 1.2% for those who did not receive RT. Most common second malignancy are meningiomas followed by high grade glioma.

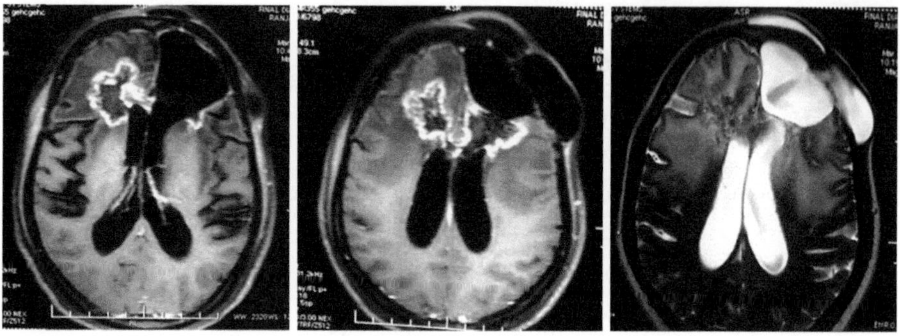

Fig. 9.4 Radiation necrosis on axial MRI images of brain

9.5 Reirradiation

Despite the multimodality curative intent treatment in the primary setting, majority of CNS tumours invariably recur during course of time. The recurrence occurs usually at the initial tumour site. Once disease recur/relapse, further salvage treatment is challenging. There is no consensus on salvage treatment for recurrent CNS tumours. Salvage options include surgery, reirradiation and systemic therapy.

Reirradiation is delivering radiotherapy for the second time to a volume that substantially overlap with previously irradiated area. Decision for reirradiation must be taken with caution, after assessing several factors and reirradiation is only given when potential benefits outweigh the toxicity and risk.

- Patient related factors
  - Age
  - Karnofsky performance status
  - Comorbidities (diabetes, CVA)
  - Duration of steroid use
- Disease related factor
  - Histology at diagnosis
  - Grade
  - Tumour volume
  - Time since previous radiotherapy
  - Tumour location
  - Proximity to critical structure or eloquent area
  - Tolerance during previous treatment
- Treatment related factors
  - Radiation dose received
  - Dose per fraction
  - Salvage surgery or chemotherapy received
  - Response to treatment

Table 9.3 Reirradiation prognostic score evaluation

Prognostic factor	Subgroup	Combs et al. [2] Value for each prognostic score	New combs prognostic score/ Kessel et al. [3] Value for each prognostic score
Histology at diagnosis	WHO I/II WHO III WHO IV	0 1 2	0 1 2
Age	<50 year ≥50 years	0 1	0 1
Time from primary RT to re-RT	>12 months ≤12 months	0 1	0 1
KPS	≥80 <80	–	0 1
Tumour volume PTV	≤47 ml >47 ml	–	0 1
Surgical resection	Yes No	–	0 1
		0—Excellent 1—Good 2—Moderate 3-4—Poor	a—0-1 b—2-3 c—4-5 d—6-7
		OS at 2 year 0-50% 1-41% 2-25% 3/4-6%	Median OS a—19.5 m b—11.3 m c—8.1 d—5.5
			MGMT methylation status and RT dose were also found significant in multivariate analysis for OS, but not included in scoring since information was available only for subgroup

- Prior radiotherapy technique
- Dose received by critical OARs

Prognostic factors which has significant correlation with survival are histology, age, time between initial radiotherapy to reirradiation, KPS, tumour volume, salvage surgical resection, extent of resection, MGMT methylation status and RT dose. The commonly used scoring systems developed by Combs et al. and Kessel et al. are summarized in Table 9.3.

9.5.1 Target Volume Delineation

GTV is contoured as the contrast enhancing lesion on T1weighted MRI.

CTV is not given usually.

PTV margin of 3–5 mm is added.

During planning, the dose to the critical structures are kept as low as possible.

9.5.2 Dose in Reirradiation

- The α/β value for brain is like late responding tissue like spinal cord.
- The α/β for normal brain is estimated between 2 and 3 [4].
- Reirradiation dose depends on:
  - Dose received during first radiotherapy
  - Time interval between first RT and reirradiation
- Fraction size should be decided depending on tumour volume and proximity to critical OARs.
  - Conventional fractionation radiotherapy (CFRT)
  - Hypo-fractionated stereotactic radiotherapy (HFSRT)
  - Stereotactic radiosurgery (SRS)

9.6 Threshold of Radiation Necrosis

- Most commonly occur for a cumulative equivalent dose in 2-Gy fractions >100 Gy, >105 Gy and >135 Gy, for patients treated with CFRT, FSRT and radiosurgery, respectively [5].
- Predicted RN 5% and 10% at a BED of 120 Gy and 150 Gy, respectively.
- In twice daily radiation, toxicity appeares at a BED of >80 Gy.
- For SRS toxicity steeply rises once the volume of brain tissue receives >12 Gy (5–10 cc).

Patient who had received a prior RT with a total dose of 60 Gy in 2 Gy fractions could receive reirradiation for an additional EQD2 of 40, 45 and 75 Gy with CFRT, FSRT and radiosurgery, respectively.

Longer the treatment interval, better is the tolerance.

Scoccianti et al. stratified tumour target volumes into small (<12.5 ml), medium (<35 ml) and large (<50 ml). The authors suggested following dose fractionation and technique:

Small tumours	EQD2 < 65 Gy	SRS
Medium tumours	EQD2 < 50 Gy	HFSRT
Large tumours	EQD2 < 50 Gy	CFRT

- Brachytherapy options for reirradiation includes interstitial brachytherapy, GliaSite, Gliadel wafers.
- Iodine 125 is used for interstitial brachytherapy.
- Median dose to the periphery of tumour used nearly 60 Gy, at a median dose rate of 40 cGy/h.

9.7 OAR Delineation

OAR	If contouring on MRI always double check on CT in case of misalignment
Brainstem	The foramen magnum to the point where the optic tract passes lateral to the midbrain (this upper limit is arbitrary but easy to define and ensures consistency). Again, for consistency, the quadrigeminal (tectal) plate should be included
Chiasm	Sits above and behind the anterior clinoids and runs backwards above the sella turcica. For consistency, the anterior and posterior "limbs" should extend 5 mm to include the start of the optic nerves anteriorly and optic tracts posteriorly. The chiasm can sometimes only be seen on a single slice as it is about 3 mm thick in cranio-caudal direction. It is often easiest to identify in the coronal plane
Cochlea	Sits just anterior to the lateral aspect of the internal auditory canal. They are most easily identified on the CT bone windows as small caves in the bone measuring 4–6 mm. Contour on 3 slices otherwise too small for dose calculation algorithms
Eyes	The whole of the outside of the globe should be contoured to include sclera and cornea. The macula lies opposite the lens
Lacrimal glands	These can be difficult (sometimes impossible!) to see, -but they lie on the superior and lateral aspect of the globe with the inferior border at the (axial) equator of the globe and wrap around superiorly about 30° (i.e. face on—left eye from 1 to 3 o'clock and right eye 9–11 o'clock). They sit anterior to the (coronal) equator of the globe. Dose limits should not be used to compromise PTV dose
Lens	Usually easy to see on the CT scan. However as cataracts are easily treatable the dose limits should never compromise PTV dose
Optic nerves	From the back of the globe to the optic chiasm passing through the optic canal to enter the skull anterior and inferior to the anterior clinoid. To help identify the exact path through the orbit change to CT bone windows. Ensure they join up with optic chiasm. It may be useful to check the structure in the sagittal plane to ensure the outlined structure is not an extra-ocular muscle
Pituitary	Within the sella turcica with chiasm lying superior and anterior to the stalk. As hypopituitarism is easily treatable the dose limits should never compromise PTV dose

References

1. Emami B, Lyman J, Brown A, et al. Tolerance of normal tissue to therapeutic irradiation. Int J Radiat Oncol Biol Phys. 1991;21(1):109–22.
2. Combs SE, Edler L, Rausch R, et al. Generation and validation of a prognostic score to predict outcome after re-irradiation of recurrent glioma. Acta Oncol. 2012;52:147–52.
3. Kessel KA, Hesse J, Straube C, Zimmer C, et al. Validation of an established prognostic score after re-irradiation of recurrent glioma. Acta Oncol. 2017;56:422–6.
4. Lee AW, Foo W, Chappell R, et al. Effect of time, dose, and fractionation on temporal lobe necrosis following radiotherapy for nasopharyngeal carcinoma. Int J Radiat Oncol Biol Phys. 1998;40:35–42.
5. Mayer R, Sminia P. Reirradiation tolerance of the human brain. Int J Radiat Oncol Biol Phys. 2008;70:1350–60.

Suggested Reading

Scoccianti S, Detti B, Gadda D, et al. Organs at risk in the brain and their dose-constraints in adults and in children: a radiation oncologist's guide for delineation in everyday practice. Radiother Oncol. 2015;114(2):230–8.

Stereotactic Radiosurgery: Planning and Evaluation

10

Sandeep Muzumder and M. G. John Sebastian

10.1 Introduction

Radiation therapy (RT) has evolved over time from manual planning to conformal therapy. The use of accurate target definition utilizing imaging, treatment planning system, radiation delivery techniques, and on-board image verification has led to contemporary stereotactic radiosurgery (SRS) practice. Traditionally, the procedures describing stereotactically guided, single-session radiation delivery would be defined as SRS and Gamma-knife (GK) alone was used. In contrast, procedures that use image-guided, multiple session radiation delivery to the entire target would be defined as stereotactic radiation therapy (SRT). SRT could be further subdivided into hypo-fractionated SRT (fractionation typically during 1 week in 3–5 sessions) or fractionated SRT (fractionation by more conventional schedules, typically during 4–6 weeks with 20–28 sessions) [1]. Presently frameless, image-guided, and single session radiation therapy delivered with a modified linear accelerator (Linac) is considered as SRS (or frameless SRS). Also, various authors refer to hypo-fractionated SRT as fractionated SRS. The present chapter deals with understanding plan evaluation of either linac- or GK-based SRS.

The SRS has transformed the management of tumors (vestibular schwannomas, meningiomas, brain metastases, pituitary adenomas, glomus jugulare, and glial tumors), arteriovenous malformations (AVM), and functional disorders (e.g., trigeminal neuralgia) of brain [2]. But the practice of SRS remains inhomogeneous for various reasons. Firstly, the various technologies used in delivery: GK (many generations), and linac-based SRS (Cone-based/micro multi-leaf collimator-based/robotic SRS or cyberknife/tomotherapy). Secondly, the diverse practice in planning:

S. Muzumder (✉) · M.G. John Sebastian
Department of Radiation Oncology, St. John's Medical College and Hospital, Bengaluru, Karnataka, India

S. Mallick et al. (eds.), *Evidence based practice in Neuro-oncology*,
https://doi.org/10.1007/978-981-16-2659-3_10

no margin to few millimeters for PTV expansion, optimal isodose for dose prescription varies from 50% to 80%, and array of metrics and indices being used for plan evaluation. Thirdly, shifting paradigms: whole brain radiation therapy to SRS for oligometastases in the brain, and treating single lesion to multiple brain lesions. Lastly, difficult interpretation of SRS literature due to many techniques being used, mixed data of benign and malignant disease, and paucity of objective long-term toxicity data. These factors make the SRS evaluation process even more complex.

10.2 Plan Evaluation of SRS

The steps involved in evaluating a conformal radiation plan are: patient selection, immobilization and simulation, contouring, beam arrangements, qualitative assay: Spatial dose distribution, quantitative assay: Dose-volume histogram (DVH), conformity index (CI), and homogeneity index (HI), and target volume (TV) and organ at risk (OARs) dose constraints evaluation [3, 4]. The evaluation process of a SRS plan is like assessing a conformal plan, albeit more intricate and usage of additional quality metrics [5, 6]. The chapter recommends considering the following steps to finalize a SRS plan.

10.2.1 Technique and Target Selection

A meta-analysis of SRS in neuro-oncologic disease showed excellent clinical outcomes (tumor stability, survival, and complications) in treating several pathologies with current stereotactic armamentarium [7]. Selecting the best SRS technique depends primarily on target size, location, shape, and its relationship to critical OARs, prescription dose, and the equipment used for radiation delivery [5]. In a multivariate analysis of the RTOG 95-05 study, maximum tumor diameter was associated with a significantly increased risk of grade 3, 4, or 5 neurotoxicity. Tumor sizes of 21–40 mm were 7.3 to 16 times more likely to develop grade 3–5 neurotoxicity compared to tumors <20 mm [8]. Larger tumors had a greater risk of unacceptable CNS toxicity than smaller ones. Hence, it is recommended to report the three maximum dimensions of the target for SRS [9]. The dose, volume, and location have complex relations with adverse radiation effects (ARE). In a study of 755 patients with AVM treated on GK with a median margin dose of 20 Gy, larger AVM volume, higher margin dose, and larger 12-Gy volume were associated with symptomatic ARE. Also, the 5-year cumulative rates of irreversible symptomatic ARE were 9.1% in thalamus, 12.1% in brainstem, and 1.4% in other locations. A target volume of 1, 4, 13, or 30 mL corresponded to an AVM with an average diameter of approximately 1.3, 2, 3, and 4 cm in diameter. In the brainstem and thalamic location, even smaller target volumes receiving a lower margin dose developed ARE. A target volume of 1 to 4 mL at the brainstem and thalamus receiving more than 16 to 20 Gy margin dose, developed 37.5% and 25% symptomatic ARE respectively, compared to none at other locations [10].

10.2.2 Dose Calculation Grid

The dose calculation grid influences all the evaluation parameters cited later, but often not considered in plan comparison. When computation of dose distribution is discretized by calculating it in a three-dimensional matrix format, lots of uncertainties are involved. When calculation points are interpolated, these discretization produces an unavoidable error especially in high dose gradient regions. Niemierko et al. recommended a grid size of 2.5 times the largest acceptable position error, when the error is in the range of 1 to a few millimeters [11]. The error of 1 mm in general is acceptable for SRS, hence the dose calculation grid can have a maximum value of 2.5 mm. Many planning systems have the option to adapt the calculation grid as required. As the number of calculation points increases, the accuracy increases at the cost of calculation time. The planner must strike a balance between dose accuracy and calculation time. The AAPM TG-101 report recommends the use of an isotropic grid size of 2 mm or finer for dose calculation grid in SRT [12].

10.2.3 Isodose Prescription

Historically, dose prescription in SRS has been using the so-called "coverage isodose" to represent the dose to the target. There is large variation in the prescription methods and there are no strict recommendations for dose prescription. The ICRU report 91 defines the dose prescription in SRS as a dose delivered to the outer border of the PTV or on an isodose surface that most optimally conforms to the outline of the PTV with optimally restricted dose to the OAR [13]. There is no ideal isodose line (IDL) for prescription in SRS and must be customized. Traditionally, the dose was prescribed to 50% IDL in GK-based SRS. The original basis for adopting this practice was built on the premise that when viewing the sigmoid curve of the dose profile of a single isocenter, the steepest dose gradient is around the half maximum, i.e., the 50% IDL. By lowering prescription IDL, the maximum dose inside the PTV increases. This may create problems for critical structure inside the PTV. The optimal prescription IDL appears to be between 50% and 75% for linac-based SRS. The larger targets tend to have higher optimal prescription IDLs. By choosing an optimal IDL, the volume of irradiated normal brain will be reduced while delivering a higher mean dose to the tumor [14]. A dosimetric study reports prescriptions at 60% and 70% IDL to provide best normal tissue sparing in linac-based SRS [15]. For robotic SRS (Cyberknife), a low IDL plan outperformed high IDL plans for all metrics in tumors >0.5 cc [16]. The local control (LC) rate of malignant lesions must be taken into consideration in IDL prescription for SRS. In a retrospective analysis of 134 patients with 374 brain metastases with a median survival of 8.7 months (range 0.2–64.8), median maximum tumor dimension of 8 mm (range 2–62 mm), and median margin dose of 20 Gy (range 5–24 Gy), the 12-month LC rate was 88.7%. The progression free survival and LC significantly improved with increasing IDL with p value of 0.003 and 0.04, respectively [17]. Hence, it is recommended to prescribe the higher IDL between 50% and 75% when feasible. The modern treatment planning system gives the option of dose prescription to cover desired percentage of target, to achieve minimum/mean/maximum dose to a structure, to receive a definite dose to a point, and to prescribe a relative or absolute IDL. The treating RO may choose a prescription method based on target coverage and dose to OARs.

10.2.4 Dose to Brain and Critical OARs

The radiographic imaging can provide a quantitative measurement of posttreatment changes like radionecrosis. However, the clinical significance of these radiation-induced transient inflammation, radionecrosis, and pseudo-progression is not clear. The focal symptoms directly attributable to radiosurgery including spinal cord myelopathy, visual deficits, hearing loss, and deficits involving specific cranial nerves can also be quantified. Though important, there is paucity of data on global problems like decline in neurocognitive function, quality of life, and hampered daily functions. The Quantitative Analyses of Normal Tissue Effects in the Clinic (QUANTEC) selected those studies that used clinical endpoints most appropriate for both the anatomic site and disease being treated (Table 10.1) [24].

For brain, receiving 12 Gy or more has been shown to correlate with both the incidence of radiation necrosis and asymptomatic radiologic changes. The risk of complications increases with the size of the target volume. Toxicity increases rapidly once the volume of the brain exposed to >12 Gy is >5–10 cm^3 [18]. For optic apparatus in single-fraction SRS, collective data from multiple studies have indicated that the incidence of radiation-induced optic neuropathy (RION) is rare for a D_{max} < 8 Gy increase in the range of 8–12 Gy, and becomes >10% in the range of

Table 10.1 OARs constraints

Sl. No.	Organ	Fractions	Endpoint	Dose-volume parameter	Rate	Evidence
1.	Brain	1	Symptomatic necrosis	V_{12} < 5–10 ml	<20%	[18]
2.	Optic apparatus	1	Optic neuropathy	D_{max} < 8 Gy D_{max} < 12 Gy	Rare <10%	[19]
3.	Brainstem	1	Permanent cranial neuropathy or necrosis	D_{max} < 12.5 Gy	<5%	[20]
4.	Spinal cord	1	Myelopathy	D_{max} < 13	1%	[21]
5.	Cochlea	1	Symptomatic hearing loss	Marginal dose ≤ 14 Gy	Rare	[22]
6.	Hippocampi	1	Neurocognitive decline	D100 ≤ 4.21 Gy, D40 ≤ 4.50 Gy, D_{max} ≤6.65 Gy	–	[23]

Dose-volume parameter of cochlea derived from SRS of vestibular schwannoma; Dose-volume constraints for Hippocampi are converted to a single-fraction scheme using equivalent doses in 2 Gy fractions with an α/β of 2

12–15 Gy [19]. For single-fraction SRS, maximum brainstem dose of 12.5 Gy is associated with a low (<5%) risk of complications. Higher doses (15–20 Gy) have been used with low reported incidence of complication in patient groups with poor prognosis for long-term survival [20]. For partial cord irradiation as part of spine radiosurgery, a maximum cord dose of 13 Gy in a single fraction or 20 Gy in three fractions appears associated with a <1% risk of injury [21]. For SRS in vestibular schwannoma (VS), the prescription dose should be limited to 12–14 Gy for hearing preservation. A hypofractionation schedule is suggested for better hearing preservation. A total prescription dose of 21–30 Gy in 3–7 Gy per fractions over 3–10 days is suggested, though data on this schedule are limited [22]. Hippocampal constraints of $D100 \le 4.21$ Gy, $D40 \le 4.50$ Gy, and $D_{max} \le 6.65$ Gy was considered successful hippocampal sparing for single-fraction GK-SRS [23]. When dose constraints of critical OARs is not achievable, the treating RO must either decrease prescription dose or transit from single-session SRS to fractionated SRS.

10.2.5 Qualitative Assessment: Spatial Dose Distribution

The result of the radiation plan is a prediction of distribution of dose deposited within the patient as seen on a planning scan. The various methods of dose display are: IDL, color wash, and isodose surface. The plan must be evaluated slice-by-slice viewing all structures and desired spectrum of isodose. A visual inspection of isodose superimposed on planning CT scan is the most valuable tool for qualitative assessment of a plan. It allows analysis of each dose level in relation to target and OARs and provides a thorough discernible spatial dose distribution. The prescription IDL must cover the target sparing critical OARs. Any inadequate coverage or excessive dose spillage outside the PTV should be identified. The slice-by-slice spatial dose distribution is the most direct and informative method of assessing a plan. All other methods of assessing dose distribution namely quantitative assay are surrogate to this and involve a loss of information to some extent. The subjective nature of visual assessment and no comparable indices for competing plans are major drawbacks.

10.2.6 Quantitative Assessment: DVH and Derived Indices

10.2.6.1 Dose-Volume Histogram (DVH)

The DVH is a graphical display of the dose distribution for each structure outlined in the planning CT scan. The X-axis represents dose in relative (%) or absolute (Grays) value, and Y-axis indicates target or OARs in relative (%) or absolute (in cc) volume. The ideal coverage is a controversial issue. Hazard et al. propose a dose prescribing to largest IDL to fulfil the following criteria: 95% of the target volume (TV) was encompassed by the prescription isodose volume (PIV) and 99% of TV was covered by 95% of the prescription dose [25]. Though variations are acceptable depending on treating radiation oncologist (RO) on a case-by-case basis. For OARs, the DVH evaluation is based on radiobiological principles that, when a percentage volume of OAR receives a certain dose, will have a relative probability of developing adverse effects of radiation [26]. In addition, DVH provides the statistical distribution of dose, such as maximum dose, minimum dose, mean dose, and median dose. The treating RO may compromise between PTV coverage and OAR constraints, to avoid unacceptable toxicity. In SRS, dose to critical OARs takes priority over target coverage.

The DVH must be used with caution. As spatial information is lost, DVH is only a secondary check. The appropriateness of target and OAR spatial coverage cannot be assessed by DVH. The DVH could report 100% coverage of the PTV by the prescription dose, but the PTV could be delineated incorrectly. There may be an excessive dose spillage in OARs, which can be seen only in a graphical dose distribution and not through a DVH. Also, only contoured structures are analyzed.

The recommended indices generated from DVH are cited in Table 10.2. These indices are used to evaluate and compare SRS plans having similar qualitative and DVH characteristics. The authors have classified the dose variation into three categories with respect to target for explanation: dose variation at edge of target, dose variation outside target, and dose variation within target.

Table 10.2 DVH indices

Sl. No.	Categories	Index	Definition	Ideal value
1.	Dose variation at edge of target	Conformity index	$\text{PITV} = \frac{\text{TV}}{\text{PIV}}$	1
		Paddick conformity index	$\frac{(\text{TV}_{\text{PIV}})^2}{\text{TV} \times \text{PIV}}$	>0.85
		New conformity index	$\frac{\text{TV} \times \text{PIV}}{(\text{TV}_{\text{PIV}})^2}$	1.18
2.	Dose variation outside target	Gradient index	$\frac{\text{PIV}_{\text{half}}}{\text{PIV}}$	<3.0
3.	Dose variation inside target	Homogeneity index	$\frac{\text{Maximum dose}}{\text{Prescription dose}}$	1
		ICRU-83 homogeneity index	$\frac{\text{D2\%} - \text{D98\%}}{\text{D50\%}}$	0

PITV prescription isodose to target volume ratio, *TV* target volume, *PIV* prescription isodose volume, TV_{PIV} volume of target covered by prescription isodose, PIV_{half} prescription isodose volume at half the prescription isodose (Fig. 10.1); D2%, D50%, and D98% are the dose received by 2%, 50%, and 98% of TV, respectively

10.2.6.2 Dose Variation at Edge of Target: Conformity Index

The dose conformity is an important parameter assessing the quality of dose distribution. It characterizes the extent to which high dose region conforms to the shape, size, and location of the target volume. This is expressed in terms of conformity index (CI) and is used to compare between different plans, evaluate a treatment plan in terms of expected normal tissue complications and tumor control. An ideal conformity index is unity. Shaw et al. proposed the term PITV (Prescription Isodose to Target Volume Ratio) as a measure of the conformity of a treatment plan [9]. The PITV is defined as

$$\text{PITV} = \frac{\text{Target Volume}(\text{TV})}{\text{Prescription Isodose Volume}(\text{PIV})}$$

A PITV ratio more than one indicates overtreatment and less than one indicates undertreatment. But it does not consider the location of PIV in relation to TV and hence even a total geographical miss can give a ratio of unity if the PIV and TV volumes are equal. These inadequacies led to the development of Paddick Conformity Index (PCI) [27]. This is the product of Coverage and Selectivity Index.

$$\text{Paddick Conformity Index} = \frac{\text{TV}_{\text{PIV}}}{\text{TV}} \times \frac{\text{TV}_{\text{PIV}}}{\text{PIV}} = \frac{(\text{TV}_{\text{PIV}})^2}{\text{TV} \times \text{PIV}}$$

The first fraction is coverage, where TV_{PIV} is the volume of the target covered by the prescription isodose (Fig. 10.1). The second part is the selectivity index and indicates how much normal tissue around the target is spared. The location of the target is also taken into consideration by PCI. A perfect plan has a score of 1, whereas plans with score less than 1 have relatively poorer conformity. Whether the compromise in conformity is due to undertreatment or overtreatment, the value is always less than or equal to 1. An ideal value for PCI conformity for practical purposes could be >0.85. PCI has been subsequently modified to its reciprocal so that it has the same scale as the PITV ratio. This is generally known as the New Conformity Index [13, 28].

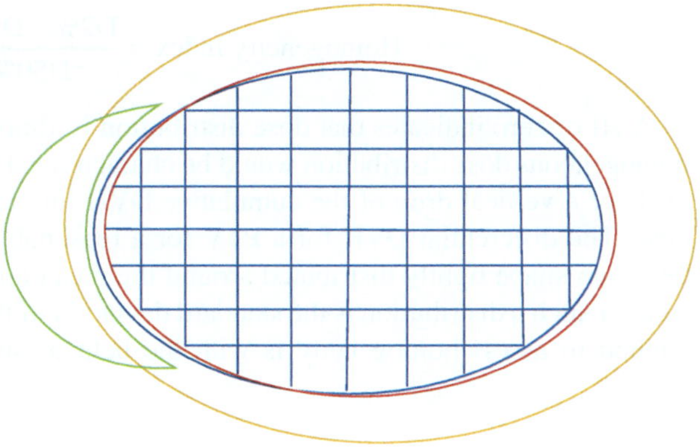

Fig. 10.1 Parameters for DVH indices. *TV* target volume (blue), TV_{PIV} volume of target covered by prescription isodose (cross-wired area), *PIV* prescription isodose volume (red), PIV_{half} prescription isodose volume at half the prescription isodose (orange), OAR (green)

$$\text{New Conformity Index} = \frac{\text{TV} \times \text{PIV}}{\left(\text{TV}_{\text{PIV}}\right)^2}$$

An ideal value of new conformity index will be less than 1.18.

10.2.6.3 Dose Variation Outside Target: Gradient Index

Though CI gives an idea of how well the distribution of radiation conforms to the size and shape of a target, the dose falloff outside the target is of equal (or even greater) importance especially in predicting normal tissue complications in SRS. Gradient Index (GI) objectively measures this dose falloff and helps to differentiate between plans of same conformity but different dose gradients. The GI is the ratio of the volume of half the prescription isodose to the volume of the prescription isodose [13, 29].

$$\text{Gradient Index} = \frac{\text{PIV}_{\text{half}}}{\text{PIV}}$$

where PIV_{half} represents the prescription isodose volume at half the prescription isodose, and PIV, the full prescription isodose volume (Fig. 10.1). The GI can be used for any prescription isodose. For a plan normalized to the 50% isodose line, it is the ratio of the 25% isodose volume ($\text{PIV}_{25\%}$) to the 50% isodose volume ($\text{PIV}_{50\%}$). An ideal value for the gradient index could be <3.0. Gradient index can help in choosing between plans of the same CI.

10.2.6.4 Dose Variation Inside Target: Homogeneity Index

The dose homogeneity characterizes the uniformity of the dose distribution within the target volume. It is the ratio of the maximum dose to the prescription dose. For a perfectly homogeneous dose distribution, this will be unity. This definition indicates only the magnitude of overdosing. It does not indicate if there is underdosing within the target volume. This definition would give a homogeneity index of unity if the maximum absorbed dose was equal to the prescription absorbed dose, but part of the target was missed. Point doses are no longer recommended, and the definition of Homogeneity Index (HI) as per ICRU report 83 is as follows [30].

$$\text{Homogeneity Index} = \frac{\text{D2\%} - \text{D98\%}}{\text{D50\%}}$$

A HI of zero indicates that dose distribution is almost homogenous. A perfectly homogeneous dose distribution would be characterized by a spike in the differential DVH or a vertical drop of the cumulative DVH line for the PTV at that absorbed dose. The differential DVH for a PTV for a reasonable treatment plan has a near Gaussian shape tightly distributed around the mean absorbed dose. A useful measure for such a distribution is the standard deviation of the mean. Therefore, the best method to assess homogeneity is with the help of standard deviation and mean absorbed dose. With this intent, the concept of sigma index was introduced [31]. It is defined as the standard deviation of the normalized differential DVH curve. With the help of statistical analysis, the standard deviation of differential DVH is used to calculate the spread or dispersion of the average dose within a structure, i.e., the degree of inhomogeneity. This gives a better evaluation of the degree of dose homogeneity. Unfortunately, most of the present treatment planning systems don't have the facility to calculate the standard deviation making sigma index calculation difficult.

10.2.6.5 Clinical Significance of DVH Indices

Though logically sound, the significance of DVH indices has not been proven clinically. In a retrospective cohort study of 925 patients with a single brain metastasis who underwent GK-based SRS, the authors reappraised whether high CI values correlate with reduced tumor progression rates, and whether low GI values correlate with reduced complications. SRS-related complications occurred in 38 patients (4.1%), with a median post-SRS interval of 11.5 (IQR 6.0–25.8) months. A multivariate analysis showed that only two clinical factors (i.e., peripheral doses and brain volume receiving ≥12 Gy) correlated with complication rates. However, neither CIs nor GIs impacted the incidences of complications. Also, cumulative incidences of local progression were significantly lower in patients with CIs < 0.65 than in those with CIs ≥ 0.65. The study clearly showed that the lower the CIs were, the lower the local progression rates were, and that the GI did not impact complication rates contrary to the majority opinion [32]. Similarly, there are conflicting reports in literature about dose inhomogeneity on LC and increased complication. Nezdi et al. reported that tumor dose inhomogeneity was associated with significantly more toxicity for intracranial SRS [33]. The RTOG 90-05 reported linac-based SRS had a 2.84 greater risk of local progression versus the GK-based SRS. The possible explanations are, 61% of GK treated patients had recurrent primary brain tumors compared to 30% of patients treated with a linac and high inherent inhomogeneity to the prescription dose in GK-based SRS [7]. Patel et al. comprehensively analyzed the evolution of planning evaluation indices and concluded that the GI emerged as a significant plan evaluation index in addition to CI, whereas homogeneity index is losing its credibility [34].

10.3 Approval of SRS Plan

The evaluation of SRS is an iterative process. Multiple plans may be generated to achieve a desired plan. The dose-volume parameters and indices are dependent on available technologies, and so is the decision of plan approval. Table 10.3 displays various parameters of a SRS plan when dose is prescribed for different IDL. The best spatial dose distribution was seen for 60% and 70% IDL. The V12 Gy of the normal brain was under 5 cc for all plans but increased with lower IDL prescription.

Table 10.3 Dose-volume parameter and DVH indices

Sl. No.	IDL (%)	Cochlea_L D_{mean} (Gy)	Brain V_{12Gy} (cc)	TV D95% (Gy)	TV D99% (Gy)	CI	GI	HI
1.	50	24.77	4.47	25.82	21.41	3.93	3.55	0.32
2.	60	20.64	2.34	21.52	17.85	3.00	2.98	0.32
3.	70	17.69	1.24	18.43	15.30	2.34	2.91	0.32
4.	80	15.48	0.59	16.13	13.39	1.81	3.08	0.32
5.	88	14.07	0.28	14.66	12.16	1.61	3.52	0.32
6.	90	13.76	0.24	14.34	11.90	1.56	3.64	0.32
7.	100	12.39	0.061	12.91	10.71	1.37	5.28	0.32

The SRS plan was generated for left sided glomus jugulare on MONACO (version 10.40.43.93) treatment planning system using APEX micro multi-leaf collimator

An ideal GI was achieved with dose prescription to 60–80% IDL. The CI improved with higher IDL values. The HI remained constant. A final approval was done by prescribing 88% IDL to achieve dose constraints of left cochlea.

10.4 Reporting of SRS

The reporting of SRS is enumerated in Table 10.4 (with an example), as recommended under ICRU report number 91: Prescribing, recording, and reporting of stereotactic treatments with small photon beams [13].

10.5 Conclusion

The chapter provides a stepwise pragmatic approach for evaluating a SRS plan. The evaluation process must be department specific and based on experience. A systematic approach will give rise to consistency in SRS implementation. DVH indices are important and not the sole criterions for finalizing a SRS plan. Special consideration must be given for prescription dose IDL, dose received by critical OARs and normal brain, and slice-by-slice spatial dose distribution. The contouring and plan evaluation must be reviewed by a second RO. If required, a RO should never hesitate to re-plan for OAR sparing and better coverage. When the desired plan is not workable, the treating RO should decrease the prescription dose or change to fractionated SRS/SRT.

Table 10.4 Reporting in SRS

Name	XXXX XXXX	Hospital ID	XXXX XXXX
Brief history	Hearing loss, tinnitus		
Examination finding	Performance status 0; profound hearing loss left ear; red polypoidal mass arising from middle ear extending to external ear		
Imaging finding	A heterogeneously enhancing lesion adjacent to left jugular foramen, in close relation to left sigmoid sinus and origin of internal jugular vein. This lesion extends antero-laterally into the middle ear cavity, suggestive of recurrent or residual lesion (glomus jugulare).		
Histopathology report	Glomus tumor		
Diagnosis with stage	Recurrent glomus jugulare (left)		
Prior treatment	Excision of tumor		
Treatment intent	Definitive		
Simulation	Supine position; thermoplastic skull cast; Simulation CT scan with 1.25 mm slice thickness; CT images were fused with contrast enhanced MRI images.		
Prescription	14 Gy in single fraction to the PTV margin		
GTV dimensions	2.70 cm (AP) × 2.62 cm (CC) × 2.20 cm (ML)		
GTV volume (cc)	4.924 cc	PTV volume (cc)	4.924 cc
Prescription dose	1500 cGy	Prescription IDL	88%
Dose in planning target volume			
Dmedian	1777 cGy		
PTV $D_{2\%}$	1921 cGy		
PTV $D_{98\%}$	1342 cGy		
Dose in OAR/PRV			
OARs	*Dose-volume*	*Constraint*	*Achieved*
Cochlea_L	D_{mean}	<14 Gy	14.07 Gy
Brain	V_{12Gy}	<10 cc	0.28 cc
Optic nerve_R	D_{max}	8 Gy	0.89 Gy
Optic nerve_L	D_{max}	8 Gy	3.35 Gy
Optic chiasm	D_{max}	8 Gy	1.40 Gy
Brainstem	D_{max}	12.5 Gy	2.32 Gy
Cochlea_R	D_{mean}	<14 Gy	0.15 Gy
DVH indices			
DVH indices	*Formula*	*Ideal value*	*Achieved*
Conformity index	$\frac{\text{TV} \times \text{PIV}}{(\text{TV}_{\text{PIV}})^2}$	1.18	1.61
Gradient index	$\frac{\text{PIV}_{\text{half}}}{\text{PIV}}$	<3.0	3.52
Homogeneity index	$\frac{\text{D2\%} - \text{D98\%}}{\text{D50\%}}$	0	0.32

References

1. Pollock BE, Lunsford LD. A call to define stereotactic radiosurgery. Neurosurgery. 2004;55(6):1371–3. https://doi.org/10.1227/01.neu.0000143613.13759.d4.
2. Kondziolka D. Current and novel practice of stereotactic radiosurgery. J Neurosurg. 2019;130(6):1789–98. https://doi.org/10.3171/2019.2.JNS181712.
3. Ventura T, Lopes MD, Ferreira BC, Khouri L. SPIDERplan: a tool to support decision-making in radiation therapy treatment plan assessment. Rep Pract Oncol Radiother. 2016;21(6):508–16. https://doi.org/10.1016/j.rpor.2016.07.002.
4. Muzumder S, John Sebastian MG. Plan evaluation in IMRT and VMAT. In: Mallick S, Rath G, Benson R, editors. Practical radiation oncology. Singapore: Springer; 2020. https://doi.org/10.1007/978-981-15-0073-2_24.
5. Piedadea PA, Radicchib LA, Peixotoc JGP. Volume-dose indexes and dose prescription descriptive review of radiosurgery planning. Braz J Rad Sci. 2019;7:1–6. https://doi.org/10.15392/bjrs.v7i3B.889.
6. Menon SV, Paramu R, Bhasi S, Nair RK. Evaluation of plan quality metrics in stereotactic radiosurgery/radiotherapy in the treatment plans of arteriovenous malformations. J Med Phys. 2018;43(4):214–20. https://doi.org/10.4103/jmp.JMP_25_18.
7. Pannullo SC, Fraser JF, Moliterno J, Cobb W, Stieg PE. Stereotactic radiosurgery: a meta-analysis of current therapeutic applications in neuro-oncologic disease. J Neuro-Oncol. 2011;103(1):1–17. https://doi.org/10.1007/s11060-010-0360-0.
8. Shaw E, Scott C, Souhami L, et al. Single dose radiosurgical treatment of recurrent previously irradiated primary brain tumors and brain metastases: final report of RTOG protocol 90-05. Int J Radiat Oncol Biol Phys. 2000;47(2):291–8. https://doi.org/10.1016/s0360-3016(99)00507-6.
9. Shaw E, Kline R, Gillin M, et al. Radiation Therapy Oncology Group: radiosurgery quality assurance guidelines. Int J Radiat Oncol Biol Phys. 1993;27(5):1231–9. https://doi.org/10.1016/0360-3016(93)90548-a.
10. Kano H, Flickinger JC, Tonetti D, et al. Estimating the risks of adverse radiation effects after gamma knife radiosurgery for arteriovenous malformations. Stroke. 2017;48(1):84–90. https://doi.org/10.1161/STROKEAHA.116.014825.
11. Niemierko A, Goitein M. The influence of the size of the grid used for dose calculation on the accuracy of dose estimation. Med Phys. 1989;16(2):239–47. https://doi.org/10.1118/1.596419.
12. Benedict SH, Yenice KM, Followill D, et al. Stereotactic body radiation therapy: the report of AAPM Task Group 101. Med Phys. 2010;37(8):4078–101. https://doi.org/10.1118/1.3438081.
13. Report 91. J Int Commiss Radiat Units Measure. 2014;14(2):1–160. https://doi.org/10.1093/jicru/ndx017.
14. Zhao B, Jin JY, Wen N, et al. Prescription to 50-75% isodose line may be optimum for linear accelerator-based radiosurgery of cranial lesions. J Radiosurg SBRT. 2014;3(2):139–47.
15. Zhang Q, Zheng D, Lei Y, et al. A new variable for SRS plan quality evaluation based on normal tissue sparing: the effect of prescription isodose levels. Br J Radiol. 2014;87(1043):20140362. https://doi.org/10.1259/bjr.20140362.
16. Xu Q, Fan J, Grimm J, et al. The dosimetric impact of the prescription isodose line (IDL) on the quality of robotic stereotactic radiosurgery (SRS) plans. Med Phys. 2017;44(12):6159–65. https://doi.org/10.1002/mp.12630.
17. Romano KD, Trifiletti DM, Garda A, et al. Choosing a prescription isodose in stereotactic radiosurgery for brain metastases: implications for local control. World Neurosurg. 2017;98:761–767.e1. https://doi.org/10.1016/j.wneu.2016.11.038.
18. Lawrence YR, Li XA, el Naqa I, et al. Radiation dose-volume effects in the brain. Int J Radiat Oncol Biol Phys. 2010;76(3 Suppl):S20–7. https://doi.org/10.1016/j.ijrobp.2009.02.091.
19. Mayo C, Martel MK, Marks LB, et al. Radiation dose-volume effects of optic nerves and chiasm. Int J Radiat Oncol Biol Phys. 2010;76(3 Suppl):S28–35. https://doi.org/10.1016/j.ijrobp.2009.07.1753.

20. Mayo C, Yorke E, Merchant TE. Radiation associated brainstem injury. Int J Radiat Oncol Biol Phys. 2010;76(3 Suppl):S36–41. https://doi.org/10.1016/j.ijrobp.2009.08.078.
21. Kirkpatrick JP, van der Kogel AJ, Schultheiss TE. Radiation dose-volume effects in the spinal cord. Int J Radiat Oncol Biol Phys. 2010;76(3 Suppl):S42–9. https://doi.org/10.1016/j.ijrobp.2009.04.095.
22. Bhandare N, Jackson A, Eisbruch A, et al. Radiation therapy and hearing loss. Int J Radiat Oncol Biol Phys. 2010;76(3 Suppl):S50–7. https://doi.org/10.1016/j.ijrobp.2009.04.096.
23. Riina MD, Stambaugh CK, Huber KE. Hippocampal dosimetry and the necessity of hippocampal-sparing in gamma knife stereotactic radiosurgery for extensive brain metastases. Adv Radiat Oncol. 2019;5(2):180–8. https://doi.org/10.1016/j.adro.2019.10.003.
24. Bentzen SM, Constine LS, Deasy JO, et al. Quantitative analyses of normal tissue effects in the clinic (QUANTEC): an introduction to the scientific issues. Int J Radiat Oncol Biol Phys. 2010;76(3 Suppl):S3–9. https://doi.org/10.1016/j.ijrobp.2009.09.040.
25. Hazard LJ, Wang B, Skidmore TB, et al. Conformity of LINAC-based stereotactic radiosurgery using dynamic conformal arcs and micro-multileaf collimator. Int J Radiat Oncol Biol Phys. 2009;73:562–70.
26. Emami B, Lyman J, Brown A, et al. Tolerance of normal tissue to therapeutic irradiation. Int J Radiat Oncol Biol Phys. 1991;21(1):109–22. https://doi.org/10.1016/0360-3016(91)90171-y.
27. Paddick I. A simple scoring ratio to index the conformity of radiosurgical treatment plans. Technical note. J Neurosurg. 2000;93(Suppl 3):219–22. https://doi.org/10.3171/jns.2000.93.supplement.
28. Van't Riet A, Mak AC, Moerland MA, Elders LH, van der Zee W. A conformation number to quantify the degree of conformality in brachytherapy and external beam irradiation: application to the prostate. Int J Radiat Oncol Biol Phys. 1997;37(3):731–6. https://doi.org/10.1016/s0360-3016(96)00601-3.
29. Paddick I, Lippitz B. A simple dose gradient measurement tool to complement the conformity index. J Neurosurg. 2006;105(Suppl):194–201. https://doi.org/10.3171/sup.2006.105.7.194.
30. Report 83. Prescribing, recording, and reporting photon-beam Intensity-Modulated Radiation Therapy. J ICRU. 2010;10(1):1–112. https://doi.org/10.1093/jicru/ndq001.
31. Yoon M, Park SY, Shin D, et al. A new homogeneity index based on statistical analysis of the dose-volume histogram. J Appl Clin Med Phys. 2007;8(2):9–17. https://doi.org/10.1120/jacmp.v8i2.2390.
32. Aiyama H, Yamamoto M, Kawabe T, et al. Clinical significance of conformity index and gradient index in patients undergoing stereotactic radiosurgery for a single metastatic tumor. J Neurosurg. 2018;129(Suppl1):103–10. https://doi.org/10.3171/2018.6.GKS181314.
33. Nedzi LA, Kooy H, Alexander E 3rd, Gelman RS, Loeffler JS. Variables associated with the development of complications from radiosurgery of intracranial tumors. Int J Radiat Oncol Biol Phys. 1991;21(3):591–9. https://doi.org/10.1016/0360-3016(91)90675-t.
34. Patel G, Mandal A, Choudhary S, Mishra R, Shende R. Plan evaluation indices: a journey of evolution. Rep Pract Oncol Radiother. 2020;25(3):336–44. https://doi.org/10.1016/j.rpor.2020.03.002.

Radiation Induced Brain Injury

11

Supriya Mallick and Aman Sharma

Radiation to the brain in therapeutic dose can cause significant injury.

Pathophysiology:

	Hypothesis	Site	Outcome
1	Vascular hypothesis	Vessels	Ischemia and secondary white matter necrosis
2	Parenchymal hypothesis	Oligodendrocyte	Oligodendrocyte type 2 astrocyte fails to maintain myelin sheaths
3	Microglia	Microglia	Production of reactive oxygen species
4	Astrocytes	Astrocytes	Acute cellular swelling and chronic hypertrophy hyperplasia
5	*Neurons*	*Neurons*	DNA damage

Broadly classified as (a) acute effects, (b) early delayed effects, (c) late effects

(a) Acute effects: occur during or shortly after the radiation exposure which is secondary to edema and disruption of the blood–brain barrier (BBB).

	Clinical manifestation	Management
1.	Fatigue characterized by lack of improvement by rest	Methylphenidate 10 mg BD, escalating to 30 mg BD [1, 2] Side effect: Anxiety and insomnia
2.	Dizziness	Fludrocortisone 0.1 mg OD All these drugs cause hypertension, hypokalemia

S. Mallick (✉)
Department of Radiation Oncology, National Cancer Institute, AIIMS, Haryana, India

A. Sharma
Department of Radiation Oncology, National Cancer Institute, AIIMS, New Delhi, India

S. Mallick et al. (eds.), *Evidence based practice in Neuro-oncology*,
https://doi.org/10.1007/978-981-16-2659-3_11

	Clinical manifestation	Management
3.	Increased intracranial pressure	Dexamethasone 8 mg TDS with PPI If not controlled mannitol infusion 1 mg/kg every 8 h Acetazolamide 250 mg BD
4.	Hair loss	Regrowth occurs
5.	Skin erythema	During radiation saline dressing, steroid skin cream may be applied Once radiation is completed topical emollients should be used

(b) Early delayed effects: Occurs 6–12 weeks after radiation exposure; due to transient demyelination.
   1. Generalized weakness.
   2. Somnolence.

   Somnolence evaluation

1. Subjective methods include.
   (a) Visual Analog Scales (VAS).
   (b) Epworth Sleepiness Scale.
2. Objective methods include.
   (a) Physician rated scale.
   (b) Littman Somnolence Syndrome (LSS) scale.
   (c) Electro-Encephalographic (EEG).

   These are reversible.

   Oral dexamethasone is used 4–8 mg TDS with PPI.

(c) Late effects: more than 1 year after radiation exposure leads to irreversible neurological consequences.
   1. Cognitive deficits.
   2. Necrosis of brain.
   3. Seizures.
   4. Cranial neuropathy.

Management policy for late delayed effects:

- Oral dexamethasone is prescribed most common: Initial dose of 4 mg BD for mild symptoms and 4 mg QID or 8 mg TDS for moderate to severe symptoms.
- If life-threatening symptoms appear dose of dexamethasone 10–25 mg bolus followed by 4–10 mg QID.
- The patients should receive PPI along with dexamethasone therapy to reduce chances of gastritis.
- After symptomatic improvement dexamethasone should be tapered by 50% dose every third day starting 5–7 days after therapy initiation.
- For prolonged steroid therapy or if symptoms arise of oral thrush anti-fungal therapy should be used. Fluconazole 200 mg for first day subsequently 100 mg daily for 6 days should be used.

- Cognitive dysfunction occurs with total doses 20 Gy in adults and 24 Gy in children (conventional fractions).

11.1 Necrosis of Brain

1. Incidence of radiation necrosis varies widely as it is difficult to estimate the true diagnosis. The authors who used histopathology as the method of diagnosis concluded incidence 5–7%, whereas by radiological modality the incidence ranges from 20% to 25%.
2. Risk factors.
   (a) Dose-volume: As the lesion size increases, deliverable dose should be reduced accordingly to limit the risk of toxicity.

Lesion size (mm)	Safe dose (Gy)
≤20	24
21–30	18
31–40	15

   (b) Previous radiation exposure increases the chances of radiation necrosis.

Modality	Risk (%)
Previous SRS	20
WBRT	4
No WBRT	3

   (c) Certain parts of brain are at higher risk of radiation necrosis. Frontal cortex has highest risk while the brainstem has least chances. Few studies have also highlighted that superficial lesions are at a lower risk.
   (d) Margin for defining Planning Target Volume should be minimized with help of image guidance.
3. Diagnosis and investigations for patients with suspected RN.
   (a) While certain enhancement patterns described in the literature as "Swiss cheese," "soap bubble" or "cut green pepper" was initially thought to favor radiation necrosis, these have only a 25% positive predictive value.
4. Management:
   - Not always an irreversible process.
   - Oral corticosteroid is the preferred first line.
   - VEGF inhibitor: bevacizumab appears to be a promising agent [3].
   - Oral pentoxifylline and vitamin E are also being attempted.
   - Laser interstitial thermal therapy (LITT) also appears to be a promising approach [4].
   - Surgical resection can be an option in selected cases [5].

References

1. Gehring K, Patwardhan SY, Collins R, et al. A randomized trial on the efficacy of methylphenidate and modafinil for improving cognitive functioning and symptoms in patients with a primary brain tumor. J Neuro-Oncol. 2012;107:165–74.
2. Meyers CA, Weitzner MA, Valentine AD, Levin VA. Methylphenidate therapy improves cognition, mood, and function of brain tumor patients. J Clin Oncol. 1998;16:2522–7.
3. Park M, Gwak HS, Lee SH, et al. Clinical experience of bevacizumab for radiation necrosis in patients with brain metastasis. Brain Tumor Res Treat. 2020;8(2):93–102.
4. Sujijantarat N, Hong CS, Owusu KA, et al. Laser interstitial thermal therapy (LITT) vs. bevacizumab for radiation necrosis in previously irradiated brain metastases. J Neuro-Oncol. 2020;148(3):641–9.
5. Vellayappan B, Tan CL, Yong C, et al. Diagnosis and management of radiation necrosis in patients with brain metastases. Front Oncol. 2018;8:395.

Part III

Management of Brain Tumours

Grade 1 Gliomas 12

Shikha Goyal and Renu Madan

12.1 Introduction

Central nervous system (CNS) gliomas are divided according to *St. Anne-Mayo* grading system into four grades based on morphology, the criteria being: cellular atypia, mitoses, microvascular (endothelial) proliferation, and necrosis. Further description is provided in the chapter on pathology of gliomas.

Grade I gliomas do not satisfy any of these parameters. They have a slow growth rate, are considered benign and have a long-term survival.

Grade II gliomas have cellular atypia but none of the other criteria. They also have a slow growth rate but may recur as high grade gliomas (HGG) later. Clinical behavior may be variable—benign or malignant.

Together, grade I and II tumors comprise low-grade gliomas (LGGs).

As per World Health Organization (WHO) 2016 CNS classification, LGGs may be either astrocytic or mixed neuronal-astrocytic tumors [1].

The various tumors described as LGGs are noted in Table 12.1.

12.2 Epidemiology

The term LGG includes tumors of varied origins but with relatively indolent behavior. They usually present in children or young adults, possess a slow growth velocity, but may be locally invasive. Occasionally they may undergo secondary transformation to HGGs. Grade 2 gliomas commonly affect the age group 35–44 years, are usually supratentorial, and may arise in the eloquent cortex. They represent nearly 10% of newly diagnosed tumors that originate in the brain.

S. Goyal (✉) · R. Madan
Department of Radiotherapy, Post Graduate Institute of Medical Education and Research, Chandigarh, India

S. Mallick et al. (eds.), *Evidence based practice in Neuro-oncology*,
https://doi.org/10.1007/978-981-16-2659-3_12

Table 12.1 WHO 2016 CNS classification of low-grade gliomas/mixed/neuronal-glial tumors

	Grade 1	Grade 2
Astrocytic tumors		
Diffuse astrocytic and oligodendroglial tumors	–	• IDH mutant diffuse astrocytoma • IDH mutant and 1p19q co-deleted ODG • IDH wild-type diffuse astrocytoma
Other astrocytic tumors	• Pilocytic astrocytoma • SEGA	• Pleomorphic xanthoastrocytoma
Other gliomas	• Angiocentric glioma	• Chordoid glioma of third ventricle
Neuronal and mixed neuronal-glial tumors	• DNET • Ganglioglioma • Gangliocytoma • Dysplastic gangliocytoma of the cerebellum • Desmoplastic infantile astrocytoma and ganglioglioma • PGNT • RGNT	• Central neurocytoma • Extraventricular neurocytoma • Cerebellar liponeurocytoma

DNET dysembryoplastic neuroepithelial tumor, *IDH* isocitrate dehydrogenase, *ODG* oligodendroglioma, *PA* pilocytic astrocytoma, *PGNT* papillary glioneuronal tumor, *PXA* pleomorphic xanthoastrocytoma, *RGNT* rosette-forming glioneuronal tumor, *SEGA* subependymal giant cell astrocytoma

The entities of pilomyxoid astrocytoma, protoplasmic astrocytoma, and oligoastrocytoma were listed as grade 2 in WHO 2007 classification but are no longer part of WHO 2016 classification

Pilocytic astrocytomas (PA) constitute the most common grade 1 LGGs, followed by other variants. In children, LGGs constitute the most common brain tumors (>20%) without any specific age predilection. The most common locations are the cerebellum, visual pathways, and diencephalon. The group encompasses several histologies with differing biological origins. Most childhood LGGs are sporadic in origin but may have an association with familial cancer predisposition syndromes, or neurocutaneous syndromes. Children with neurofibromatosis 1 (NF-1) have a higher risk of developing gliomas involving optic pathways and other locations [2]. Children with tuberous sclerosis may have a high risk of developing subependymal giant cell astrocytoma (SEGA) [3].

12.3 Pathology

Gliomas were earlier categorized according to cell lineage, histology, differentiation, and proliferation. Of late, improved understanding of tumor genetics has changed the landscape of how these tumors are classified, prognosticated, and treated. WHO 2007 classification introduced immunohistochemistry (IHC) to confirm cell lineage and differentiate astrocytic versus oligodendroglial tumors. The

WHO classification in 2016 introduced a system of molecular classification whereby isocitrate dehydrogenase (IDH) mutation, alpha-thalassemia/mental retardation syndrome X-liked (ATRX) loss, and 1p19q codeletion assessment became key to establishing the diagnosis and prognostic group of a tumor. The newer classification aims to group together tumors with similar biology and clinical behavior, fine-tuning both the prognostic assignment and treatment intensity, in addition to making the diagnostic process more objective by giving more importance to genotype than histopathologic diagnosis. IDH mutations are rare among grade 1 gliomas. However, individual tumor types may have specific mutations such as BRAF mutations or fusions.

12.4 Clinical Presentation

Grade 1 gliomas may cause symptoms due to cortical irritation (seizures) or mass effect over ventricles or optic pathways, although they may also be incidentally diagnosed on imaging done for other indications such as head trauma. Neurologic deficits are remarkably less common than in HGGs. In children, the presentation will also depend on age. Focal symptoms may include seizures, ataxia, limb weakness, focal sensory changes, cranial nerve palsies, aphasia, vision loss or field defects, nystagmus, Parinaud's syndrome (pupillary abnormalities, paralysis of upward gaze, or convergence-retraction nystagmus, eyelid retraction), diabetes insipidus, or growth retardation. Diffuse symptoms may include headache, neck stiffness, vomiting, lethargy, behavior change, change in cognition, or increased irritability. In infants, raised intracranial pressure may manifest as fontanelle bulge, split sutures, increase in head circumference, and "setting sun" sign. The specific presentation, histopathologic appearance, and management differ by tumor type and are discussed under specific tumor types.

12.4.1 Other Astrocytic Tumors

12.4.1.1 Pilocytic Astrocytoma (WHO Grade I)

PAs constitute the most common gliomas in childhood and the second most common brain tumors after embryonal tumors. Cerebellum is the most common location followed by midline structures such as hypothalamus and optic pathways. Other locations such as cerebral hemispheres, brainstem, and spinal cord are less common. Presentation is dependent on site of origin. Cerebellar lesions present with ataxia or features of raised intracranial tension (ICT). Optic pathway tumors present with proptosis, diplopia, nystagmus, or gradual vision loss. Tumors close to fourth or third ventricle may cause hydrocephalus, causing headache, vomiting, increased head size, delayed milestones, or behavior changes. Hypothalamic lesions may cause short stature, precocious puberty or symptoms related to hydrocephalus. Tumors involving both optic chiasm and hypothalamus may produce the diencephalic syndrome consisting of severe weight loss, failure to thrive while having normal or accelerated growth, hyperalertness, hyperkinesis, and mood elation. Tectal plate location is usually considered characteristically pilocytic histology and these tumors are treated without biopsy confirmation. These patients may have papilledema on examination or have Parinaud's syndrome.

Children having NF-1 are prone to develop optic pathway gliomas, nearly 15–20% develop the tumor before the age of 7 years [4]. Nearly 2% of brainstem lesions may have NF-1 association; patients with other stigmata of NF-1 should be identified.

PAs are well-circumscribed tumors, often consisting of a cystic lesion with a solid component, usually present as a characteristic mural nodule. On histology, they have a biphasic architecture with variable proportion of compact fibrillary component (bipolar piloid cells, eosinophilic Rosenthal fibers), intermixed with spongy microcystic element (loose stellate cells, microcysts, granular bodies). The piloid cells are glial fibrillary acid protein (GFAP) positive while the granular bodies show periodic acid–Schiff (PAS), alpha1-antitrypsin and alpha1-antichymotrypsin positivity. MIB-1 labeling index (MIB-LI) varies from 1% to 3%.

Magnetic resonance imaging (MRI) shows a well-defined cyst wall, with T2 hyperintensity but minimal edema. The mural nodule or solid component shows brilliant enhancement while the cyst wall may show variable enhancement. The cyst fluid is more proteinaceous than cerebrospinal fluid (CSF) and is T1 iso- to hypointense but T2/FLAIR hyperintense. Peripherally located lesions may occasionally cause skull bone thinning. These tumors are usually unifocal. Visual pathway tumors have less defined boundaries. Optic pathway tumors are T1 iso- or hypointense and T2 hyperintense. Only 50% show contrast enhancement. Variable degree of optic pathway involvement may be seen; sporadic lesions usually arise from chiasm while NF-1 related lesions more commonly arise from optic nerve.

Propensity for CSF dissemination is rare even if located close to the ventricles; it has been reported occasionally in de novo but largely in aggressive and recurrent tumors (3–5%). Even with leptomeningeal dissemination, median survival exceeding 5 years has been reported [5].

Several genetic pathways are associated with disease causation and progression. Causative factors include NF1 gene inactivation in neurofibromatosis 1 and BRAF gene overexpression/fusion/BRAF V600 E mutation in sporadic tumors, cause upregulation of MAPK/Erk pathway. KIAA1549-BRAF fusions occur more often in optic pathway lesions and cerebellum while BRAFV600E point mutation in supratentorial and diencephalic lesions [6].

Molecular alterations such as mammalian target of rapamycin (mTOR) or phosphoinositol 3 kinase (PI3K) pathway deregulation, or Matrilin 2 overexpression may lead to transformation into aggressive or recurrent disease [7]. Recurrent tumors treated with multiple lines of chemotherapy may have high CD133 and MDR1 expression that may act via PI3K-Akt-NF-κB signaling pathway and contribute to drug resistance [8].

Spontaneous regression is occasionally seen in these tumors. Surgical excision forms the mainstay of treatment and is curative if gross total resection (GTR) is achieved. A multicenter study on grade 1 gliomas showed better 5-year progression free survival (PFS) with greater extent of resection in PA but not for ganglioglioma or dysembryoplastic neuroepithelial tumor (DNET) [9]. Temporal lobe tumors had better PFS than occipital/parietal lobe tumors. Intraoperative MRI may help improve surgical extent but not necessarily impact survival.

However, optic pathway gliomas are not resectable, even biopsy may cause vision loss. If associated with NF-1, they remain largely asymptomatic and do not need to be treated. Sporadic optic pathway gliomas in the absence of NF-1 may be more aggressive and require treatment.

Adjuvant radiotherapy (RT) after incomplete resection or biopsy in brainstem or spinal lesions may offer durable long-term control. It may also be offered in symptomatic sporadic visual pathway tumors and tumors refractory to chemotherapy or recurrent tumors that are not amenable to re-excision. Residual tumor seen on a recent T1 contrast MRI is marked as gross tumor volume (GTV). Clinical target volume (CTV) margin of 5–10 mm is generated around GTV respecting anatomic boundaries. A 3–5 mm planning target volume (PTV) is generated around CTV (depending on setup technique, imaging verification frequency, and institutional data) and a dose of 50.4 gray (Gy) in 28 fractions is planned using conformal techniques such as three-dimensional conformal radiation therapy. Other techniques such as intensity modulated radiation therapy with or without image guidance, volumetric modulated arc therapy, and proton therapy may be offered depending on proximity of tumor to critical normal structures. With adjuvant RT, long-term survival over 95% and event free survival between 70% and 90% is seen [10]. Not all patients with optic pathway PAs will have vision recovery. Children with NF-1 may be more prone to second cancers and vasculopathies with RT, where it should be avoided. However, in young children, chemotherapy consisting of vincristine-carboplatin may slow down progression and delay RT in residual, unresectable lesions [11].

Figure 12.1 shows the T1 contrast image (a) and adjuvant radiotherapy plan (b) of a large residual pilocytic astrocytoma involving brainstem in a young patient. A dose of 50.4 gray was planned to planning target volume.

Since these tumors affect young children and have an excellent prognosis, adjuvant radiation is not offered even with incomplete resection unless there is neurological compromise due to mass effect from large residual component in locations such as brainstem. The risk of malignant transformation should be explained before offering RT for any low-grade tumors in children and young adults. In very young children, chemotherapy may have a role in slowing down disease progression in unresectable tumors. Adult PAs are highly vascular, show high vascular endothelial growth factor (VEGF) expression and may be responsive to antiangiogenic therapies such as bevacizumab in recurrent/unresectable/disseminated settings [12]. mTOR inhibitors, MEK inhibitors, and farnesyl transferase inhibitors may have a therapeutic role. Systemic agents such as dabrafenib and trametinib are being investigated for treatment of such aggressive tumors through inhibition of mitogen activated protein kinase (MAPK) pathway.

Survival is good and depends on age as well as tumor location. Overall survival (OS) at 10 years for children exceeds 95% while for adults >40 years, it is around 70%. Tumors that have a chance of GTR have a better survival than

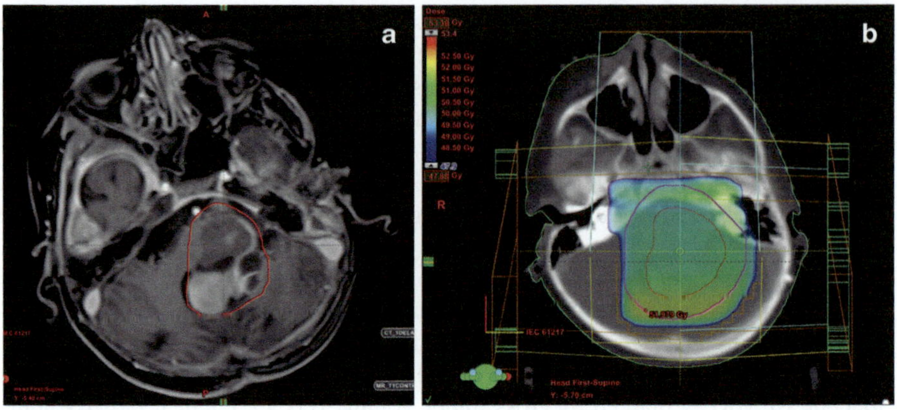

Fig. 12.1 (**a**) T1 Contrast enhanced axial section of magnetic resonance imaging showing a large residual tumor involving left half of brainstem and left cerebellar peduncle with solid and cystic areas, patchy enhancement, and mass effect over normal brainstem, fourth ventricle, and cerebellum. (**b**) Three-dimensional conformal radiotherapy plan for the same patient. A dose of 50.4 gray in 28 fractions is planned to planning target volume

unresectable tumors (brainstem, spine). Recurrence rates after GTR are under 5%, while nearly 50% after incomplete resection. Even with incomplete resection, it is reasonable to follow up and offer re-excision at progression for non-critical locations.

12.4.1.2 Subependymal Giant Cell Astrocytomas

SEGAs are slow-growing benign tumors arising primarily in periventricular region [13]. They are strongly associated with tuberous sclerosis, which is an autosomal dominant neurocutaneous disorder with TSC or TSC2 gene mutations, and the classic triad of *mental retardation, seizures, and multiple organ hamartomas*. SEGA was historically considered pathognomonic for tuberous sclerosis but rarely, instances of SEGA in absence of tuberous sclerosis are also reported although clinical features of the same may develop later in life [14]. There is a male predominance with a mean age of 11 years at diagnosis.

They typically originate from subependymal nodules in the region of foramen of Monro, and can be unilateral or bilateral. The only differentiation from subependymal nodules can be made with serial imaging showing growth or development of obstructive symptoms; even histopathology may be unable to distinguish SEGA from subependymal nodules. Rate of growth is slow, over weeks to months, although parenchymal invasion, perilesional edema, or atypical locations within hypothalamus or pineal are occasionally seen. For asymptomatic tumors, neuroimaging every 1–3 years is recommended. Symptoms may arise from acute or chronic obstructive hydrocephalus: vomiting, positional headache, diplopia or blurring of vision, new or worse seizures, and personality changes.

MRI shows T1 hypointense and T2/FLAIR hyperintense periventricular lesions, with only 10% showing contrast enhancement. Multiple cortical tubers/

subependymal nodules may be noted; only a serial change in size identifies a SEGA with certainty [15].

A symptomatic (intractable seizures) or growing SEGA should be surgically excised to avoid complications from raised ICT. Early surgery is recommended to avoid hydrocephalus-related sequelae [16]. GTR is curative. If resection is incomplete, tumor growth may still continue. For larger tumors, surgery may have its attendant risk of hemiparesis, ventriculoperitoneal shunt dependence, stroke, transient memory impairment, etc., and in these situations, medical management may be considered to shrink or stabilize the tumor. mTOR is a protein kinase that is constitutively activated in tuberous sclerosis. mTOR inhibitors (rapamycin, everolimus) have been used successfully with significant tumor shrinkage and resolution of hydrocephalus; tumor regrowth may happen following interruption of the drug. There are complex interactions with anticonvulsants, in addition to the significant adverse effects such as thrombocytopenia, delayed wound healing, and immunosuppression, hence judicious use is recommended [17]. RT is usually not recommended; however, gamma knife (doses of 11–13 Gy at periphery) has been used for growing but asymptomatic tumors who were not candidates for surgery or medical management, with variable results [18].

A Surveillance, Epidemiology, and End Results (SEER)-18 database analysis of 226 cases over 10 years places the incidences at 0.027 per 100,000 person-years. Age younger than 18 years and surgery were both prognostic for survival while gender, tumor size, location, extent of surgery, and receipt of RT had no impact [19].

12.4.2 Other Gliomas

12.4.2.1 Angiocentric Glioma

Angiocentric glioma is a rare and new distinct tumor type of glial and possibly partial ependymal origin first defined in 2005, that forms one of the tumors constituting the group of drug-resistant long-term epilepsy associated tumors (LEAT). There is a slight male preponderance, with median age of 13 years at diagnosis [20]. Seizures start in childhood. Disease location is supratentorial, unilobar (most commonly in frontal lobe) with involvement of superficial cerebral cortex and extension to subcortical white matter. MRI shows rim-like T1 hyperintensities in cerebral cortex with stalk-like extension to ventricle on T2/FLAIR [21]. Lesions are usually well defined though diffuse change is also reported. Contrast enhancement is variable. Radiologic differential diagnoses comprise other LEAT associated lesions such as focal cortical dysplasias (FCD), ganglioglioma, or cortical tubers. On histology, typical features include monomorphous, bipolar tumor cells showing angiocentric arrangements and diffuse infiltration of cerebral cortex and focal subpial growth pattern. Proliferative activity is rare. IHC for GFAP, vimentin, S100B, and epithelial membrane antigen (EMA) is positive; p53, epidermal growth factor receptor (EGFR), neurospecific nucleoprotein (NeuN), smooth muscle actin (SMA), chromogranin A (CgA), microtubule associated protein 2 (MAP 2) and CD34 and CD99 are negative. Ki-67 is <5%, usually ~1%. Differential diagnoses include PAs,

Table 12.2 Differential diagnosis between AG and similar grade 1–2 gliomas

Tumor type (WHO 2016)	Age	Symptoms (in addition to epilepsy)	Location	MRI	Histopathology	IHC (in addition to GFAP)
Angiocentric glioma	Children and young adults	–	Supratentorial (frontal, temporal)	Small non-enhancing mass	Angiocentric growth pattern, monomorphous bipolar cells	S-100, vimentin, EMA
Ganglioglioma	Children and young adults	Raised ICT	Temporal lobe, cerebellum	Nodular, cystic, rim-like or solid, contrast enhancing	Both ganglion and glial cells	NF, NeuN
Diffuse grade 2 gliomas	Young adults	Focal neurologic deficits, neurocognitive dysfunction, raised ICT	Cerebral hemispheres	No or minimal enhancement	Diffusely infiltrating	S-100, Olig2, EMA, nestin

AG angiocentric glioma, *EMA* epithelial membrane antigen, *GFAP* glial fibrillary acidic protein, *ICT* intracranial tension, *IHC* immunohistochemistry, *MRI* magnetic resonance imaging, *NF* neurofilament, *NeuN* neurospecific nucleoprotein, *NSE* neurospecific enolase, *Olig2* oligodendrocyte transcription factor-2, *WHO* World Health Organization

gangliogliomas, and ependymomas, but they are differentiated based on morphology and IHC. Table 12.2 lists important differences between angiocentric gliomas and other LGGs. Some angiocentric gliomas have loss of chromosomal bands 6q24-q25 (PLAGL1/ZAC1 gene) and gain on 11p11.2 (PTPRJ gene, CD148) and these may have pathobiological relevance. GTR is the treatment of choice, with most patients relieved of epilepsy. Clinical course is long and indolent. There are occasional reports of recurrence after incomplete resection with histologic signs of malignant change at progression. Since this is a relatively new entity with limited reports, long-term follow-up information is not available.

12.4.3 Neuronal and Mixed Neuronal-Glial Tumors

12.4.3.1 Dysembryoplastic Neuroepithelial Tumor (WHO Grade I)

The entity of DNET was first described in 1988. It comprises benign cortical tumors, mostly in supratentorial location. Presentation is with refractory seizures; other neurologic signs are usually absent. DNET is one of several LEATs; others being angiocentric glioma, ganglioglioma, pleomorphic xanthoastrocytoma (PXA), and isomorphic astrocytoma.

Both DNET and ganglioglioma have common characteristics of young age of presentation (childhood and adolescence), tumor location in temporal lobe, benign behavior and chronic epilepsy (starting before the age of 20 years) due to cortical involvement. Both are also tumors related to abnormal proliferation of neuronal and glial elements and are associated with FCD and hemimegalencephaly [22].

On imaging, DNETs appear as intracortical tumors without any mass effect or perilesional edema. Nearly half the patients have overlying skull deformities. Computed tomography (CT) shows a hyperdense lesion, occasionally with calcific changes. On MRI, lesions are T1 hypointense and T2 hyperintense, well demarcated, multilobulated or gyriform. Presence of contrast enhancement as multiple rings is seen in only one-third, but when present, it should raise suspicion of DNET (vs. HGGs) especially when mass effect and edema are absent. High choline peaks on MR spectroscopy, high apparent diffusion coefficient (ADC) on diffusion weighted MRI sequences, lack of high methionine uptake on PET with ^{11}C-methionine may help distinguish DNET or ganglioglioma from FCD [23].

On histopathology, glial nodules, cortical dysplasia, and glioneuronal elements are seen. Nearly 40% DNETs have cystic change; other forms may include single or multiple nodules or diffuse change. Nearly 30% have FCD. Oligodendroglioma (ODG)-like areas may also occur with corresponding myelin oligodendrocyte glycoprotein (MOG) expression on IHC and need to be distinguished from pure ODGs by presence of neuronal elements with a columnar appearance perpendicular to cortical surface, FCD, and a myxoid matrix. It may be difficult to distinguish from extraventricular neurocytoma, other LGGs and ODG. 1p19q codeletion is absent in DNET.

Genome sequencing shows an increased propensity to alteration of FGFR1 gene (duplication or point mutation), but rarely BRAF alterations [24].

With small tumors, complete excision may be possible (often with a rim of surrounding dysplastic tissue) and is curative, also treating the epilepsy. However, cortical location may preclude complete excision in others. Usually, rate of regrowth is slow, and most tumors do not need re-treatment even after incomplete excision. Due to frequent diffuse brain involvement, there may be relapse of epilepsy despite no disease recurrence. DNET does not adversely impact survival. RT and chemotherapy have no role.

12.4.3.2 Ganglioglioma (WHO Grade I)

Grade 1 ganglioglioma consists of mixed neuronal and glial elements. They represent less than 1% of CNS tumors. Median age of diagnosis is 12 years with a male preponderance [25]. Temporal lobe is the most common location and seizures the most common presentation.

Histopathology shows mature neoplastic astrocytes admixed with variable proportions of atypical large binucleate neurons with prominent nucleoli. Nearly 75% have alterations of BRAF gene (BRAF V600E or other mutations, BRAF fusion) [25]. Other mutations include RAF, NF1, KRAS, or FGFR1/2 alterations. MRI shows a T1 iso- to hypointense and T2 hyperintense solid-cystic or solid mass, most commonly in temporal lobe. Contrast enhancement, if present, may be solid or nodular rim. Extensive calcification may be seen. Mass effect and perilesional edema are usually not seen. Surrounding cortical dysplasia may be noted in the form of cortical thickening, abnormal gray and white matter signals, localized cortical volume loss, etc.

Complete resection is usually possible in more than two-third cases, and is curative. They have a favorable prognosis. Age less than 1 year at diagnosis and tumor origin in brainstem are associated with poor OS. SEER database results from pediatric ganglioglioma/gangliocytoma suggest that nearly 2/3 have complete resection; non-brainstem lesions had an excellent 5-year OS of 96.6%. Gangliogliomas have some malignant potential especially in recurrent situations, though rarely in children [26]. RT is rarely needed, for inoperable primary or recurrent tumors, largely in brainstem location.

12.4.3.3 Gangliocytoma

Gangliocytomas are clinically, radiologically, and histologically similar to gangliogliomas, but represent just 1% of gangliogliomas. They present in children and young adults with prolonged epilepsy. Most tumors are cerebral in location, most common location being temporal lobe. They may occasionally occur in pituitary gland in association with pituitary adenomas [27]. They are not distinguishable from ganglioglioma on imaging but sometimes have a dural tail. They are purely neuronal differentiated neoplasms. They are CD34 negative, comprise primarily of dysplastic neurons in a background of non-neoplastic glial cells, without almost any proliferation activity. Complete surgery is curative. These tumors are not known to have any malignant change on follow-up. No long-term follow-up information is available since they are largely clubbed with gangliogliomas [28]. It is not certain if these constitute a distinct tumor entity despite inclusion in all WHO CNS classifications till date. They may possibly represent dysplastic or malformative lesions [29].

12.4.3.4 Dysplastic Gangliocytoma of the Cerebellum

Dysplastic gangliocytoma of the cerebellum (Lhermitte–Duclos disease) was first described in 1920. There is no predilection for any gender. Presentation is most often in young adults (third to fourth decade). The mean age of presentation is 4.3 years in children and 42.5 years in adults.

It manifests as a slowly growing cerebellar cortical mass, and may have a syndromic association and association with congenital malformations like polydactyly, cutaneous or mucosal lesions, gigantism, megalencephaly, and hemihypertrophy [30]. Patients have a long history of vague symptoms related to raised ICT, brainstem compression or cerebellar involvement, including headache, gait ataxia, and cranial nerve palsies. Sudden neurologic deterioration from acute or decompensation of chronic hydrocephalus may happen. An association with Cowden disease (an autosomal dominant neoplastic-hamartomatous disorder) is also reported, with multiple hamartomas and high incidence of malignancies of breast, endometrium, thyroid, and genitourinary systems. PTEN mutations have been identified in these patients [31].

On histology, the granular, Purkinje, and glial cells are present, loosely mimicking the architecture of cerebellar cortex but in a disorganized manner. Ki67 is low. Histopathology shows widened molecular layer, absent Purkinje cell layer and hypertrophic granule cell layer. Whether the origin is hamartomatous or neoplastic is still not clear.

MRI may show enlargement of cerebellar cortical folia with non-enhancing gyriform pattern. Lesions are well circumscribed, T1 hypointense, T2 hyperintense with a striated pattern (interspersed bands of isointensity, also known as tiger stripes), with minimal or no contrast enhancement.

Tumor growth is slow but there may be rapid deterioration with mass effect in some. Surgical excision is the treatment of choice. Residual or recurrent lesions have not demonstrated malignant change in any reports. RT and chemotherapy have no role. Due to the association with Cowden disease, these patients need long-term follow-up after treatment to help early detection of other associated cancers and genetic counseling.

12.4.3.5 Desmoplastic Infantile Astrocytoma and Ganglioglioma

These are tumors of newborns or infants, usually presenting in first 18 months of life. Presentation may be with increased head size, seizures, and focal neurologic signs. MRI shows a typically supratentorial tumor with more than one lobe involved, with T2 heterogeneity (hyperintense cystic component and hypointense solid component). The solid component enhances with contrast. Superficial cerebral cortical and leptomeningeal involvement with dural attachment with a desmoplastic reaction are characteristic. MRS shows high choline/N-acetylaspartate (NAA) ratio, perfusion studies show increased perfusion, and diffusion studies show low ADC values [32]. Histology shows mixed ganglionic and astrocytic cells. Presence of primitive cells may be indicated by a high number of mitosis and these may mimic malignant astrocytomas in appearance.

BRAF mutations are occasionally found. Complete surgical resection is curative although possible in only 1/3 cases. Challenges to surgery include young age with low weight, large size of tumor with hypervascularity. Even after partial resection, these tumors merit observation only. Chemotherapy may be appropriate for recurrent disease after a second look surgery. Long-term prognosis is excellent. Midline (hypothalamic) tumors have more aggressive course [33].

12.4.3.6 Papillary Glioneuronal Tumor

Papillary glioneuronal tumor (PGNT) are rare tumors of younger population, with mean age of presentation at 27 years. Clinical features include symptoms due to raised ICT, seizures, focal neurologic changes as per tumor location (hemiparesis, dysphasia, vision changes). MRI shows a cystic lesion (hypointense on T1, iso- to hyperintense on T2) with a solid component or mural nodule in supratentorial location, often frontal and occasionally temporal or parietal [34]. The solid component or nodule is contrast enhancing. A possible origin from subependymal plate explains their close proximity to ventricles. On histology, the tumor is low grade, biphasic with papillary and solid parts containing neurocytic and glial elements. Sometimes, ODG-like cells or mini-gemistocytes may be seen. GTR is curative, and extent of resection and Ki-67 are prognostic factors for local control (5-year PFS 91.9% with GTR vs 46.7% with partial resection) [35]. Recurrences may happen following partial resections especially in parietal location (due to proximity to eloquent cortex, basal ganglia, or internal capsule). Efforts should be made to remove as much of solid part as possible. RT is reserved for aggressive lesions that are not candidates for re-surgery.

12.4.3.7 Rosette-Forming Glioneuronal Tumor

Rosette-forming glioneuronal tumor (RGNT) is a rare tumor predominantly affecting females with a mean age of 31.8 years at detection. The tumor is most often located in relation to fourth ventricle (~60%), extending to cerebellar vermis or thalamus [36]. Presenting symptoms may include headache, gait imbalance, vertigo, or vision problems. MRI may show a mass (average size 3 cm) spanning cortex and underlying white matter with solid, cystic or mixed features. The solid areas show T1 iso- to hypointensity and T2 hyperintensity with focal contrast enhancement. Hydrocephalus may be seen in nearly half of them. Nearly one-fourth may show calcification. Histopathology shows a biphasic tumor with neuronal (neurocytic rosettes and perivascular pseudorosettes) and glial (spindled astrocytic cells, elongated nuclei, slender bipolar processes, similar to PA) elements without mitotic activity, necrosis, or vascular endothelial proliferation [34]. IHC shows synaptophysin, neuron specific enolase (NSE), neurofilament protein (NFP) positivity in neuronal component and GFAP in glial component, with low Ki-67 index. PIK3CA gene mutations are seen. GTR is the goal; however, often only partial resection is possible due to intimate relationship with neural structures of cerebellum. Most patients do well despite partial resection. Risk of postoperative neurologic deficits (ataxia, cranial nerve palsies) may be considerable.

References

1. Louis DN, Perry A, Reifenberger G, et al. The 2016 World Health Organization classification of tumors of the central nervous system: a summary. Acta Neuropathol. 2016;131(6):803–20.
2. Listernick R, Louis DN, Packer RJ, et al. Optic pathway gliomas in children with neurofibromatosis 1: consensus statement from the NF1 Optic Pathway Glioma Task Force. Ann Neurol. 1997;41:143–9.
3. DiMario FJ, Sahin M, Ebrahimi-Fakhari D. Tuberous sclerosis complex. Pediatr Clin N Am. 2015;62:633–48.
4. Bornhorst M, Frappaz D, Packer RJ. Pilocytic astrocytomas. Handb Clin Neurol. 2016;134:329–44.
5. Mazloom A, Hodges JC, Teh BS, Chintagumpala M, Paulino AC. Outcome of patients with pilocytic astrocytoma and leptomeningeal dissemination. Int J Radiat Oncol Biol Phys. 2012;84(2):350–4.
6. Collins VP, Jones DT, Giannini C. Pilocytic astrocytoma: pathology, molecular mechanisms and markers. Acta Neuropathol. 2015;129(6):775–88.
7. Bikowska-Opalach B, Szlufik S, Grajkowska W, Jozwiak J. Pilocytic astrocytoma: a review of genetic and molecular factors, diagnostic and prognostic markers. Histol Histopathol. 2014;29(10):1235–48.
8. Xi G, Li YD, Grahovac G, et al. Targeting CD133 improves chemotherapeutic efficacy of recurrent pediatric pilocytic astrocytoma following prolonged chemotherapy. Mol Cancer. 2017;16(1):21.
9. Yahanda AT, Patel B, Sutherland G, et al. A multi-institutional analysis of factors influencing surgical outcomes for patients with newly diagnosed grade I gliomas. World Neurosurg. 2020;135:e754–64.
10. Merchant TE, Kun LE, Wu S, Xiong X, Sanford RA, Boop FA. Phase II trial of conformal radiation therapy for pediatric low-grade glioma. J Clin Oncol. 2009;27(22):3598–604.
11. Packer RJ, Ater J, Allen J, et al. Carboplatin and vincristine chemotherapy for children with newly diagnosed progressive low-grade gliomas. J Neurosurg. 1997;86(5):747–54.
12. Wasilewski A, Mohile N. Durable response to bevacizumab in adults with recurrent pilocytic astrocytoma. CNS Oncol. 2018;7(3):CNS26.
13. David N, Louis HO, Wiestler OD, Cavenee WK, editors. WHO classification of tumours of the central nervous system (revised). 4th ed. Lyon; 2016.
14. Cuccia V, Zuccaro G, Sosa F, et al. Subependymal giant cell astrocytoma in children with tuberous sclerosis. Childs Nerv Syst. 2003;19(4):232–43.
15. Evans JC, Curtis J. The radiological appearances of tuberous sclerosis. Br J Radiol. 2000;73(865):91–8.
16. de Ribaupierre S, Dorfmuller G, Bulteau C, et al. Subependymal giant-cell astrocytomas in pediatric tuberous sclerosis disease: when should we operate? Neurosurgery. 2007;60(1):83–9. discussion 89–90
17. Campen CJ, Porter BE. Subependymal giant cell astrocytoma (SEGA) treatment update. Curr Treat Options Neurol. 2011;13:380–5.
18. Park KJ, Kano H, Kondziolka D, et al. Gamma Knife surgery for subependymal giant cell astrocytomas. J Neurosurg. 2011;114(3):808–13.
19. Nguyen HS, Doan NB, Gelsomino M, et al. Subependymal giant cell astrocytoma: a surveillance, epidemiology, and end results program-based analysis from 2004 to 2013. World Neurosurg. 2018;118:e263–8.
20. Han G, Zhang J, Ma Y, Gui Q, Yin S. Clinical characteristics, treatment and prognosis of angiocentric glioma. Oncol Lett. 2020;20(2):1641–8.
21. Preusser M, Hoischen A, Novak K, et al. Angiocentric glioma: report of clinico-pathologic and genetic findings in 8 cases. Am J Surg Pathol. 2007;31(11):1709–18.
22. Phi JH, Kim SK. Clinical pearls and advances in molecular researches of epilepsy-associated tumors. J Korean Neurosurg Soc. 2019;62(3):313–20.

23. O'Brien DF, Farrell M, Delanty N, et al. The Children's Cancer and Leukaemia Group guidelines for the diagnosis and management of dysembryoplastic neuroepithelial tumours. Br J Neurosurg. 2007;21(6):539–49.
24. Rivera B, Gayden T, Carrot-Zhang J, Nadaf J, Boshari T, Faury D, et al. Germline and somatic FGFR1 abnormalities in dysembryoplastic neuroepithelial tumors. Acta Neuropathol. 2016;131:847–63.
25. Dudley RW, Torok MR, Gallegos DR, et al. Pediatric low-grade ganglioglioma: epidemiology, treatments, and outcome analysis on 348 children from the surveillance, epidemiology, and end results database. Neurosurgery. 2015;76(3):313–20.
26. Majores M, von Lehe M, Fassunke J, Schramm J, Becker AJ, Simon M. Tumor recurrence and malignant progression of gangliogliomas. Cancer. 2008;113:3355–63.
27. Cossu G, Daniel RT, Messerer M. Gangliocytomas of the sellar region: a challenging diagnosis. Clin Neurol Neurosurg. 2016;149:122–35.
28. Schramm J, Kristof R. Glioneuronal tumors. In: Tonn JC, Grossman SA, Rutka JT, Westphal M, editors. Neuro-oncology of CNS tumors. Berlin, Heidelberg: Springer; 2006.
29. Blümcke I, Wiestler OD. Gangliogliomas: an intriguing tumor entity associated with focal epilepsies. J Neuropathol Exp Neurol. 2002;61(7):575–84.
30. Ma J, Jia G, Chen S, Jia W. Clinical perspective on dysplastic gangliocytoma of the cerebellum (Lhermitte-Duclos disease). World Neurosurg. 2019;122:16–23.
31. Nowak DA, Trost HA, Porr A, Stölzle A, Lumenta CB. Lhermitte-Duclos disease (dysplastic gangliocytoma of the cerebellum). Clin Neurol Neurosurg. 2001;103(2):105–10.
32. Demir MK, Yapıcıer O, Yılmaz B, Kılıç T. Magnetic resonance imaging findings of mixed neuronal–glial tumors with pathologic correlation: a review. Acta Neurol Belg. 2018;118(3):379–86.
33. Bianchi F, Tamburrini G, Massimi L, Caldarelli M. Supratentorial tumors typical of the infantile age: desmoplastic infantile ganglioglioma (DIG) and astrocytoma (DIA). A review. Childs Nerv Syst. 2016;32(10):1833–8.
34. Louis DN, Ohgaki H, Wiestler OD, et al. The 2007 WHO classification of tumours of the central nervous system [published correction appears in Acta Neuropathol. 2007 Nov;114(5):547]. Acta Neuropathol. 2007;114(2):97–109.
35. Ahmed AK, Dawood HY, Gerard J, Smith TR. Surgical resection and cellular proliferation index predict prognosis for patients with papillary glioneuronal tumor: systematic review and pooled analysis. World Neurosurg. 2017;107:534–41.
36. Hsu C, Kwan G, Lau Q, Bhuta S. Rosette-forming glioneuronal tumour: imaging features, histopathological correlation and a comprehensive review of literature. Br J Neurosurg. 2012;26(5):668–73.

Grade 2 Gliomas

13

Shikha Goyal and Renu Madan

13.1 Introduction

This chapter deals with grade 2 diffuse astrocytic and oligodendroglial tumors, "other" astrocytic tumors and mixed neuronal and neuronal-glial tumors, as defined by World Health Organization (WHO) 2016 Central nervous system (CNS) classification. For the ease of discussion, low grade gliomas (LGGs) mentioned in this chapter would essentially refer to grade 2 tumors. Grade 1 tumors have been discussed in a separate chapter.

13.2 Epidemiology

LGGs comprise about 2000 cases per annum and represent nearly 10% of primary brain tumors. They have a slight male preponderance, with a mean age of diagnosis at 37 years (range 20–50 years), the incidence decreasing with advancing age. The presenting complaints for a patient with LGG may depend on the location of disease and may span several years before they are reported; these include seizures (65%, higher in LGG than glioblastoma, GBM), headache (35%), focal weakness (30%), visual disturbances (15%), personality changes (14%), and nausea or vomiting (10%).

13.3 Pathology and Molecular Characteristics

Traditionally, they were defined as astrocytic, oligodendroglial, and mixed tumors, but now, with advent and incorporation of molecular characteristics in WHO 2016 CNS classification, the "mixed" group is no longer valid. Apart from infiltrative

S. Goyal (✉) · R. Madan
Department of Radiotherapy, Post Graduate Institute of Medical Education and Research, Chandigarh, India

S. Mallick et al. (eds.), *Evidence based practice in Neuro-oncology*,
https://doi.org/10.1007/978-981-16-2659-3_13

histopathology, low mitotic rate and absence of necrosis/microvascular proliferation, molecular characteristics are now crucial for diagnosis, treatment decisions, and prognostication. Astrocytomas and oligodendrogliomas (ODG) are clubbed together based on growth pattern, behavior, and isocitrate dehydrogenase (IDH) mutation. Grade 2 astrocytic/oligodendroglial tumors are divided into three groups: (a) IDH-mutant diffuse astrocytoma, (b) IDH-mutant and 1p19q codeleted ODG, (c) IDH wild-type diffuse astrocytoma. "Other" astrocytic grade 2 tumors with a more circumscribed growth pattern, lack of IDH gene changes, and BRAF mutations include pleomorphic xanthoastrocytoma (PXA); grade 2 glioneuronal tumors include central neurocytoma, extraventricular neurocytoma, and cerebellar liponeurocytoma, which will also be discussed briefly.

Incorporation of both phenotypic and genotypic characteristics makes the new classification more objective and integrates histologic information, grade, and molecular information into one diagnosis [1].

Diffuse astrocytomas are lesions having tumor cells with wide infiltration into brain parenchyma; cellular density and anaplasia increase with the grade of tumor. Grade 2 diffuse astrocytomas have moderate increase in cellularity with highly fibrillar processes extending from neoplastic cells. There is normal CNS parenchymal architectural distortion with splaying of neuropil and microcysts. Nuclear enlargement, oblong shape, heterogeneous hyperchromatism, and irregular contours differentiate them from normal astrocytes. In contrast, classic ODGs have round regular nuclei with small nucleoli, clear perinuclear haloes, and bland chromatin. Glial fibrillary acid protein (GFAP) is positive. IDH-mutant diffuse astrocytoma (grade 2), IDH-mutant anaplastic astrocytoma (grade 3), and IDH-mutant GBM (grade 4) all lie on a continuum of increasing aggressiveness of similar origin cells. Similarly, grade 2 and grade 3 ODGs line on the same continuum with no grade 4 counterpart.

The main molecular markers used in the diagnostic workup are isocitrate dehydrogenase (IDH-1 and IDH-2), alpha-thalassemia/mental retardation syndrome X-linked (ATRX), Tumor protein p53 (p53), 1p19q codeletion, O6-methylguanine-DNA methyltransferase (MGMT), and telomerase reverse transcriptase (TERT), of which the first three are essential.

IDH can help distinguish gliomas from gliosis and may be present in astrocytomas, ODG, and 10% GBM. IDH-1 mutations (R132H) are more common and usually detectable by immunohistochemistry (IHC). If IHC is negative, IDH-1/IDH-2 genotyping may be needed if there is clinical suspicion. IDH wild-type tumors are those without IDH-1 or IDH-2 mutations. IDH-1 mutation confers a good prognosis in gliomas but is also a driver mutation promoting tumorigenesis. If IDH testing is inconclusive or cannot be performed, it is called IDH NOS (not otherwise specified).

ATRX: This gene has several important roles in cell division. Mutations will lead to loss of ATRX staining and is seen in several high grade gliomas (HGG).

p53: This is a tumor suppressor gene located on 17p chromosome. Loss of p53 promotes tumorigenesis in several CNS tumors, including GBM, medulloblastoma, and >50% IDH-mutant astrocytomas.

1p19q codeletion detectable by fluorescent in situ hybridization (FISH) is a diagnostic hallmark for ODGs. It has a strong prognostic value as well as predictive for response to chemotherapy and radiotherapy (RT).

MGMT gene encodes a DNA repair enzyme that may counter effect of alkylating agents such as temozolomide (TMZ). Methylation of MGMT promoter silences the gene. This promoter methylation is an established prognostic and predictive factor in GBM but has emerging value in LGGs as well.

TERT promoter mutations: They predict poor survival and treatment resistance in gliomas. This may be found in GBM and ODG.

IDH-mutant astrocytoma is characterized by a combination of IDH mutation, TP53 mutation, and ATRX loss.

If IDH mutation is absent by both IHC and genotyping, it qualifies as IDH wild-type diffuse astrocytoma, but only once IDH-1 and IDH-2 both have been deemed non-mutated. This group is an uncommon diagnosis and other alternative diagnoses such as ganglioglioma and IDH wild-type GBM should be ruled out.

IDH mutation with retained ATRX, limited p53 positivity and 1p19q codeletion define ODGs.

Consequent to WHO 2016 classification, the boundary between grade 2 and grade 3 gliomas has become less well defined in terms of survival outcomes. Instead of grade, IDH-1 mutation and 1p19q codeletion have become major drivers of disease categorization and eventual prognosis. Many tumors earlier defined as grade 3, such as anaplastic ODG, may have a good prognosis if IDH mutant and 1p19q codeleted, while a tumor previously defined as a diffuse astrocytoma (grade 2) may turn out to be IDH wild type and have an aggressive course akin to GBM. Epidermal growth factor receptor (EGFR) amplification, chromosome 7 gain, chromosome 10 loss, TERT promoter mutations, if present, would qualify grade 4 behavior and though not officially part of WHO classification, this entity has a proposed designation of "diffuse astrocytic glioma, IDH wild type, with molecular features of GBM, WHO grade 4."

Hence the term "high risk lower grade gliomas" is sometimes used to encompass the IDH-mutant anaplastic gliomas along with the traditional definition of age, tumor, and surgical extent. Since most phase 2 and 3 trials did not use these molecular characteristics for inclusion in trials until recently, we do not have clear-cut data for decision-making based on these molecular groupings. However, general principles of prognostic categorization and post hoc analysis for some of the earlier trials may assist in treatment choices.

Malignant transformation can occur in up to 70% of grade 2 glioma patients, the main risk factors being initial contrast enhancement, large tumor size, and subtotal resection (STR). In EORTC 22845 trial, early RT after surgery did not confer a higher risk for malignant transformation in LGGs compared to observation. Patients with malignant transformation have a shortened lifespan.

Astrocytic and oligodendroglial tumors may occur anywhere in the neuraxis. IDH-mutant gliomas are more common in cerebral subcortical white matter of frontal lobe. IDH wild-type astrocytomas arise in supratentorial region, and may involve multiple lobes or deep grey matter. Astrocytomas which appear grade 2 on histology but are in midline/thalamic location should be tested for H3 K27M mutation to rule out grade 4 diffuse midline glioma, which is particularly aggressive.

13.4 Imaging in LGGs

13.4.1 Computed Tomography (CT) Scan

Grade 2 gliomas are slow growing tumors and may produce calvarial erosion, expansion, or remodeling on plain CT scan. Tumors are seen as ill-defined homogenous hypodense masses. Most of them show no to minimal contrast enhancement and do not have significant perilesional edema. ODGs may have characteristic heterogeneous density with calcification in nearly 70%. Enhancement in previously known LGGs may raise suspicion of secondary transformation to HGG.

13.4.2 Multiparametric Magnetic Resonance Imaging (MRI)

This is now the standard and preferred imaging modality for brain tumors. It helps distinguish LGGs from HGGs better and gives fine anatomic detail of tumor location and extent that helps to plan local therapies such as surgery and RT, and also monitor treatment response. Diffuse astrocytomas are hypointense on T1 weighted sequences and hyperintense on T2/FLAIR, with no to minimal enhancement with gadolinium. ODGs may sometimes be isointense on T1 and may show heterogeneous enhancement. Calcification in ODGs appears as areas of low intensity blooming on GRE, SWI, or T2 sequences. MR spectroscopy may help distinguish brain tumors from non-neoplastic lesions and also to determine grade. *N*-acetylaspartate (NAA) is a surrogate marker of neuronal viability; it is present in higher levels in LGGs than HGGs due to preservation of neuronal integrity and astrocyte function. Choline (Cho) is a marker of cell membrane density and is raised in higher grade tumors and tumors with high Ki-67. Creatine (Cr) denotes energy metabolism and is largely stable. LGGs have higher NAA/Cr ratios and lower Cho/Cr ratios than HGGs. Cho peak may be higher in LGGs compared to background normal brain. Lactate and lipid peaks usually denote necrosis and myelin sheath disruption and are almost never seen in LGGs. Diffusion tensor imaging (DTI) helps visualize the path of white matter tracts owing to physiologic water molecule anisotropy within them. LGGs have more discrete boundaries and rarely invade along or into white matter tracts, unlike HGG. Surgeries are often performed with MRI guidance where preoperative MRI is co-registered with the operative field to help define planes for resection between normal brain and tumor tissue; T2 and FLAIR sequences are particularly useful in LGGs. CT guidance may be useful in some ODGs with calcification or CT angiography for navigation of critical vascular structures in tumor vicinity. Ultrasound, on the other hand, may be a useful real-time navigation tool, especially in the setting of brain shift following tumor removal. Intraoperative MRI in the operating suite itself also helps ensure the completeness of resection, though more useful in HGGs.

13.4.3 Positron Emission Tomography (PET)-CT Scans

They have a limited role in brain tumors but C-11 methionine and 18F-fluoro-ethyl-L-tyrosine (FET)-PET may have some value in distinguishing gliomas from non-neoplastic lesions and in stereotactic biopsy planning.

13.5 Treatment

Most evidence on LGG management derives from seminal trials that did not segregate patients based on molecular characteristics. Additionally, incompletely resected pilocytic astrocytomas (PA) which are grade 1, and some aggressive forms of diffuse astrocytomas (IDH wild, H3K27M mutant) were included, diluting the results somewhat. However, post hoc analyses of several trials suggest that majority of patients do belong to the recently defined grade 2 category and the trial interpretations would still be valid.

13.5.1 Surgery

This is the first-line treatment for LGGs, even when detected incidentally and neurologic signs are absent. This relieves the mass effect over surrounding normal brain, relieves seizures if they were the presenting symptom, gives tissue for histopathology and molecular characterization and prognostication, reduces risk of malignant transformation, and even gives information for planning adjuvant therapy. Early resection is favored over watchful waiting as it shows a survival advantage [2]. A greater extent of resection (mostly based on volumetric studies comparing the pre- and postoperative tumor extent) correlates with better overall survival (OS). Gross total resection (GTR, defined as >90% removal based on postoperative MRI taken within 48 hours) improves survival. In a retrospective series, 8-year OS was 91% for patients with >90% resection versus (vs.) 60% for STR [3, 4]. A maximal safe resection is advocated to avoid short-term and long-term neurologic deficits; GTR is the goal but may not always be possible due to large size, proximity to eloquent brain and vascular involvement or encasement. Intraoperative electrostimulation mapping/functional mapping on awake patients helps in improving resection extent (and possibly long-term control and survival) without increased risk of neurologic deficits.

13.5.2 Adjuvant Radiotherapy and Chemotherapy

Indications for RT traditionally include all high risk patients—diffuse astrocytoma or ODGs occurring in patients over 40 years of age or those with STR. Additional factors that may have a role in decision-making include tumor size, midline extension and baseline neurologic deficits; importance of these shall be discussed in detail citing individual studies incorporating these factors. As evidence discussed subsequently will show, RT is no longer supported as monotherapy and should be followed up with adjuvant chemotherapy. Most of the available evidence does not take into account the molecular characteristics of the tumor which may change treatment recommendations in coming years. At present, the evidence is agnostic to molecular subtype, and observation after STR should be offered only to the subset of young patients with extremely favorable tumor characteristics willing for regular imaging.

13.5.2.1 Timing of Radiotherapy

One randomized controlled trial (RCT) determined that early RT may delay tumor growth and control seizures better but does not improve survival, hence there is an option for deferring RT in patients with good general condition and compliance for follow-up.

The EORTC 22845 ("Non-Believers Trial," 1986–1997) recruited supratentorial LGGs (excluded completely resected PA, brainstem and optic pathway gliomas) between the ages of 16 and 65 years, post-surgery for randomization to early RT (54 gray, Gy) vs. delayed RT at progression, with follow-up contrast CT every 4 months. At a median follow-up of 7.8 years, the study concluded that early RT delayed local progression (median progression free survival, PFS, 5.3 vs. 3.4 years; $p < 0.0001$) and better seizure control (1 year, 25% vs. 41%) but did not improve OS (median OS, 7.4 vs. 7.2 years; $p = 0.872$). Nearly 65% patients in delayed RT group eventually received salvage RT at progression, and this probably contributed to lack of OS advantage. Both groups had similar cognitive function, performance status (PS), and focal deficits on follow-up [5].

13.5.2.2 Dose of Radiotherapy

Two randomized studies established that adjuvant RT doses of 45–50.4 Gy at 1.8 Gy per fraction (Fr) may be sufficient for LGGs, and contrary to popular belief, focal RT at given doses does not cause any major cognitive decline in adults.

EORTC 22844 trial ("Believers Trial," 1985–1991) investigated the adjuvant RT dose for LGG (excluding completely resected PA) patients aged 16–65 years, comparing two schedules: 45 Gy/25 Fr vs. 59.4 Gy/33 Fr. Both groups had similar OS (58% vs. 59%) and PFS (47% vs. 50%) at a median follow-up of 6.2 years, suggesting no dose response above 45 Gy [6]. The high dose group also had worse quality of life (QOL) and symptom burden (fatigue, malaise, and insomnia at RT completion and leisure time and emotional functioning after 7–15 months of randomization) [7]. This trial proved a shorter RT schedule was more convenient and comfortable while not compromising on disease control. Tumor size was deemed prognostic for survival.

The Intergoup (NCCTG/RTOG/ECOG) trial (1986–1994) also randomized patients adults with supratentorial LGGs (other than PA) to adjuvant RT doses of 50.4 Gy vs. 64.8 Gy. At a median follow-up of 6.43 years, this trial also reported no survival advantage with high RT doses (5-year OS, 64% vs. 72%; $p = 0.48$). In addition to tumor size (<5 cm), this trial also noted histologic subtype (oligo component) and age (under 40 years) as significant prognostic factors for better survival (5-year OS, 82% for <40 years and ODG vs. 32% for >40 years and astrocytoma).

High dose arm had higher incidence of grade 3–5 RT necrosis (5% vs. 2.5%) [8]. Cognitive testing of these patients and follow-up for >7 years discovered that most patients had a stable neurocognitive status after RT irrespective of dose or baseline patient or tumor characteristics. Some patients with abnormal cognition at baseline showed improvement after RT [9].

13.5.2.3 Radiotherapy Target Volume and Technique

Limited RT fields are used for treatment. Gross tumor volume (GTV) is the postoperative tumor bed and any residual tumor usually defined by the T2 or FLAIR abnormality on postoperative MRI. Clinical tumor volume (CTV) is generated by 1.5 cm anatomically constrained expansion around GTV. Planning target volume (PTV) is dependent on the expected setup errors with a treatment technique as well as frequency of verification imaging, and is usually 0.3–1.0 cm. Organs at risk would include optic structures, brainstem, cochlea, temporal lobes, and possibly hippocampi. The dose is variable across regions. European institutions prefer a dose of 45 Gy @ 1.8 Gy per fraction due to lack of benefit at higher doses in EORTC trials. American institutions typically use 54 Gy at same dose per fraction. Three-dimensional conformal radiation therapy suffices for most patients; however, when the target is close to critical structures or hippocampal avoidance is desired, intensity modulation with image guidance may help reduce high dose spillage with smaller PTV margins.

Figure 13.1 shows (a) the postoperative magnetic resonance image with outlined tumor bed (gross tumor volume), and (b) the three-dimensional conformal radiotherapy plan for a patient with grade 2 oligodendroglioma, post gross total resection.

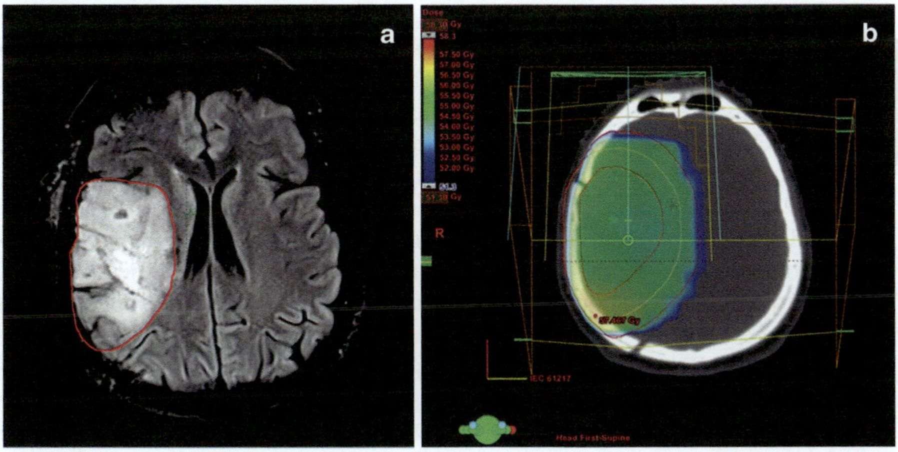

Fig. 13.1 A 50-year-old male who underwent gross total resection for grade 2 oligodendroglioma of right parietal lobe was advised adjuvant radiotherapy (RT) followed by adjuvant chemotherapy. (**a**) RT planning computed tomography (CT) scan co-registered with magnetic resonance imaging (MRI) of brain shows the T2 hyperintense tumor bed with minimal residual tumor. (**b**) Three-dimensional conformal radiotherapy plan for the same patient. A total dose of 54 gray (Gy) in 27 fractions was planned in two phases (46 Gy in phase 1 followed by 8 Gy in phase 2 with a shrinking field technique)

Protons have been advocated to reduce long-term morbidity, especially neurocognitive decline in pediatric patients, by reducing dose spillage beyond the target, but large clinical data or direct comparisons with photons are not available. Additionally, limited availability and cost of proton therapy preclude wider use, especially in developing countries. A randomized trial (NRG BN005) comparing proton beam with intensity modulated radiation therapy (IMRT) to preserve brain function in IDH-mutant grade 2–3 gliomas opened in 2017. Primary outcome is change in cognition and secondary outcomes include QOL, treatment toxicity, OS, and PFS.

13.5.2.4 Addition of Chemotherapy Following Adjuvant RT

Experience with chemotherapy agents procarbazine/lomustine (CCNU)/ vincristine combination (PCV) and TMZ in recurrent LGG gave encouraging response rates though without a demonstrable survival advantage [10].

One of the initial studies evaluating addition of chemotherapy to RT was by the Southwest Oncology Group (SWOG, 1980–1985). They randomized adult LGGs following STR or biopsy to receive RT (55 Gy) with or without chemotherapy (CCNU 100 mg/m^2 every 6 weeks, starting 2 days before RT and continuing for 2 years). Median OS was 4.45 years for RT alone vs. 7.4 years for RT + CCNU ($p = 0.7$). However, the 10-year OS of the chemotherapy arm was much lower (40% for RT and 20% for CCNU). Age and PS had the maximum impact on OS [11].

RTOG 9802 (1992–2002) randomized adults with high risk supratentorial LGGs to receive RT alone (54 Gy/30 Fr) or RT followed by chemotherapy (6 cycles PCV). Eligible patients were those older than 40 years with any surgical extent, or those younger than 40 years with STR or biopsy. Initial results of this trial reported PFS advantage with PCV (5-year PFS, 63% vs. 46%; $p = 0.005$) but no OS advantage (5-year OS, 72% vs. 63%; $p = 0.13$). 2-year survivors had a survival advantage with PCV on post hoc analysis suggesting a delayed benefit with chemotherapy [12]. Updated results with 12 year follow-up indeed proved the delayed survival advantage with PCV (10-year OS, 60% vs. 40%; median OS 13.3 vs. 7.8 years; HR 0.59; $P = 0.003$, no difference in initial 4 years) and delayed progression (10-year PFS, 51% vs. 21%, no difference in initial 2 years). This difference of 5.5 years in OS happened despite the 77% cross-over rate to receive PCV in patients who progressed in RT alone arm. Favorable prognostic factors for OS were age under 40 years, ODG histology, and chemotherapy addition [13]. A post hoc study done on 106 patients to see impact of molecular subgroups on prognosis showed that 35% patients had both IDH-1 mutation and 1p19q codeletions and 41% had IDH-1 mutation without 1p19q codeletion. PCV was associated with longer PFS (HR 0.32 for IDH-mutant/non-codeleted, HR 0.13 for IDH-mutant with 1p19q codeletion) and OS (HR 0.38 for IDH-mutant/non-codeleted; HR 0.21 for IDH-mutant with 1p19q codeletion) in both these groups, thus confirming that PCV had a benefit for all IDH-mutant patients [14]. TERT promoter and Capicua transcriptional repressor (CIC) mutations also have a significant association with OS, while IDH-1/2 mutations have a significant association with PFS [15]. MGMT promoter methylation has emerged as an important prognostic marker for PFS in this group of high risk LGG treated with RT ± PCV [16].

RTOG 0424 (2005–2009) was a single phase 2 study that attempted to replace PCV with TMZ as adjuvant therapy for high risk LGGs using Pignatti criteria (described later, in prognostic factors), and compared the data with a historical cohort. Patients who had at least 3 of the 5 high risk factors were treated with RT (54 Gy/30 Fr) along with concurrent (75 mg/m^2) and adjuvant TMZ (150–200 mg/m^2 D1–5 q4 weeks for 12 cycles) with pneumocystis carinii pneumonia (PCP) prophylaxis. Long-term results from this study with a median follow-up of 9 years show a median OS of 8.2 years, higher than historical controls treated with RT alone (median OS for intermediate risk, 7.2–7.6 years; high risk, 4.8–5.5 years) with acceptable toxicity [17]. MGMT methylation status was available for nearly 60% of patients in this study; 76% of these had methylated MGMT. Unmethylated MGMT promoter had a statistically worse OS and PFS association, independent of IDH1/2 mutation status, signifying MGMT promoter methylation as an important prognostic biomarker for grade 2 gliomas treated with RT and TMZ, possibly highlighting the effect of TMZ as radiosensitizer in this group as well [18].

Eastern Cooperative Oncology Group (ECOG) had started a randomized study (E3F05) in 2009 comparing RT with or without concurrent and adjuvant TMZ for 12 cycles for symptomatic or progressive LGGs. However, the study recruitment was suspended after RTOG 9802 results were published, assuming that RT alone would be an inferior arm.

EORTC 26951 and RTOG 9402 trials compared adjuvant RT with or without 6 cycles of PCV in patients with anaplastic ODGs and demonstrated higher OS in chemotherapy arm. Although traditionally grade 3 tumors, the new WHO 2016 classification places them in the better prognostic category of "high risk lower grade gliomas." Post hoc analysis on these trials suggested that the OS benefit of PCV was limited to patients with 1p19q codeletion group and IDH mutation group, potentially identifying the true molecularly defined ODG subgroup [19–22].

13.5.2.5 Radiotherapy Versus Chemotherapy

Though RT has been established as an adjuvant therapy choice in LGGs with unfavorable factors, some patients who are not candidates for RT or unwilling for the same may be offered adjuvant TMZ alone at the risk of higher hematologic toxicity. This option may not be valid in high risk LGGs since combinations of both RT and chemo have shown better survival in that cohort.

EORTC 22033–26033 (2005–2010) was a phase 3 trial that compared RT (50.4 Gy/28 Fr) with TMZ (75 mg/m^2 OD X 21 days q4 weeks for 12 cycles) in adult LGG patients with at least one of the five unfavorable factors mentioned by Pignatti et al. Primary endpoint was PFS, and secondary endpoints included OS, toxicity, neurocognition, and correlation of PFS with several biomarkers. At a median follow-up of 4 years, both arms had similar PFS (3.8 years with RT vs. 3.25 years with TMZ; $p = 0.22$). Median OS was not reached. Patients with IDH mutation without 1p19q codeletion had longer PFS with RT (HR 1.86) while the two arms had similar PFS in patients with (a) IDH mutation with 1p19q codeletion, and (b) IDH wild-type tumors. TMZ arm had higher grade 3–4 hematologic toxicity, infections, and fatigue [23]. There was no difference between the two groups in neurocognition or QOL in first 3 years of follow-up [24]. Genomic studies have tried to define an MGMT methylation score in IDH-mutant tumors with 1p19q codeletion that may help identify patients who merit TMZ as first-line adjuvant therapy for PFS benefit [25].

A retrospective National cancer database (NCDB) analysis of 1054 patients with high risk LGGs (age ≥40 years or STR as per RTOG 9802 criteria) who had received chemotherapy or chemoradiation also revealed no significant OS difference between the two groups (8-year OS, 52.8% vs. 56.7%) but tumor size >6 cm, older age, and astrocytoma histology predicted worse OS. Younger patients with ODG histology or 1p19q codeletion were offered chemotherapy more often [26].

13.5.2.6 Choice of Adjuvant Chemotherapy: PCV Versus TMZ

There are no direct comparisons between the two treatment regimes. RTOG 9802 used PCV in a randomized study while RTOG 0424 used TMZ in a single arm study. Median PFS in the former study with PCV was 10.4 years (3 years PFS 75–80%) while it was 4.5 years in the latter (3 years PFS 59%) giving some indication that PCV may be superior to TMZ, notwithstanding the inherent bias and challenges of such cross-trial comparisons.

The CODEL trial (NCT00887146) is a randomized phase 3 study that initially started to compare addition of TMZ vs. PCV to RT in adjuvant management of anaplastic ODGs with 1p19q codeletion and IDH mutation, but has now extended its inclusion to grade 2 gliomas as well; the trial is still recruiting and results will take a long time to mature.

13.5.3 Which Patients with LGG may Do Well with Observation Only

RTOG 9802 also had a low risk arm (111 patients) which consisted of patients younger than 40 years having neurosurgeon-determined GTR; these patients were followed up without any adjuvant therapy after surgery. This group had an excellent 5-year OS of 93% but 48% PFS rate over this period. However, it was discovered on review of postoperative MRI that almost all these patients had residual disease; nearly 31% had residual more than 1 cm. Preoperative tumor diameter ≥4 cm, astrocytoma component and residual ≥1 cm predicted poorer PFS [27].

NCDB data for RTOG 9802-defined low risk LGG patients between 2010 and 2013 were analyzed. Within this cohort, patients with tumor size >5 cm and non-ODG histology had an inferior OS compared to other low risk patients (80.5% vs. 94.0%). There are recommendations to include these patients into an “intermediate risk” group and investigate appropriate adjuvant therapy strategies for them [28].

13.5.4 Prognostic Factors

The following factors have prognostic value as seen in several randomized studies:

1. Patient factors
   (a) Patient age ≥40 years—unfavorable

   (b) Baseline neurologic deficits—unfavorable
   (c) Seizures at diagnosis—favorable
2. Tumor factors
   (a) Size ≥6 cm—unfavorable
   (b) Astrocytoma histology—unfavorable
   (c) Tumors crossing midline—unfavorable
   (d) Molecular studies
      IDH mutation—favorable
      1p19q codeletion—favorable
      MGMT promoter methylation—favorable
      p53 mutation—unfavorable
      TERT mutation—unfavorable
      CIC mutation—favorable
3. Extent of surgery: subtotal resection or biopsy—unfavorable

Based on the patient data in EORTC trials 22844 and 22845, Pignatti et al. proposed the following five factors as unfavorable for survival: age ≥40 years, neurologic deficit present before surgery, preoperative tumor diameter ≥6 cm, tumor crossing midline, and astrocytoma histology. They grouped the patients into categories of low risk if up to two factors were present (median OS, 7.8 years), and high risk if three or more factors (median OS, 3.4 years) were present, known as Pignatti criteria for risk grouping [29]. RTOG 9802, however, defined only two factors: age above 40 years and STR, presence of either of which was considered a high risk disease.

The data from EORTC trials 22844 and 22845 were reanalyzed by only including patients who had undergone a central pathology review and confirmed to be grade 2 gliomas; this excluded 21% patients. This was believed to be a more homogenous group after excluding HGGs, and had median PFS and OS of 4.6 years and 7.2 years, respectively. These were further modeled based on several factors including some of Pignatti criteria along with time since LGG symptoms, to generate three risk groups: low, intermediate, and high, which were later validated on patient data from intergroup and RTOG 9802 trials. The median OS stratified by risk groups was 11.6–12.7 years for low risk, 7.2–7.6 years for intermediate risk, and 4.8–5.4 years for high risk groups [30].

13.5.5 Reirradiation/Salvage

Recurrence in LGGs may manifest as symptomatic worsening or progression on MRI. Nearly 70% may have malignant transformation on long-term follow-up. Salvage treatment option should include surgery, if feasible—to verify histology and transformation, reduce mass effect and thereby relieve symptoms. Reirradiation, either with conventional fractionation or SRS to small residual disease may be decided on the merit of an individual case; risks of critical structure damage and radiation necrosis should be evaluated carefully. Systemic therapy may include PCV, TMZ, lomustine alone or bevacizumab, with variable results, depending on type of prior therapy given and resultant response/toxicity. The longer the time since prior therapy, the better is the expected outcome from retreatment.

13.6 "Other" Astrocytic Tumors

13.6.1 Pleomorphic Xanthoastrocytoma Grade 2

These rare tumors account for 1% of astrocytic neoplasms, first incorporated in 1993 in WHO classification. They may affect children or young adults, with median age at diagnosis of 29 years, without any gender predilection. Most common location is supratentorial in cerebral hemispheres (usually temporal lobe) but other sites such as sella, ventricles, spine, retina, etc., have also been reported. On CT, they appear as well-circumscribed superficial solid-cystic masses, usually hypo- or isodense, with areas of calcification in solid component. The solid component and cyst wall enhance with contrast. On MRI, the lesions are T1 hypointense, T2 hyperintense and heterogeneously enhancing on T1 contrast with rim enhancement of the cyst wall as well. Minimal perilesional edema can be seen. The lesion is often in contact with leptomeninges and there may occasionally be leptomeningeal enhancement [31]. In histology, pleomorphic lipidized cells may be seen with multinucleated giant cells, eosinophilic granular bodies in an inflammatory background. Necrosis is absent and mitoses are rare. The histologic appearance is often confused with high grade astrocytomas such as GBM. On IHC, S-100, GFAP, and oligodendrocyte transcription factor-2 (Olig2) are positive [32]. Neuronal markers such as synaptophysin and neurofilament (NF) are occasionally seen. Ki-67 is low. Reticulin fiber staining is commonly seen, suggesting an origin potentially from subpial fibrous astrocytes. Nearly 60–70% PXA have BRAF V600E mutation and this confers a better outcome [33]. Nearly 25% have p53 mutations. Grade 2 PXAs have a favorable prognosis with 5-year survival >75% and 5-year PFS >60%. Treatment of choice is surgery and extent of surgery (GTR vs. STR) has prognostic value. Adjuvant RT has been tried more often for salvage in recurrent settings than in adjuvant setting after STR at a dose of 45–54 Gy with or without concurrent TMZ with a possible PFS benefit but no OS advantage [34]. Systemic therapy (TMZ, carmustine, lapatinib, bevacizumab) has been used in salvage settings but with inconsistent outcomes.

13.7 Other Gliomas

13.7.1 Chordoid Glioma of Third Ventricle

These are rare neuroepithelial tumors with a slight female preponderance and median age of presentation at 48 years, originating in midline in suprasellar region or anterior part of third ventricle, and containing both glial and chordoid elements [35]. Due to their location, presenting features are due to raised intracranial tension (ICT) with obstructive hydrocephalus (headache, nausea, vomiting, ataxia), hypothalamic dysfunction, memory deficits, mental status changes, and visual impairment. On CT, they appear most commonly as well-circumscribed, solid, hyperdense lesions with homogenous contrast enhancement. MRI shows a mass that is T1 iso- or hypointense, T2 hyperintense and homogenously enhancing on T1 contrast sequences [36].

On histology, they appear as clusters and cords of ovoid or polygonal epithelial cells in a mucinous background. GFAP is strongly positive; other positive IHC include epithelial membrane antigen (EMA), CD34, cytokeratin (CK), S100, and vimentin. Mitoses are low or absent with MIB-LI $\leq 5\%$, and necrosis is absent [37].

Surgery in hypothalamic location is tricky, with higher risk of neuroendocrine dysfunction (including diabetes insipidus) when attempting GTR. GTR is feasible in less than half of patients although it is the most crucial factor for local control. Hence, for most tumors, STR only is feasible but there are reports where the surgical approach (transcortical and transcallosal vs. trans-lamina terminalis) is linked to increased complications rather than extent of surgery [38]. RT has been used occasionally for residual tumors in the form of stereotactic radiosurgery (SRS) with marginal doses of 12–20 Gy or fractionated RT with 54 Gy in 30 fractions, with variable results [39, 40]. Adjuvant chemotherapy has no demonstrable role. The disease carries high postoperative morbidity and mortality as well as high recurrence rates following STR, hence a lot of attention has to be directed to surgical planning and postoperative care.

13.8 Mixed Neuronal and Neuronal-Glial Tumors

13.8.1 Central Neurocytoma

Central neurocytomas (CN), described first in 1982, are uncommon neuroepithelial tumors with neuronal differentiation, possibly originating from neuroglial precursor cells. They constitute less than 0.5% of all adult intracranial tumors, most commonly presenting in the third decade. Presenting features are commonly due to raised ICT (headache, vomiting) but vision changes, gait imbalance, and memory disturbances may also occur. The lesions originate in subependymal plate of ventricles. CT demonstrated a well-circumscribed enhancing ventricular mass, most commonly in lateral ventricles. Calcification is seen in 50%. The lesion is iso- to hyperintense on T2 MRI with heterogeneous contrast enhancement on T1 contrast sequences. Histology demonstrates small round cells, rounded nuclei, and scant cytoplasm. IHC is positive for synaptophysin. Some cases may show features such as high mitotic activity, microvascular proliferation, necrosis, and Ki-67 >3%, and are referred as atypical CN with higher recurrence potential [41].

Most cases have a favorable prognosis with benign biological behavior. Maximal safe resection is the dictum of management. GTR gives 5-year survival of 93% and needs no further adjuvant management. For patients with STR or high MIB-LI, adjuvant RT (dose ~56 Gy) may help improve local control as well as OS [42, 43]. For small recurrent lesions, SRS has also been used with peripheral doses of 10.5–17.0 Gy, with local control rates exceeding 90% [44]. TMZ and PCV have been tried as salvage therapies in recurrent inoperable tumors with variable outcomes [45].

13.8.2 Extraventricular Neurocytoma

Extraventricular neurocytoma (EVN) was first described as an entity in 1997, having the same histopathological characteristics as CN but located outside the ventricular system. It is a rare glioneuronal tumor, with two incidence peaks in the second and fifth decades [46]. Most common presenting complaints are seizure and headache. There are no specific imaging characteristics. Typically it manifests as a large circumscribed mass, most commonly frontal, with a solid portion and variable cystic regions, calcification, hemorrhage, and perilesional edema. More than 80% show contrast enhancement of solid portion, mostly heterogeneous [47].

The standard therapy is surgery; GTR is curative. In case of STR when tumors lie in proximity to critical structures or eloquent cortex, or when histopathology shows atypical features, recurrence risk is high, often 50% or more; in these situations, adjuvant RT (50.4–54.0 Gy) may help improve control. Salvage RT may also be considered in inoperable recurrences or after incomplete removal of recurrent tumor. Overall outcome is poorer compared to CN. This may be due to low GTR frequency and difficulty of salvage surgeries in the same critical locations. There is no role of adjuvant chemotherapy but salvage TMZ has been used anecdotally for recurrences [48].

13.8.3 Cerebellar Liponeurocytoma

This rare tumor was first described by Bechtel et al. in 1978. It was earlier regarded as grade 1 but upgraded to grade 2 in WHO 2007 classification due to reported atypia and recurrence patterns. The mean age of presentation is 49 years with a female preponderance [49]. Most common location is hemispheric, though direct intraspinal extension to cervical spine and lumbar spine metastases have also been reported [50]. Presenting features are due to cerebellar involvement (gait ataxia, frequent falls, dizziness) or obstructive hydrocephalus (headache, vomiting, nausea). On CT imaging, the fatty part is detected as a hypodense lesion. MRI shows well-demarcated T1 hyperintense lesion with inversion on fat suppressed images and minimal contrast enhancement; T2/FLAIR imaging shows heterogeneous intensity and diffusion weighted imaging (DWI) shows high intensity with minimal perilesional edema. There is a possible familial association with autosomal dominant inheritance [51]. Germline mutations have also been postulated. Pathology shows variable neuronal/neurocytic differentiation and focal lipomatous and astrocytic differentiation. Histologic features include isomorphic oval or round cells with focal accumulation of lipid-filled neuroepithelial cells. IHC shows synaptophysin, neuron-specific enolase (NSE), neurospecific nucleoprotein (NeuN) and microtubule associated protein 2 (MAP 2) positivity and IDH-1, EMA, CK and nuclear Olig2 negativity. The lipid vacuole membranes of lipomatous cells are positive for synaptophysin, MAP 2, vimentin, and S100. GFAP is limited to reactive astrocytes. 1p19q codeletion is absent, differentiating them from ODG. MIB-LI is low (<5%) [52].

Surgery is the primary treatment of choice. Adjuvant RT (dose 54 Gy) reduces recurrence risk by nearly 50%, even in the setting of GTR. Nearly 30% patients recur even after long follow-up periods of 8–10 years. For recurrent tumors, resurgery followed by RT should be considered [52].

13.9 Response to Treatment

Since LGGs have a long follow-up with potential for progression as well as conversion to HGG, changes in tumor characteristics over time are monitored using several criteria. Uniformity of these criteria is necessary to ensure comparison of results across institutions and treatment strategies. Macdonald criteria were developed for supratentorial gliomas in 1990. They were initially based on CT and later, adapted to MRI [53]. They defined four groups: Complete response (CR): Complete disappearance of tumor on consecutive imaging, at least a month apart, without steroids, Partial response (PR): ≥50% decrease in the enhancing area of tumor without neurologic deterioration or increasing steroids; Progressive disease (PD): ≥25% increase in size of enhancing lesion; Stable disease (SD): Not fitting any of the previous three criteria. This system is unable to account for changes such as radiation necrosis, pseudoprogression, response to bevacizumab, TMZ or immune changes, and is largely not applicable in non-enhancing LGGs. The Response Assessment in Neuro-Oncology (RANO) Working Group defined MRI-based criteria for LGGs in 2011 with more emphasis on T2/FLAIR sequences which have more relevance in LGGs. The new criteria also incorporate new lesions, enhancement pattern changes, and clinical changes [54]. Five groups as per RANO criteria include CR: Full resolution of T2/FLAIR changes, no new radiographic findings, clinically improved or stable, and off steroids, sustained for 4 weeks; PR: >50% decrease in product of perpendicular diameters of pre-treatment T2 or FLAIR change, no new lesions, sustained for 4 weeks, clinically improved or stable, and stable steroid dose; Minor response: >25% but <50% reduction in product of perpendicular diameters of pre-treatment T2 or FLAIR change, no new lesions, sustained for 4 weeks, clinically improved or stable, and stable steroid dose; Stable response: Imaging unchanged, clinically stable, stable steroid dose; PD: New lesions, new enhancement (evidence of progression), or >25% increase in T2/FLAIR change, clinical decline not due to causes other than tumor.

13.10 Treatment-Related Morbidity

13.10.1 Early/Acute Effects

The acute effects of cranial RT include nausea, appetite loss, and somnolence. Patchy alopecia and dry desquamation, dryness of eyes, and fatigue may be present, but these problems are usually self-limiting. PCV or TMZ may cause hematologic toxicity and increased risk of infections such as pneumocystis, for which appropriate monitoring and prophylaxis are recommended.

13.10.2 Delayed Effects: Hearing, Neurocognition, and Neuroendocrine Dysfunction

Patients with temporal lobe lesions may experience conductive or sensorineural hearing loss in the long term if the middle ear or cochlea receive mean doses exceeding 45 Gy. Due to the prolonged survival in LGGs, neurocognition is an important issue that may be affected and impacts function and QOL. With the RT doses used in EORTC 22033–26033 trial, neurocognitive deficits and QOL issues did not differ between the treatment arms of RT vs. chemotherapy till 3 years follow-up; they rather improved in some people with baseline deficits, pointing to the fact that largely the disease itself contributes to this issue. Reducing the volume of RT has helped preservation of intelligence quotient (IQ) and neuroendocrine function in children [55]. Hippocampal sparing is now being attempted for most low grade tumors and even for whole brain RT for brain metastases to preserve memory [56]. Theoretically, proton therapy may contribute further to this goal.

13.10.3 Pseudoprogression and Radiation Necrosis

Pseudoprogression refers to increase in size or enhancement of a lesion in response to treatment; it is more common after combined RT and chemotherapy and in ODGs, and less often seen after RT alone. Median time to appearance varies from 8–12 months but delayed presentations are also seen. In LGGs, pseudoprogression is more often seen in volumes receiving doses over 45 Gy, and in 1p19q intact and IDH wild-type tumors [57].

Radionecrosis refers to coagulative necrosis of brain tissue in response to focal radiation. It is less common in LGGs than in HGGs, but may occasionally be seen after reirradiation for recurrent or progressive disease.

Both pseudoprogression and radionecrosis may be confused with disease progression and may need to be differentiated for management decisions. Pseudoprogression may have mismatch between worsening radiologic appearance and clinical features which may be stable or improved. Radiation necrosis may be suspected based on the history of radiation dose, treatment technique, and volume. Multiparametric MRI with perfusion and diffusion weighted sequences, and occasionally PET scan may help distinguish them from recurrent disease.

13.11 Current Status and Future Directions

The group LGG encompasses several heterogeneous primary CNS tumors with different natural history and treatments. Surgery (maximal safe resection) is the standard treatment with need for adjuvant therapy defined by extent of resection, tumor characteristics, and patient age. For most astrocytic and oligodendroglial tumors, both adjuvant radiation and chemotherapy are recommended. RT alone is not recommended after RTOG 9802 findings. RT delay or observation can only be justified in very small IDH-mutant completely resected tumors in young patients provided they continue rigorous follow-up. Molecular characterization and inclusion of these factors into clinical trial stratification and inclusion criteria will help fine tune treatment intensity to meet the dual goals of maximum control with minimum long-term morbidity.

References

1. Gupta A, Dwivedi T. A simplified overview of world health organization classification update of central nervous system tumors 2016. J Neurosci Rural Pract. 2017;8(4):629–41.
2. Jakola AS, Skjulsvik AJ, Myrmel KS, et al. Surgical resection versus watchful waiting in low-grade gliomas. Ann Oncol. 2017;28(8):1942–8.
3. Aghi MK, Nahed BV, Sloan AE, Ryken TC, Kalkanis SN, Olson JJ. The role of surgery in the management of patients with diffuse low grade glioma: a systematic review and evidence-based clinical practice guideline. J Neuro-Oncol. 2015;125(3):503–30.
4. Smith JS, Chang EF, Lamborn KR, et al. Role of extent of resection in the long-term outcome of low-grade hemispheric gliomas. J Clin Oncol. 2008;26(8):1338–45.
5. van den Bent MJ, Afra D, de Witte O, et al. Long-term efficacy of early versus delayed radiotherapy for low-grade astrocytoma and oligodendroglioma in adults: the EORTC 22845 randomised trial [published correction appears in Lancet. 2006 Jun 3;367(9525):1818]. Lancet. 2005;366(9490):985–90.
6. Karim AB, Maat B, Hatlevoll R, et al. A randomized trial on dose-response in radiation therapy of low-grade cerebral glioma: European Organization for Research and Treatment of Cancer (EORTC) Study 22844. Int J Radiat Oncol Biol Phys. 1996;36(3):549–56.
7. Kiebert GM, Curran D, Aaronson NK, et al. Quality of life after radiation therapy of cerebral low-grade gliomas of the adult: results of a randomised phase III trial on dose response (EORTC trial 22844). EORTC Radiotherapy Co-operative Group. Eur J Cancer. 1998;34(12):1902–9.
8. Shaw E, Arusell R, Scheithauer B, et al. Prospective randomized trial of low- versus high-dose radiation therapy in adults with supratentorial low-grade glioma: initial report of a North Central Cancer Treatment Group/Radiation Therapy Oncology Group/Eastern Cooperative Oncology Group study. J Clin Oncol. 2002;20(9):2267–76.
9. Brown PD, Buckner JC, O'Fallon JR, et al. Effects of radiotherapy on cognitive function in patients with low-grade glioma measured by the folstein mini-mental state examination. J Clin Oncol. 2003;21(13):2519–24.
10. Soffietti R, Rudà R, Bradac GB, Schiffer D. PCV chemotherapy for recurrent oligodendrogliomas and oligoastrocytomas. Neurosurgery. 1998;43(5):1066–73.
11. Eyre HJ, Crowley JJ, Townsend JJ, et al. A randomized trial of radiotherapy versus radiotherapy plus CCNU for incompletely resected low-grade gliomas: a Southwest Oncology Group study. J Neurosurg. 1993;78:909–14.
12. Shaw EG, Wang M, Coons SW, et al. Randomized trial of radiation therapy plus procarbazine, lomustine, and vincristine chemotherapy for supratentorial adult low-grade glioma: initial results of RTOG 9802. J Clin Oncol. 2012;30(25):3065–70.
13. Buckner JC, Shaw EG, Pugh SL, et al. Radiation plus procarbazine, CCNU, and vincristine in low-grade glioma. N Engl J Med. 2016;374(14):1344–55.
14. Bell EH, Zhang P, Shaw EG, et al. Comprehensive genomic analysis in NRG oncology/RTOG 9802: a phase III trial of radiation versus radiation plus procarbazine, lomustine (CCNU), and vincristine in high-risk low-grade glioma. J Clin Oncol. 2020;2020:JCO1902983.
15. Bell EH, McElroy JP, Fleming J, et al. Comprehensive mutation analysis in NRG oncology/RTOG 9802: a phase III study of RT vs RT + PCV in high-risk low-grade gliomas (LGGs). J Clin Oncol. 2016;34(15_suppl)
16. Bell EH, Zhang P, Aldape K, et al. OS01.7 MGMT promoter methylation status independently predicts progression-free survival in NRG oncology/RTOG 9802: a phase III trial of RT vs RT + PCV in high-risk low-grade gliomas. Neuro-Oncology. 2017;19(Suppl 3):iii2–3.

17. Fisher BJ, Pugh SL, Macdonald DR, et al. Phase 2 study of a temozolomide-based chemoradiation therapy regimen for high-risk, low-grade gliomas: long-term results of radiation therapy oncology group 0424. Int J Radiat Oncol Biol Phys. 2020;107(4):720–5.
18. Bell EH, Zhang P, Fisher BJ, et al. Association of MGMT promoter methylation status with survival outcomes in patients with high-risk glioma treated with radiotherapy and temozolomide: an analysis from the NRG oncology/RTOG 0424 trial. JAMA Oncol. 2018;4(10):1405–9.
19. van den Bent MJ, Brandes AA, Taphoorn MJB, et al. Adjuvant procarbazine, lomustine, and vincristine chemotherapy in newly diagnosed anaplastic oligodendroglioma: long-term follow-up of EORTC brain tumor group study 26951. J Clin Oncol. 2013;31(3):344–50.
20. Cairncross G, Wang M, Shaw E, et al. Phase III trial of chemoradiotherapy for anaplastic oligodendroglioma: long-term results of RTOG 9402. J Clin Oncol. 2013;31(3):337–43.
21. van den Bent MJ, Carpentier AF, Brandes AA, et al. Adjuvant procarbazine, lomustine, and vincristine improves progression-free survival but not overall survival in newly diagnosed anaplastic oligodendrogliomas and oligoastrocytomas: a randomized European Organisation for Research and Treatment of Cancer P. J Clin Oncol. 2006;24(18):2715–22.
22. Cairncross JG, Wang M, Jenkins RB, et al. Benefit from procarbazine, lomustine and vincristine in oligodendroglial tumors is associated with mutation of IDH. J Clin Oncol. 2014;32(8):783–90.
23. Baumert BG, Hegi ME, van den Bent MJ, et al. Temozolomide chemotherapy versus radiotherapy in high-risk low-grade glioma (EORTC 22033-26033): a randomised, open-label, phase 3 intergroup study. Lancet Oncol. 2016;17(11):1521–32.
24. Reijneveld JC, Taphoorn MJB, Coens C, et al. Health-related quality of life in patients with high-risk low-grade glioma (EORTC 22033-26033): a randomised, open-label, phase 3 intergroup study. Lancet Oncol. 2016;17(11):1533–42.
25. Bady P, Kurscheid S, Delorenzi M, et al. The DNA methylome of DDR genes and benefit from RT or TMZ in IDH mutant low-grade glioma treated in EORTC 22033. Acta Neuropathol. 2018;135(4):601–15.
26. Jhaveri J, Liu Y, Chowdhary M, et al. Is less more? Comparing chemotherapy alone with chemotherapy and radiation for high-risk grade 2 glioma: an analysis of the National Cancer Data Base. Cancer. 2018;124(6):1169–78.
27. Shaw EG, Berkey B, Coons SW, et al. Recurrence following neurosurgeon-determined gross-total resection of adult supratentorial low-grade glioma: results of a prospective clinical trial. J Neurosurg. 2008;109(5):835–41.
28. Jairam V, Kann BH, Park HS, et al. Defining an intermediate-risk group for low-grade glioma: a national cancer database analysis. Anticancer Res. 2019;39(6):2911–8.
29. Pignatti F, van den Bent M, Curran D, et al. Prognostic factors for survival in adult patients with cerebral low-grade glioma. J Clin Oncol. 2002;20(8):2076–84.
30. Gorlia T, Wu W, Wang M, et al. New validated prognostice models and prognostic calculators in patients with low-grade gliomas diagnosed by central pathology review: a pooled analysis of EORTC/RTOG/NCCTG phase III clinical trials. Neuro Oncol. 2013;15:1568–79.
31. Shaikh N, Brahmbhatt N, Kruser TJ, et al. Pleomorphic xanthoastrocytoma: a brief review. CNS Oncol. 2019;8(3):CNS39.
32. Liu XF, Du X, Zhang XT, Yang M, Han YM, Lin XY. Pleomorphic xanthoastrocytoma inside lateral ventricle: a rare case report and literature review. Int J Clin Exp Pathol. 2019;12(4):1118–23.
33. Dias-Santagata D, Lam Q, Vernovsky K, et al. BRAF V600E mutations are common in pleomorphic xanthoastrocytoma: diagnostic and therapeutic implications. PLoS One. 2011;6:e1794.
34. Mallick S, Benson R, Melgandi W, Giridhar P, Rath GK. Grade II pleomorphic Xanthoastrocytoma; a meta-analysis of data from previously reported 167 cases. J Clin Neurosci. 2018;54:57–62.
35. Brat DJ, Scheithauer BW, Staugaitis SM, Cortez SC, Brecher K, Burger PC. Third ventricular chordoid glioma: a distinct clinicopathologic entity. J Neuropathol Exp Neurol. 1998;57:283–90.

36. Vanhauwaert DJ, Clement F, Van Dorpe J, Deruytter MJ. Chordoid glioma of the third ventricle. Acta Neurochir. 2008;150(11):1183–91.
37. Reifenberger G, Weber T, Weber RG, et al. Chordoid glioma of the third ventricle: immunohistochemical and molecular genetic characterization of a novel tumor entity. Brain Pathol. 1999;9:617–26.
38. Ampie L, Choy W, Lamano JB, et al. Prognostic factors for recurrence and complications in the surgical management of primary chordoid gliomas: a systematic review of literature. Clin Neurol Neurosurg. 2015;138:129–36.
39. Kobayashi T, Tsugawa T, Hashizume C, et al. Therapeutic approach to chordoid glioma of the third ventricle. Neurol Med Chir. 2013;53:249–55.
40. Hanbali F, Fuller GN, Leeds NE, Sawaya R. Choroid plexus cyst and chordoid glioma. Report of two cases. Neurosurg Focus. 2001;10:E5.
41. Goyal S, Kataria T, Gupta D, et al. Atypical central neurocytoma with leptomeningeal dissemination: a case report. J Egypt Natl Canc Inst. 2020;32:23.
42. Imber BS, Braunstein SE, Wu FY, et al. Clinical outcome and prognostic factors for central neurocytoma: twenty year institutional experience. J Neuro-Oncol. 2016;126(1):193–200.
43. Mallick S, Roy S, Das S, et al. Role of adjuvant radiation in the management of central neurocytoma: experience from a tertiary cancer care center of India. Indian J Cancer. 2015;52(4):590–7.
44. Bui TT, Lagman C, Chung LK, et al. Systematic analysis of clinical outcomes following stereotactic radiosurgery for central neurocytoma. Brain Tumor Res Treat. 2017;5(1):10–5.
45. Johnson MO, Kirkpatrick JP, Patel MP, et al. The role of chemotherapy in the treatment of central neurocytoma. CNS Oncol. 2019;8(3):CNS41.
46. Mallick S, Benson R, Rath GK. Patterns of care and survival outcomes in patients with an extraventricular neurocytoma: an individual patient data analysis of 201 cases. Neurol India. 2018;66(2):362–7.
47. Romano N, Federici M, Castaldi A. Imaging of extraventricular neurocytoma: a systematic literature review [published online ahead of print, 2020 Apr 25]. Radiol Med. 2020; https://doi.org/10.1007/s11547-020-01198-8.
48. El Demellawy D, Sur M, Ahmed AD, Provias J. Hemispheric extra-ventricular glioneurocytoma: a clinicopathological review with detailed immunohistochemical profile. Pathol Res Pract. 2012;208:444–51.
49. Oudrhiri MY, Raouzi N, El Kacemi I, et al. Understanding cerebellar liponeurocytomas: case report and literature review. Case Rep Neurol Med. 2014;2014:186826.
50. Deora H, Prabhuraj AR, Saini J, Yasha TC, Arimappamagan A. Cerebellar liponeurocytoma: a rare fatty tumor and its literature review. J Neurosci Rural Pract. 2019;10(2):360–3.
51. Wolf A, Alghefari H, Krivosheya D, et al. Cerebellar liponeurocytoma: a rare intracranial tumor with possible familial predisposition. Case report J Neurosurg. 2016;125:57–61.
52. Gembruch O, Junker A, Mönninghoff C, et al. Liponeurocytoma: systematic review of a rare entity. World Neurosurg. 2018;120:214–33.
53. Macdonald DR, Cascino TL, Schold SC Jr, Cairncross JG. Response criteria for phase II studies of supratentorial malignant glioma. J Clin Oncol. 1990;8(7):1277–80.
54. van den Bent MJ, Wefel JS, Schiff D, et al. Response assessment in neuro-oncology (a report of the RANO group): assessment of outcome in trials of diffuse low-grade gliomas. Lancet Oncol. 2011;12(6):583–93.
55. Jalali R, Mallick I, Dutta D, et al. Factors influencing neurocognitive outcomes in young patients with benign and low-grade brain tumors treated with stereotactic conformal radiotherapy. Int J Radiat Oncol Biol Phys. 2010;77(4):974–9.
56. Farjam R, Pramanik P, Aryal MP, et al. A radiation-induced hippocampal vascular injury surrogate marker predicts late neurocognitive dysfunction. Int J Radiat Oncol Biol Phys. 2015;93(4):908–15.
57. Lin AL, White M, Miller-Thomas MM, et al. Molecular and histologic characteristics of pseudoprogression in diffuse gliomas. J Neuro-Oncol. 2016;130(3):529–33.

14 Oligodendroglioma

Supriya Mallick

14.1 Introduction

Oligodendrogliomas (ODG) are a group of biologically distinct brain tumors which arise from the oligodendrocytes. These oligodendrocytes arise from oligodendrocyte progenitor cells (OPCs). These OPCs are present in abundance in the subventricular zone (SVZ) lining the walls of the lateral ventricle as well as in white matter regions. These tumors compared to other glial tumors show GFAP negativity and exclusiveness in 1p19q co- or mono-deletion. Prognosis of these tumors is always better than other glial tumors.

14.2 Epidemiology

Incidence of central nervous system tumors taken together is increasing slowly. According to the GLOBOCAN data, central nervous system tumors have an ASR of 3.5 globally and 2.1 in India. In India, according to PBCR data, incidence of CNS tumors is less and listed in top ten only in Bengaluru registry. Oligodendroglioma is considered the third most common glioma accounting for 2–5% of primary brain tumors and 5%–18% of all glial neoplasms.

The ODGs are most found in the fourth and fifth decades of life with mild higher incidence in male (M: F 1.3:1). ODGs are most found to involve the cerebral hemispheres with frontal lobe being the commonest site.

As these predominantly involve the cortical gray matter, most of the patients present with seizures.

S. Mallick (✉)
Department of Radiation Oncology, National Cancer Institute, AIIMS, Haryana, India

S. Mallick et al. (eds.), *Evidence based practice in Neuro-oncology*,
https://doi.org/10.1007/978-981-16-2659-3_14

14.2.1 Risk Factors

These are no different than that of glioblastoma.

1. Radiation exposure: rare.
2. Use of cell phone.
3. Syndromic or associated with family history, viz. Neurofibromatosis type 1, Li–Fraumeni.

14.3 Diagnostic Workup

1. Biopsy is confirmatory: Pathology report reveals atypia, endothelial proliferation, mitotic figure, nuclear pleomorphism, necrosis (not seen)
2. History and physical examination
3. Fasting and PP blood sugar, Complete blood count, Liver function test, Renal Function test
4. Non-contrast CT brain within 24-h post-surgery—To assess completeness, bleeding
5. Contrat Enhanced MRI (CEMRI) Brain—Investigation of Choice after 3 weeks of surgery
6. 11C-Methionine PET—Investigational
7. Cerebrospinal Fluid Cytology (CSF) cytology: Not indicated unless clinical suspicion
8. WHO classification of CNS tumors in 2016 described four variants:
   (a) Oligodendroglioma (WHO grade II): IDH-mt, 1p19q co-deleted
   (b) Anaplastic oligodendroglioma (WHO grade III): IDH-mt, 1p19q co-deleted
   (c) Oligodendroglioma NOS (WHO grade II): IDH and 1p19q indeterminate
   (d) Anaplastic oligodendroglioma NOS (WHO grade III): IDH and 1p19q indeterminate

14.3.1 Radiology

14.3.1.1 CT

- Tumors are hypodense to iso-dense on non-contrast CT.
- 70–90% of oligodendrogliomas are calcified.
- Variable post contrast enhancement.

14.3.1.2 MRI

- T1: hypointense.
- T2: hyperintense (calcified areas are exception); calcium appears as "blooming."
- T1 post contrast: variable degree of heterogeneous contrast enhancement.
- DWI does not show restriction of diffusion.

- DWI may help to distinguish lower grade ODG from astrocytomas as astrocytomas have high Apparent diffusion coefficient (ADC) value.
- Perfusion weighted images reflect variation in relative cerebral blood volume (rCBV) and help to differentiate lower from higher grade tumors.

14.3.2 Pathology

- Well-circumscribed mass.
- Nearly 70–90% oligodendrogliomas are calcified.
- Frequently demonstrate focal hemorrhage.
- Frequently cystic change also seen.
- After 2016 classification, a tumor must have Isocitrate dehydrogenase1 (IDH1) mutation and 1p19q co-deletion to qualify as ODG.
  (a) *Oligodendroglioma (WHO grade II)*
  Neoplastic oligodendrocytes appear as regular cells with spherical nuclei containing finely granular chromatin surrounded by a halo of cytoplasm "fried egg" appearance under the light microscope.
  Typically contains a delicate network of anastomosing capillaries giving it a so-called "chicken wire" appearance. These tumors are slowly growing.
  (b) *Anaplastic oligodendroglioma (WHO grade III)*
  Demonstrate increased cellular density, increased mitotic activity, microvascular proliferation, and necrosis. Nuclear anaplasia is also common.
  In contrast to astrocytomas, oligodendrogliomas with necrosis and microvascular proliferation are considered only WHO grade III anaplastic oligodendrogliomas and not WHO grade IV GBM.
  (c) *Oligoastrocytoma*
  These are tumors of mixed oligodendroglioma and astrocytoma cell populations, that have historically been variably reported depending on local practice.

14.4 Prognostic Factors (Favorable) [1]

1. Pediatric patients (<18 years), among adults age <50 years
2. Location (frontal lobe)
3. Size of the tumor (<3 cm or <6 cm)
4. Gross total resection of tumor
5. Performance status (PS)
6. Use of radiotherapy
7. Use of chemotherapy

14.5 Surgery

1. Surgical options:
   (a) Biopsy only
   (b) Decompression/Sub total resection (STR)
   (c) Gross total resection (GTR)
2. Maximal safe surgical resection—Is the aim of surgery. It delay progression and improve survival, symptom relief and allows access to tissue sample. GTR – associated with decreased mortality, lower risk of disease progression and improved seizure control. Brown et al in a systematic review and meta analysis reported significantly reduced risk of disease progression and mortality associated with a gross total resection (GTR) compared to sub-total resection (STR). Relative Risk (RR) of death at 2, 5, and 10 years for GTR vs STR was 0.29, 0.39 and 0.50 (SS). RR of progression at 2, 5, and 10 years was 0.37, 0.50 and 0.67 (GTR vs STR).
3. Extent of surgery has significant prognostic value. Roelz et al established that OS benefit of maximal safe resection was limited to patients with residual tumour volume <15 mm^3.

14.6 Radiation

14.6.1 Radiation for ODG Grade II

- Early radiation vs Observation

EORTC 22845, Phase III randomized study, evaluated early radiation compared to observation in Grade II glioma

- 14% oligodendroglioma,13% mixed oligoastrocytoma.
- Radiation dose: 54 Gy (Preop tumor volume).
- At progression 65% patients of observation arm received radiation.
- Progression free survival (PFS) 5.3 years versus 3.4 years. This translates to a 5-year PFS of 55 versus 35%, respectively ($p < 0.0001$) favoring early radiation.
- Seizure control at 1 year: 41 vs 25%.
- OS not different. Hence, Pignatti and UCSF criteria emerged to find out the patient at higher risk of recurrence and should receive early radiation.
- Pignatti score/UCSF score for early Radiation (RT).

Pignatti Score	UCSF criteria
Unfavorable factors	*Unfavorable factors*
Age > or = 40 years	Age > or = 50 years
Astrocytoma histology subtype	Largest diameter of the tumor > or = 4 cm
Largest diameter of the tumor > or = 6 cm	KPS < 80
Tumor crossing the midline	Eloquent tumor location
Presence of neurologic deficit before surgery	Low risk: 0–1, medium risk: 2, high risk: 3–4
Low risk: 0–2, high risk: 3–5	

14.6.1.1 Optimum Timing

Available data suggest early radiation compared to delayed radiation improves seizure control and provides significantly better PFS without improving the OS. Concern of neuro cognitive compromise persists with early radiation. Therefore, patients with high risk of recurrence according to Pignatti criteria or UCSF criteria should receive immediate radiation, preferably within four to six weeks of surgery. The patients at lower risk may receive at progression. Elderly patients are at higher risk and should be considered for early radiation.

14.6.1.2 Dose

- EORTC 22844 Phase III trial compared 59.4 Gy/33 fractions to 45 Gy in 25 fractions.
  - Oligodendroglioma (22%) and mixed oligoastrocytoma (9%).
  - Gross total resection (GTR) was in 25%.
  - 5-year overall survival (OS) of 58 vs 59% or 5-year PFS of 47 vs 50%.
- NCCTG/RTOG/ECOG phase III trial compared 50.4 Gy/28 fractions to 64.8 Gy/36 fractions.
  - Oligodendroglioma (68%).
  - 2- and 5-year OS was 94% and 72%, with low-dose RT and 85% and 64%, with high-dose RT (log rank $P = 0.48$)
  - 2-year actuarial grade 3 to 5 radiation necrosis 2.5% with low-dose RT and 5% with high-dose RT.
- Altered fractionation schemes has been used without much advantage.

Adult AODG, PS 0–1: 60 Gy in 30 fractions in 6 weeks.
Elderly AODG, PS 0–1: 40 Gy in 15 fractions over 3 weeks
Elderly AODG PS 2–3: 34 Gy in 10 fractions or 25 Gy in 5 fractions.
Pediatric AODG: 60 Gy in 30 fractions in 6 weeks (standard).

14.6.1.3 Target Volume

Target volume grade II ODG	Target volume grade III ODG (Single Phase Plan)	Target volume grade III ODG (Two phase plan)
Single phase Gross Tumor Volume (GTV): Resection cavity + residual enhancing tumor as visible in T1 post contrast MRI image Clinical Target Volume (CTV): GTV +1 cm anatomical constrained isotropic expansion Planning Target Volume (PTV): CTV +3–5 mm margin Dose: 45–54 Gy in 25–30 fractions Trials have used T2 or fluid-attenuated inversion recovery (FLAIR) MRI sequence	Single phase GTV: Resection cavity + residual enhancing tumor as visible in T1 post contrast MRI image CTV: GTV +2 cm anatomical constrained isotropic expansion PTV: CTV +3–5 mm margin Dose: 60 Gy in 30 fractions	Phase I: GTV1: Resection cavity + residual enhancing tumor+ edema as visible in T2 FLAIR image CTV1: GTV +2 cm anatomical constrained isotropic expansion (in case edema not visible add margin of 2.5 cm) PTV1: CTV +3–5 mm margin Dose: 50.4 Gy in 28 fractions Phase II GTV2: Resection cavity + residual enhancing tumor as visible in T1 post contrast MRI image CTV2: GTV +1 cm anatomical constrained isotropic expansion PTV2: CTV +3–5 mm margin Dose: 9 Gy in 5 fractions

14.6.2 Radiation Technique

- Radiation planning should be done at least with a conformal technique.
- Target volume delineation is based on post-operative contrast enhanced MRI brain.
- 11C Methionine PET may help in target delineation.
- White matter abnormalities are better seen on diffusion tensor imaging (DTI) and have been called image-based high-risk volumes (IHV).
- Efforts must be made to spare all possible brain tissue.
- For lateralized tumors intensity modulated raduiotherapy (IMRT) from ipsilateral fields or partial arc should be preferred.

14.6.3 Supportive Treatment During Radiation

- Patients who are off steroid should be started on low-dose steroid for first 3–5 days and subsequently tapered.
- Prolonged use of steroid may lead to opportunistic infection, and compromised survival outcome.
- Radiation may lead to nausea and hence an oral serotonin 5-hydroxytryptamine (5-HT3) antagonist should be administered 30 min before each dose.
- A proton pump inhibitor should also be started.
- Anti-epileptics should be continued. Seizure is the most common symptom in LGG; 25–80% at onset referred for seizures are IDH mut and location of tumour in fronto-temporal, superficial cortical and insular location Seizure arising from brain tumour is classified as symptomatic seizure with a focal onset; therefore AED effective in focal seizures should be used for treatment Levetiracetam is the first choice in glioma for lack of interaction with other drugs, good tolerability and easy titration, rate of seizure freedom 60–100% Treatment can be initiated after first episode of seizures; prophylactic treatment not supported When add on therapy is needed, levetiracetam + valproate is most effective.

14.6.4 Radiation Alone Vs Radiation with Chemotherapy Vs Chemotherapy Alone for Grade II ODG

Role of chemotherapy not well established [2–5].

PCV and Temozolomide widely used in different approach.

Trial/ author	Risk stratification	Design	Result			Conclusion
RTOG 9802 Buckner et al., Phase III, n:251 [2]	Age 18–39; if STR or biopsy Age >40: Biopsy or resection	Arm A: RT 54 Gy Arm B: RT 54 Gy → 6 PCV		RT (years)	RT + PCV (years)	• 44% grade 3 or higher neutropenia • 51% grade 3 or higher anemia • OS not significantly better for astrocytoma • IDH mutation strong predictor for CTRT
			Total cohort PFS	4 (10-year PFS 21%)	10.4 (10-year PFS 51%)	
			Total cohort OS	7.8 (10-year OS 40%)	13.3 (10-year OS 60%)	

Trial/ author	Risk stratification	Design	Result	Conclusion
RTOG 0424. Fisher et al., Phase II; n:129 [3]	High-risk low-grade glioma (LGG) Definition of high risk: 3 or more risk factors as Pignatti score	Single arm: RT 54 Gy with concurrent TMZ → 12 TMZ D-D5 q 4 weeks	3-year OS rate was 73.1% Three-year PFS 59.2% Updated results: TMZ improves outcome	• Significantly improved compared to that of prespecified historical control values ($P < 0.001$) • Grades 3 and 4 adverse events 43% and 10%
EORTC 22033-26033 *Baumert* et al. *Phase III;* *n = 477* [4]	High-risk LGG At least ONE of the following 1. *Aged >40 years* 2. *Progressive disease* 3. *Tumor size >5 cm* 4. *Tumor crossing the midline* 5. *Neurological symptoms*	Arm A: RT 50.4 Gy Arm B: Dose dense TMZ 75 mg/m² D1-D21 q 4 weeks for 12 cycles	RT / TMZ PFS: 46 months / 39 months OS: NR / NR	• IDH-mt/ non-codel tumors longer PFS with RT • IDH-mt/codel and IDHwt tumors—No difference • Grade 3–4 hematological toxicity in TMZ arm: 14%

- PCV Regimen:
  - Procarbazine 60 mg/m² PO D8–D21
  - CCNU 110 mg/m² PO D1
  - Vincristine 1.4 mg/m² [maximum dose, 2.0 mg] intravenously on D8, D29
  - Cycle length 8 weeks

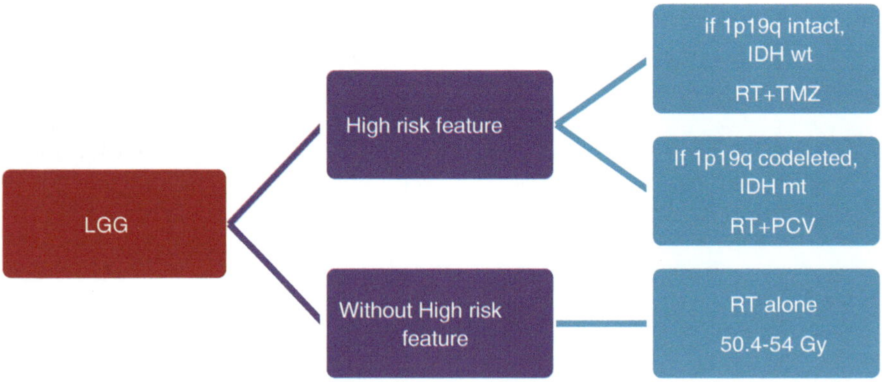

14.7 Radiation Alone Vs Radiation with Chemotherapy Vs Chemotherapy Alone for Anaplastic Oligodendroglioma

Trial	Feature	Design	Result
NOA 04 Wick et al. [6]	Phase III trial $N = 318$ (78 ODG)	Randomization 2:1:1 (A: B1: B2) Conventional radiotherapy (arm A) Procarbazine, lomustine (CCNU), and vincristine (PCV; arm B1) Temozolomide (arm B2) Crossover allowed at progression or toxicity	PFS and OS similar in A, B1, B2 arm Higher grade ¾ hematological toxicity in B1 TTF:54 months; PFS: 52 months
EORTC 26951 Van den Bent et al. [7]	Phase III trial $N = 318$	Arm A: Conventional radiation 59.4 Gy Arm B: Conventional radiation 59.4 Gy followed by 6 cycles PCV	ITT population Median PFS: 13.2 vs 24.3 months Median OS: 30.6 vs 42.3 months Confirmed ODG Median PFS: 10.4 vs 19.1 months Median OS: 28.7 vs 35 months 1p/19q co-deleted Median PFS: 49.9 vs 156.8 months Median OS: 111.1 vs NR months 20-year survival OS 13.6% vs 37.1% PFS 10.8% vs 31.3%
RTOG 9402 Cairncross G et al. [8]	Phase III trial $N = 289$	Arm A: Intensive PCV followed by immediate RT Arm B: RT alone	OS entire cohort: 4.6 vs 4.7 years OS co-deleted vs non co-deleted tumors: 14.7 vs 2.6 years OS co-deleted tumors: 14.7 vs 7.3 years

14.8 Evolving Approaches

- Hypofractionation may be the next approach
  - Hypofractionation radiation is evovlving for many tumors. Recurrence within the previously treated area or primary location of tye tumor is most common. Modern radiation techniques enables delivery of higher fraction size without compromising the organ at risk.
  - Hypofractionation may result in higher BED and shorter OTT

- Additional benefit
  - No increase in toxicity
  - May reduce machine time
  - Reduce cost of care

14.8.1 Proton Therapy

Proton delivers radiation precisely to the target and may help to reduce lingterm morbidity and improve disease control. However, cost and limited availability limits the use of proton radiation in such tumors, particulraly in developing world.

References

1. Puduvalli VK, Hashmi M, McAllister LD, Levin VA, Hess KR, Prados M, et al. Anaplastic oligodendrogliomas: prognostic factors for tumor recurrence and survival. Oncology. 2003;65(3):259–66.
2. Buckner JC, Shaw EG, Pugh SL, Chakravarti A, Gilbert MR, Barger GR, et al. Radiation plus procarbazine, CCNU, and vincristine in low-grade glioma. N Engl J Med. 2016;374(14):1344–55.
3. Fisher BJ, Hu C, Macdonald DR, Lesser GJ, Coons SW, Brachman DG, et al. Phase 2 study of temozolomide-based chemoradiation therapy for high-risk low-grade gliomas: preliminary results of Radiation Therapy Oncology Group 0424. Int J Radiat Oncol Biol Phys. 2015;91(3):497–504.
4. Baumert BG, Hegi ME, van den Bent MJ, von Deimling A, Gorlia T, Hoang-Xuan K, et al. Temozolomide chemotherapy versus radiotherapy in highrisk low-grade glioma (EORTC 22033-26033): a randomised, open-label, phase 3 intergroup study. The Lancet Oncology. 2016;17(11):1521–1532.
5. Hafazalla K, Sahgal A, Jaja B, Perry J, Das S. Procarbazine, CCNU and vincristine (PCV) versus temozolomide chemotherapy for patients with low-grade glioma: a systematic review. Oncotarget. 2018;9(72):33623–33.
6. Wick W, Hartmann C, Engel C, Stoffels M, Felsberg J, Stockhammer F, et al. NOA-04 Randomized phase III trial of sequential radiochemotherapy of anaplastic glioma with procarbazine, lomustine, and vincristine or temozolomide. J Clin Oncol. 2009;27(35):5874–5880.
7. van den Bent MJ, Brandes AA, Taphoorn MJ, Kros JM, Kouwenhoven MC, Delattre JY, et al. Adjuvant procarbazine, lomustine, and vincristine chemotherapy in newly diagnosed anaplastic oligodendroglioma: longterm follow-up of EORTC brain tumor group study 26951. J Clin Oncol. 2013;20;31(3):344–50. https://doi.org/10.1200/JCO.2012.43.2229. Epub 2012 Oct 15. PMID: 23071237.
8. Cairncross G, Wang M, Shaw E, Jenkins R, Brachman D, Buckner J, et al. Phase III Trial of chemoradiotherapy for anaplastic oligodendroglioma: long-term results of RTOG 9402. J Clin Oncol. 2013;31(3):337–343.

Anaplastic Astrocytoma

15

Dodul Mondal and Neelam Singh

15.1 Introduction and Epidemiology

- Anaplastic gliomas are constituted of anaplastic astrocytoma (AA) and anaplastic oligoastrocytoma. Morphologically they are WHO grade III and account for 25% of high-grade gliomas, majority are anaplastic astrocytoma [1].
- Almost 1.6% and 5.8% of newly diagnosed gliomas are either anaplastic oligodendroglioma or anaplastic astrocytoma, respectively [2].
- Median age of anaplastic oligodendroglioma (ODG) usually 45–50 years and that of anaplastic astrocytoma is 41 years [3].
- Median age of anaplastic ODG probably reflects its natural history of progression from a lower grade to higher grade which occurs at an earlier age.

15.2 Anatomy

- Mostly arise from cerebral hemispheres.
- Frontal lobe is more common than parietal/temporal followed by occipital. Cerebellum is a rare site [1, 2].

15.3 Risk Factors

- Almost 5% of gliomas can be associated with different genetic syndromes like NF1, NF2, tuberous sclerosis, Li–Fraumeni syndrome, von Hippel–Lindau syndrome (hemangioblastoma), etc. [1].
- Previous ionizing radiation [4].

D. Mondal (✉) · N. Singh
Department of Radiation Oncology, Indraprastha Apollo Hospitals, New Delhi, India

S. Mallick et al. (eds.), *Evidence based practice in Neuro-oncology*,
https://doi.org/10.1007/978-981-16-2659-3_15

15.4 Clinical Presentation

- Headache, seizures most common.
- Memory loss, motor weakness, visual symptoms, language deficit, cognitive and personality changes depending on location [1].

15.5 Evaluation

- Good clinical examination including neurological function and memory evaluation.
- Blood and biochemistry profile.
- *Imaging*:
  - MRI with gadolinium contrast. Anaplastic gliomas are typically hypointense on T1-weighted images and show heterogeneous enhancement.
  - T1, T2, FLAIR, diffusion weighted images or ADC sequence.
  - Immediate postoperative MRI (within 72 h) helps to determine the extent of surgical resection and residual disease.
  - MRI is also helpful during radiation planning to delineate surgical cavity, residual tumor, and edema wherever necessary.
  - MRI is also crucial in follow-up to detect recurrence and differentiate radionecrosis.
  - PET scan with 11C-Methionine may be useful to evaluate early recurrence and differentiate with radionecrosis [5].
  - Figure 15.1a–f shows different sequences and findings on MRI in a proven anaplastic astrocytoma.
- *Pathology*: Either a tissue diagnosis by stereotactic biopsy or surgical excision should be considered whenever possible to know morphology and molecular marker status.
  - Radiology should be considered as diagnostic for tumor located at eloquent areas of brain.
  - Routine microscopy along with molecular marker is highly recommended.
  - According to morphology, anaplastic gliomas can be:
    Anaplastic Oligodendroglioma (AODG).
    Anaplastic Astrocytoma (AA).
    - Morphologically AA typically shares some common features with that of glioblastoma (GBM) like nuclear atypia, increased cellularity, and proliferation activity identified by increased mitosis. Glial markers are usually positive such as GFAP and neuronal markers are usually negative. It also lacks endothelial proliferation or microvascular activity including necrosis, a hallmark of GBM [6].
- *Molecular classification*: All gliomas including AA are increasingly being diagnosed with molecular markers although there is no single specific marker for any of them. A combination of markers helps identifying the subtype. Only few selected markers can classify gliomas into different subtypes and include IDH1/2

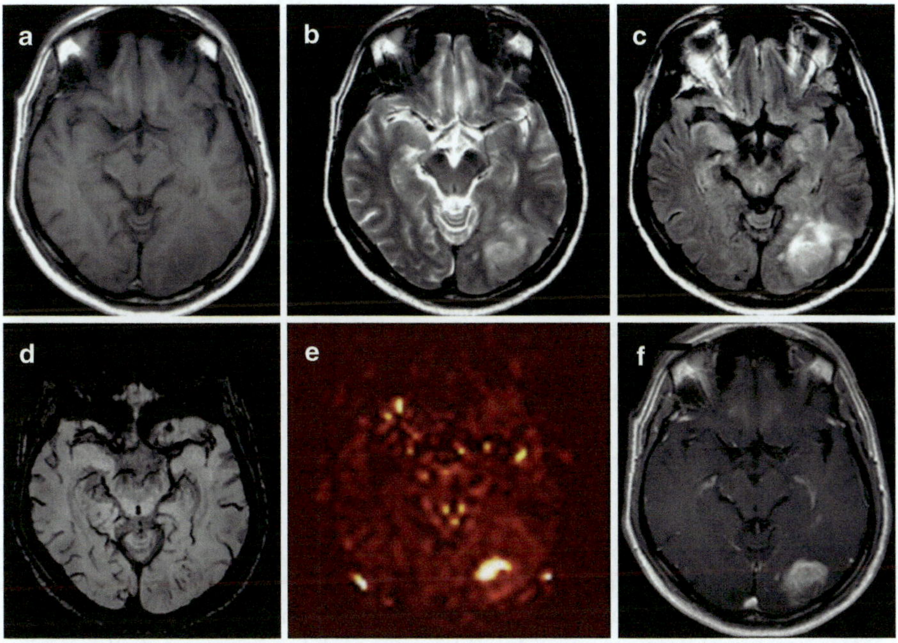

Fig. 15.1 (**a**–**f**) Anaplastic astrocytoma of left occipital lobe. The mass is hypointense on T1WI (**a**), hyperintense on T2W (**b**), and FLAIR (**c**); shows foci of punctate hemorrhage on SWI (d), high rCBV on ASL perfusion image (**e**), and avid contrast enhancement (**f**) (Courtesy: Dr. Leve Joseph)

mutation, 1p19q codeletion, and ATRX mutation. However, other mutations can also help like TERT promoter mutation, EGFR amplification, H3K27M, etc.

– Using only IDH1/2 and 1p19q codeletion, it is possible to differentiate between different grades of gliomas:
  Grade I gliomas do not have IDH mutation.
  ODG: IDH mutation +1p19q codeletion; if ATRX is mutated then it is unlikely to be ODG.
  AODG: 1p19q codeleted.
  AA: 1p19q nondeleted; IDH mutated. TP53 and ATRX are most common mutation [7].
  - Presence of IDH mutation is a good prognostic factor.
  - IDH wild type is usually seen at a median age of 50 years with a median survival of 1.7 years.
  - Implication of MGMT promoter methylation is unclear for AA.
  - AA can be further subclassified into four different types based on their molcular pattern: [8].
    – Molecular equivalent of conventional GB
    – GB-H3F3A mutated (GBM-H3)
    – GB-H3-K27
    – Lower grade gliomas

- 1p19q codeletion almost always coexists with IDH mutation.
- ATRX mutation is mutually exclusive of 1p19q codeletion and TERT promoter methylation.
- GBM has been discussed elsewhere in this book.

Prognostic Factors Following factors are associated with better prognosis:

- RPA Class I (Includes both age and performance status)
- Extent of surgical resection
- Anaplastic astrocytoma without 1p19q codeletion but with IDH mutation carries intermediate prognosis
- IDH wild type AA carries, worst prognosis

15.6 Treatment

- The current standard of care for patients is maximal safe resection followed by post-op radiotherapy and chemotherapy.
- *Surgery*: Primary treatment is maximal safe resection with neurologic preservation (maximal safe resection feasible with goal for image verified complete resection).
- *Chemotherapy*: Concurrent Temozolomide at a dose of 75 mg/m^2 daily during radiation followed by adjuvant temozolomide at a dose of 150–200 mg/m^2 daily for 5 days to be repeated every 28 days. Adjuvant temozolomide can be continued for six to 12 cycles depending on patient tolerance.

In a randomized trial by UK-MRC, 117 patients with anaplastic astrocytoma received either radiation alone or radiation followed by adjuvant PCV chemotherapy. The trial failed to show any benefit of adjuvant PCV for anaplastic astrocytoma [9].

CATNON trial is a multicenter randomized study to evaluate efficacy of temozolomide in patients with anaplastic glioma with intact 1p19q. In this 2x2 factorial design trial, patients were randomized into one of the four arms, namely radiotherapy alone, radiotherapy with concurrent temozolomide, radiotherapy followed by adjuvant temozolomide, or radiotherapy with concurrent temozolomide followed by adjuvant temozolomide. Radiation dose was 59.4 Gy delivered in 33 fractions and adjuvant TMZ was given for 12 months. First interim analysis showed significant survival improvement with TMZ in noncodeleted patients [10]. A second interim analysis published in 2019 ASCO meeting, found no significant OS benefit across the group. IDH mutated group showed a trend towards improved survival with concurrent TMZ and a survival advantage with adjuvant TMZ. More understanding about molecular aspect is awaited [11].

In another trial, Chang et al. compared concurrent and adjuvant TMZ versus nitrosourea in patients with AA (RTOG 9813). Almost 200 patients were randomized between two arms. The dose of radiation was 59.4Gy in 33 fractions. Dose and schedule of chemotherapy was somehow different (Readers are requested to refer to original article for details of chemotherapy). After a median follow-up of almost 10 years, no difference was found between two arms. IDH1-R132H mutation showed improved survival [12].

15.7 Radiotherapy Technical Details

Patient is simulated with appropriate head mask or other immobilization device and intravenous contrast. A delayed scan is preferred to have adequate penetration of contrast agent inside brain. Preferably planning CT scan is done at 3 mm interval. Planning MRI with contrast and additional sequences including T1, T2, and FLAIR is helpful in delineating residual tumor, cavity, and edema component. CT-MR fusion helps in better delineation of target and few OARs. Although a deformable registration software is ideal, rigid registration also provides accurate image fusion for all practical purpose. Figure 15.2a–c shows typical planning of AA in sequential phased manner. Because of proximity to brainstem, patient was planned in a typical sequential manner rather than a single phase plan.

Target volumes and dose prescription can be defined as follows: [13].

Component	Ph I	Ph II
GTV	Postoperative T1-weighted MRI contrast enhancement or T2/FLAIR signal change	Cavity and enhancing tumor
CTV	GTV + 1.5 to 2 cm	Cavity/enhancing tumor + 1 cm
PTV	3-5 mm based on institutional protocol	
Dose	50.4 Gy in 28 fractions	9 Gy in 5 fractions

15.8 Toxicity

- *Acute effects*: These can include but are not limited to fatigue, headache, alopecia, skin erythema, nausea, memory loss, cerebral edema, increased intracranial pressure, seizures, loss of consciousness, etc.
- *Late effects*: Depend on tumor location, irradiated volume, pre-existing neurocognitive function; may include radiation necrosis, long-term memory/cognitive changes, hearing loss, optic neuritis, neurological damage, cataracts, hypopituitarism, etc.

15.9 Follow-up

- For evaluating tumor control and toxicity.
- A good clinical examination including neurological examination, memory and cognitive assessment is important.
- MRI brain with contrast every 3–4 months till 3 years and then every 4–6 months for an indefinite period.

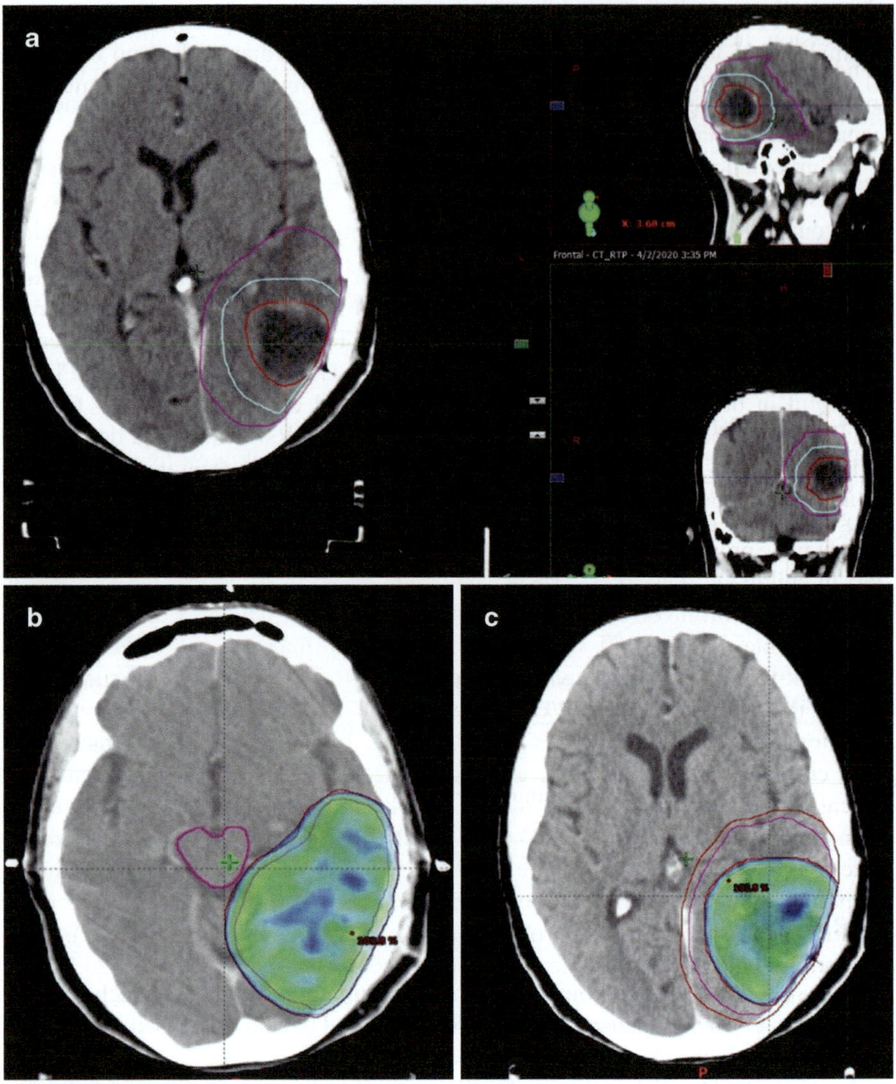

Fig. 15.2 (**a**) Postoperative cavity, first phase CTV, boost CTV (red, magenta, and color, respectively). Parts (**b**, **c**) shows excellent coverage of target volume with prescribed dose

15.10 Treatment of Recurrence

- Recurrences are common.
- No standard treatment guidelines.
- Options include surgical resection, re-irradiation with or without chemotherapy, systemic therapy including TMZ, PCV, anti VEGF therapy (Bevacizumab), immunotherapy etc. Data are sparse, mostly retrospective in nature or phase I–II.
- Treatment of recurrence depends on many factors

- Interval from last radiation
- Volume of recurrence
- Resectability
- If unresectable, then expected morbidity from re-irradiation
- Dose to OARs from previous irradiation
- Expertise of the treating physician and the center in treating recurrences

Proton Beam Therapy Owing to its particle nature and Bragg peak, proton beam therapy is supposed to reduce neurocognitive impairment in patients receiving radiation for AA. Also, it gives the opportunity to dose escalation trials.

Funding Nil.

Conflict of Interest Nil.

Acknowledgement Figure 15.1 Copyright and courtesy: Dr. Leve Joseph, Associate Professor, Department of Neuroradiology, AIIMS Delhi, India.

References

1. Louis DN, Perry A, Reifenberger G, von Deimling A, et al. The 2016 World Health Organization classification of tumors of the central nervous system: a summary. Acta Neuropathol. 2016;131(6):803–20. https://doi.org/10.1007/s00401-016-1545-1.
2. Ostrom QT, Gittleman H, Fulop J, Liu M, et al. CBTRUS statistical report: primary brain and central nervous system tumors diagnosed in the United States in 2008–2012. Neuro-oncol. 2015;17(Suppl 4):iv1–iv62.
3. Grimm SA, Chamberlain MC. Anaplastic astrocytoma. CNS Oncol. 2016;5(3):145–57. https://doi.org/10.2217/cns-2016-0002.
4. Marquet G, Dameron O, Saikali S, et al. Grading glioma tumors using OWL-DL and NCI thesaurus. AMIA Annu Symp Proc. 2007;2007:508–12.
5. Garcia JR, Cozar M, Baquero M, Fernández Barrionuevo JM, et al. The value of ^{11}C-methionine PET in the early differentiation between tumour recurrence and radionecrosis in patients treated for a high-grade glioma and indeterminate MRI. Rev Esp Med Nucl Imagen Mol. 2017;36(2):85–90.
6. Louis DN, Ohgaki H, Wiestler OD, Cavenee WK, et al. The 2007 WHO classification of tumours of the central nervous system. Acta Neuropathol. 2007;114(2):97–109. https://doi.org/10.1007/s00401-007-0243-4.
7. Brennan C. Genomic profiles of glioma. Curr Neurol Neurosci Rep. 2011;11(3):291–7. https://doi.org/10.1007/s11910-011-0198-7.
8. Ellison DW. Multiple molecular data sets and the classification of adult diffuse gliomas. N Engl J Med. 2015;372(26):2555–7. https://doi.org/10.1056/NEJMe1506813.
9. Medical Research Council Brain Tumor Working Party. Randomized trial of procarbazine, lomustine, and vincristine in the adjuvant treatment of high-grade astrocytoma: a Medical Research Council trial. J Clin Oncol. 2001;19(2):509–18. https://doi.org/10.1200/JCO.2001.19.2.509.
10. van den Bent MJ, Baumert B, Erridge SC, Vogelbaum MA, et al. Interim results from the CATNON trial (EORTC study 26053-22054) of treatment with concurrent and adjuvant temozolomide for 1p/19q noncodeleted anaplastic glioma: a phase 3, randomised, open-label intergroup study. Lancet. 2017;390(10103):1645–53.

11. Van Den Bent MJ, Erridge S, Vogelbaum MA, et al. Second interim and first molecular analysis of the EORTC randomized phase III intergroup CATNON trial on concurrent and adjuvant temozolomide in anaplastic glioma without 1p/19q codeletion. J Clin Oncol. 2019;37(15_suppl):2000.
12. Chang S, Zhang P, Cairncross JG, Gilbert MR, et al. Phase III randomized study of radiation and temozolomide versus radiation and nitrosourea therapy for anaplastic astrocytoma: results of NRG oncology RTOG 9813. Neuro-Oncology. 2017;19(2):252–8.
13. Weller M, van den Bent M, Hopkins K, Tonn JC, et al. EANO guideline for the diagnosis and treatment of anaplastic gliomas and glioblastoma. Lancet Oncol. 2014;15(9):e395–403. https://doi.org/10.1016/S1470-2045(14)70011-7.

Mixed Gliomas or Oligoastrocytoma

16

BhanuPrasad Venkatesulu

16.1 Introduction

Mixed gliomas are glial cell neoplasms with more than one type of glial cell in their phenotype. The WHO 2016 classification of CNS brain tumors classifies mixed gliomas as oligoastrocytoma, NOS or anaplastic oligoastrocytoma, NOS. These are very rare neoplasms in view of better molecular diagnostic testing stratifying tumors into astrocytic (IDH-mutant, ATRX-mutant, 1p/19q-intact) and oligodendroglial (IDH-mutant, ATRX-wildtype, and 1p/19q-codeleted) tumor populations. The mixed histology typically has astrocytic and oligodendroglial components intermingled with one another or existing separately. There is lack of clearly defined criteria to classify as mixed gliomas but the general consensus is that glial neoplasms that exhibit phenotypic heterogeneity are classified as oligoastrocytomas. Oligoastrocytomas can be grade 2 or anaplastic oligoastrocytoma are grade 3 tumors [1]. In this chapter, we will discuss the clinical presentation and management of mixed gliomas.

16.2 Clinical Presentation

The clinical presentation is similar to other low-grade gliomas with symptoms that include headache, nausea or vomiting, confusion, memory loss, personality changes, difficulty with balance, vision problems, speech difficulties, and seizures.

B. Venkatesulu (✉)
Department of Radiation Oncology, Stritch School of Medicine Loyola University Chicago, Loyola University Medical Center, Maywood, IL, USA

Edward Hines VA Hospital, Maywood, IL, USA

S. Mallick et al. (eds.), *Evidence based practice in Neuro-oncology*,
https://doi.org/10.1007/978-981-16-2659-3_16

16.3 Management

Investigations include history and physical examination with MRI brain. The treatment is on similar lines of grade 2 or grade 3 tumors. There is lack of prospective data to support any particular treatment modality for oligoastrocytoma. They are managed on similar lines of grade 2 or grade 3 oligodendroglioma.

16.4 Surgery

Surgery is considered the cornerstone of therapy.

16.5 Adjuvant Therapy

Grade 2 oligoastrocytoma management is based on extrapolation of management of grade 2 oligodendroglioma with high risk features should be assessed for IDH-mutant as well as 1p/19q codeleted status.

If they are IDH-mutant should receive upfront radiotherapy with adjuvant PCV, if they are 1p/19q codeleted the usual approach is to administer upfront chemotherapy and reserve radiotherapy for salvage in case of future progression.

Dose: radiation dose 54–56 Gy may be optimum.

Technique: A conformal technique should be advised.

Target should be single volume, CTV should include post-operative cavity, any residual enhancing volume, with 1 cm isotropic expansion for low-grade tumors and 1.5 cm for high-grade tumors. PTV should be based on the institutional standard [2, 3].

16.6 Follow-up

First follow-up after 1 month of surgery or radiation with clinical examination.

Follow-up every 3 months with contrast enhanced MRI brain for first 3 years.

Subsequently continue follow-up every 6 months with contrast enhanced MRI.

References

1. Louis DN, Perry A, Reifenberger G, von Deimling A, Figarella-Branger D, Cavenee WK, et al. The 2016 World Health Organization classification of tumors of the central nervous system: a summary. Acta Neuropathol. 2016;131(6):803–20. https://doi.org/10.1007/s00401-016-1545-1.

2. Shaw EG, Scheithauer BW, O'Fallon JR, Davis DH. Mixed oligoastrocytomas: a survival and prognostic factor analysis. Neurosurgery. 1994;34(4):577–82.; discussion 82. https://doi.org/10.1227/00006123-199404000-00002.
3. Walker DG, Kaye AH. Diagnosis and management of astrocytomas, oligodendrogliomas and mixed gliomas: a review. Australas Radiol. 2001;45(4):472–82. https://doi.org/10.1046/j.1440-1673.2001.00959.x.

Glioblastoma

17

Supriya Mallick

17.1 Introduction

Glioblastoma (GBM) is the most aggressive brain tumor of glial cell origin making up a fifth of all intracranial tumors [1]. These tumors are characterized by marked mitotic activity, substantial tendency toward neo-angiogenesis (microvascular proliferation) and necrosis. GBM has threefold to fivefold higher proliferative rate compared to anaplastic astrocytoma [1].

17.2 Epidemiology

Incidence of central nervous system tumors is slowly increasing. According to the GLOBOCAN data central nervous system tumors has an age standardized rate (ASR) of 3.5 globally and 2.1 in India. GBM has a prevalence of 26,000 cases, with a mortality rate of 15,000 cases yearly in the USA. According to data from National Cancer Institute, GBM accounts for 52% of all primary brain tumors in the USA. In India, according to the population based cancer registry (PBCR) data, the incidence of CNS tumors is less. Only in Bengaluru registry it is among the top ten most common malignancies.

17.2.1 Risk Factors

1. Radiation exposure: rare.
2. Use of Cell phone: Cell phones do not emit ionizing radiation and hence unlikely to cause cancer. However, some studies show higher incidence of benign brain

S. Mallick (✉)
Department of Radiation Oncology, National Cancer Institute, AIIMS, Haryana, India

S. Mallick et al. (eds.), *Evidence based practice in Neuro-oncology*,
https://doi.org/10.1007/978-981-16-2659-3_17

tumors like meningioma, vestibular schwannomas as well as glioma with cell phone use [2]. Judicious use of cell phones and use of hands-free phones or land phone may be advised for minimizing risk.

3. Syndromic or Associated with family history
   (a) Neurofibromatosis type 1 also known as von Recklinghausen disease: Other than neurofibroma, it is also associated with increased incidence of gliomas, schwannomas, and meningiomas.
   (b) Li–Fraumeni syndrome shows increased incidence of gliomas, breast, adrenal cancer, and sarcoma.
   (c) Gorlin syndrome (basal cell nevus syndrome).
   (d) Turcot syndrome.
   (e) Cowden syndrome.

17.3 Diagnostic Workup

1. Contrast Enhanced MRI (CEMRI) Brain is the initial investigation of choice (Fig. 17.1).
2. Cerebrospinal fluid cytology indicated in *Epithelioid and Small Cell GBM.*
3. History and Physical Examination.
4. Fasting and postprandial blood sugar, Complete blood count, Liver Function Test, Renal Function Test.
5. Non-contrast CT brain within 24-h post-surgery—To assess completeness, bleeding.
6. CEMRI Brain—Investigation of Choice after 3 weeks of surgery.
7. 11C-Methionine PET—Investigational.

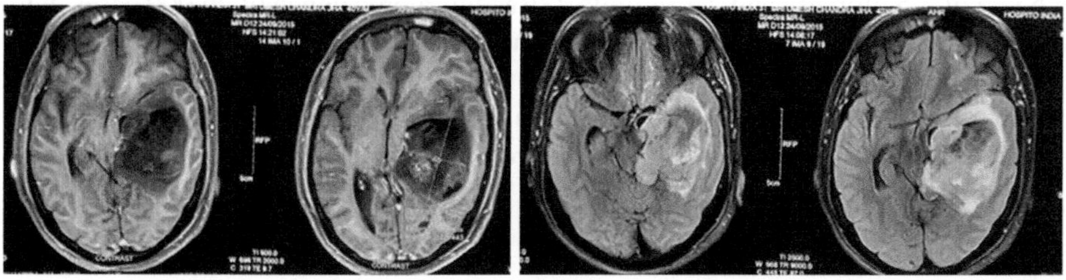

Fig. 17.1 T1 post contrast axial section shows large hypointense mass with heterogenous contrast enhancement arising from left temporal lobe with perilesional edema and mass effect in T2 FLAIR

17.4 Prognostic Factors (Good)

1. Younger age (<40 years), more extensive tumor resection, KPS
2. O^6-methylguanine-DNA methyltransferase (MGMT) methylation status [5]—Predictive and Prognostic
3. Mini-Mental State Examination (MMSE) score of 27 or higher

4. No corticosteroid treatment at baseline
5. Isocitrate dehydrogenase 1 (IDH-1) mutation in secondary GBM
6. Epidermal Growth Factor Receptor (EGFR) amplification, maintenance of Phosphatase And Tensin Homolog (PTEN), wild-type p53 and p16 all appear to be associated with improved survival in patients treated with CRT

17.5 Surgery

1. Surgical options
   (a) Biopsy only
   (b) Decompression
   (c) Complete resection
2. Maximal safe surgical resection is the aim of surgery. Extent of surgery has significant prognostic value. Gross total resection of high grade gliomas increases median survival rate by 200% compared to subtotal resection.
3. In subtotal resection at least 70% tumor resection shows improvement in overall survival.
4. Intraoperative magnetic resonance imaging decreases the risk of damaging principal areas in the brain during surgery.
5. Development in surgical techniques
   (a) Awake craniotomy performed under local sedation allows the identification of eloquent areas in tumors involving cortical and subcortical regions.
   (b) Laser Interstitial Thermal Therapy for GBM Ablation is a thermocoagulative therapy. Laser rays are delivered percutaneously by an optical fiber into tissue to cause heat-generated necrosis and ablation.
   (c) Fluorescence-Guided Surgery using 5-ALA, which is a photosensitive precursor of protoporphyrin IX. It helps to identify tumor in eloquent areas with promising results.
   (d) Near infrared confocal endomicroscopy with indocyanine green also improves tumor visualization.
   (e) Optical coherence tomography helps in differentiating tumor and non-tumor tissue.

17.6 Adjuvant Therapy

Nearly all patients of GBM require adjuvant therapy. The standard adjuvant therapy consists of radiation and concurrent Temozolomide (TMZ) followed by adjuvant TMZ. Chemotherapy may be omitted for elderly patients with PS 2–3. Radiation (RT) may be omitted for MGMT methylated, elderly GBM patients with good performance status and they may be treated with only TMZ. Further explanation regarding adjuvant therapy is explained in subsequent sections.

17.6.1 Radiotherapy

1. BTCG 69-01 and 1981 SGSG studies initially demonstrated significant (doubled) survival with adjuvant RT over supportive care after surgery, and RT became a standard component of adjuvant treatment.
   - A systematic review and pooled analysis by Laperriere et al. in 2002 also demonstrated improvement in survival with adjuvant radiation. Pooling of 6 randomized trials showed significant survival benefit with adjuvant radiation.
   - Relative risk (RR) for 1-year mortality 0.81 ($p < 0.0001$).

17.6.2 Optimum Timing

Delay in adjuvant radiotherapy after surgery decreases survival of patients with high grade glioma. Retrospective studies have shown, 9 weeks loss of survival if radiation is started 6 weeks after surgery (nearly equal the benefit of adding temozolomide).

17.6.3 Dose

Adult GBM, PS 0–1: 60 Gy in 30 fractions in 6 weeks (standard) with concurrent and adjuvant TMZ

Elderly GBM, PS 0–1: 40 Gy in 15 fractions over 3 weeks with TMZ

Elderly GBM PS 2–3: 34 Gy in 10 fractions or 25 Gy in 5 fractions. No TMZ

Pediatric GBM: 60 Gy in 30 fractions in 6 weeks (standard). Role of TMZ debateable

17.6.4 Target Volume

Partial brain radiation is standard

Central or in-field failure very common

	In-field recurrence (%)	Distant (%)
Methylated	64	31
Unmethylated	91	5.4

Target volume delineation:

Target volume as EORTC	Target volume as RTOG
Phase I GTV: Resection cavity + residual enhancing tumor as visible in T1 post-contrast MRI image CTV: GTV +2 cm anatomical constrained isotropic expansion PTV: CTV +3–5 mm margin Dose: 60 Gy in 30 fractions	Phase I: GTV1: Resection cavity + residual enhancing tumor + edema as visible in T2 FLAIR image CTV1: GTV +2 cm anatomical constrained isotropic expansion (in case edema not visible add margin of 2.5 cm) PTV1: CTV +3–5 mm margin Dose: 46 Gy in 23 fractions Phase II GTV2: Resection cavity + residual enhancing tumor as visible in T1 post-contrast MRI image CTV2: GTV +2 cm anatomical constrained isotropic expansion PTV2: CTV +3–5 mm margin Dose: 14 Gy in 7 fractions

17.6.5 Radiation Technique

- Radiation planning should be done with a conformal technique.
- Target volume delineation is based on postoperative contrast enhanced MRI brain.
- ^{11}C Methionine PET may help in target delineation.
- White-matter abnormalities are better seen on diffusion tensor imaging (DTI) and have been called image-based high-risk volumes (IHV).
- Efforts must be made to spare all possible brain tissue.
- For lateralized tumors IMRT from ipsilateral fields or partial arc should be preferred.
- GBM proximity to the subventricular zone (SVZ) impacts recurrence patterns and prognosis. It has been hypothesized that irradiation of the cancer stem cells niche in the SVZ may provide therapeutic benefit in selected patients.

17.6.6 Supportive Treatment During Radiation

- Patients who are off steroid should be started on low dose steroid 2 mg TDS for first 3–5 days and subsequently stopped.
- Prolonged use of steroid may lead to opportunistic infection, and also compromise survival outcome.
- Radiation may lead to nausea and hence an oral serotonin 5-hydroxytryptamine (5-HT3) antagonist may be administered 30 minutes before each dose.
- A proton pump inhibitor should also be started.
- Anti-epileptics should be continued.

17.7 Chemotherapy

17.7.1 Temozolomide Forms the Backbone of Chemotherapy in GBM and Is Given Concurrently with RT Followed by Maintenance Phase

- Concurrent and maintenance TMZ both are important as the pivotal trial [3, 4] was not designed to separately evaluate the impact of each strategy.
- Dose during concurrent phase is 75 mg/m^2 daily including weekend. After 1 month of completion of radiation 150–200 mg/m^2 for maximum six such cycles repeated every 4 weeks is the maintenance regimen.
- First cycle should be started with a lower dose (150 mg/m^2) to assess tolerance. Subsequent cycle's dose may be increased to 200 mg/m^2.
- Standard versus dose dense TMZ: RTOG 0525 randomly assigned 833 patients to receive standard radiation and standard maintenance temozolomide (150–200 mg/m^2 D1–D5 q 4 weekly) or dose dense temozolomide (75–100 mg/m^2 D1–D21 q 4 weekly) for 6–12 cycles [6]. The trial found no improvement either in median OS (16.6 vs 14.9 months) or PFS (5.5 vs 6.7 months) in dose dense TMZ but it was associated with higher grade 3/4 toxicity. Subset analysis suggested that a greater number of temozolomide cycles may be better for MGMT methylated, IDH mutated and young patients.
- Temozolomide should be stopped if platelets fall below 100,000/microL or absolute neutrophil count (ANC) falls below 1500/microL.
- Supportive care during adjuvant chemotherapy: Temozolomide is moderately emetogenic, requires premedication with an oral serotonin 5-hydroxytryptamine (5-HT3) antagonist, 30 min before each dose.
- As ondansetron causes constipation and temozolomide adds to it, laxatives may be required for few patients.
- Antimicrobial prophylaxis for pneumocystis pneumonia is required during the concomitant radiation phase of treatment, due to selective CD4+ T cell depletion. Risk is increased in patients receiving glucocorticoids.

17.7.2 Bevacizumab Has Been Extensively Evaluated in GBM. It Is a Humanized Monoclonal Antibody Targeted Against Vascular Endothelial Growth Factor (VEGF-A)

- Bevacizumab was evaluated in RTOG 0825 [7] and AVAglio [8] trials.
- Overall survival (OS) was not improved with bevacizumab in both trials.
- RTOG 0825 noted increased symptom burden, a worse quality of life, and a decline in neurocognitive function in patients treated with bevacizumab.
- Bevacizumab for the first-line treatment of newly diagnosed GBM has limited effectiveness and is NOT standard of care.
- Bevacizumab is effective in recurrent GBM and considered as one of the options in treatment.

- Dose of bevacizumab used is 10 mg/kg every 2 weeks till progression.
- Common adverse effects: hypertension and proteinuria.

17.7.3 Novel Anti-Angiogenic Therapy

17.7.3.1 Integrins

Cilengitide is a selective inhibitor of αvβ3 and αvβ5 integrins. Stupp et al. in a phase III trial randomly assigned 545 newly diagnosed MGMT methylated GBM cases to receive either standard chemoradiotherapy or chemoradiotherapy with added cilengitide (standard dose of 2000 mg intravenously twice weekly on days 1 and 4, beginning 1 week before starting temozolomide and radiotherapy) [9]. Median overall survival was identical 26.3 months in both the arms. Median PFS was 13.5 months in the experimental arm and 10.7 months in the standard arm but could not reach statistical significance.

17.7.3.2 Anti-EGFR Monoclonal Antibody

EGFR overexpression is found in 60% cases of GBM. Nearly 50% of these are due to a mutant form of the receptor, EGFRvIII, which has a constitutively active kinase domain. Nimotuzumab is a monoclonal antibody against EGFR without intrinsic stimulating activity. A phase III trial randomly assigned 149 patients with newly diagnosed GBM to receive either intravenous nimotuzumab 400 mg weekly added to standard radio-chemotherapy followed by 400 mg biweekly after 12 weeks or standard radio-chemotherapy only [10]. Median overall survival for the experimental arm was 22.3 months vs 19.6 months in the standard arm and difference was not statistically significant. Interestingly, EGFR amplification did not correlate with clinical efficacy of nimotuzumab.

17.7.3.3 Non-Conventional Therapy (Novo TTF)

1. The NovoTTF-100A system (Novocure Ltd., Haifa, Israel) is a novel portable device.
2. It delivers low intensity, intermediate frequency, alternating electric fields (Tumor Treating Fields—TTF) with non-invasive, disposable transducer arrays.
3. These fields lead to misalignment of the microtubule subunits in the mitotic spindle during the metaphase to anaphase transition and dielectrophoretic movement of intracellular micro-molecules and organelle during telophase and thereby interfere with cell division and cell death.
4. In a Phase III trial of 695 patients [11].

	TTFields-temozolomide	Temozolomide-alone
Median PFS	6.7 months	4 months ($P < 0.001$)
Median OS	20.9 months	16 months ($P < 0.001$)

17.8 Outcome

Treatment options	PFS (months)	OS (months)	Remark
Radiation alone vs RT + TMZ [4]	5 vs 6.9 ($P < 0.001$)	12.1 vs 14.6	TMZ adds modest 2.5 months survival benefit, 2-year survival 10.9% vs 27.2%
Radiation alone vs RT + TMZ, MGMT methylated [5]	5.9 vs 10.3 ($p = 0.001$)	15.3 vs 21.7 ($p = 0.007$)	MGMTm patients far better when treated with RT-TMZ
Radiation alone vs RT + TMZ, MGMT unmethylated [5]	4.4 vs 5.3 ($p = 0.02$)	11.8 vs 12.7 ($p = 0.06$)	MGMTu patients derive significant benefit when treated with RT-TMZ
RT + TMZ vs RT + TMZ + Novo TTF [11]	4 vs 6.7 ($P < 0.001$)	16 vs 20.9 ($P < 0.001$)	Addition of TTFields resulted in statistically improved PFS, OS
RT + TMZ vs RT + TMZ + Bevacizumab [7]	7.3 vs 10.7 ($P < 0.001$)	16.1 vs 15.7 ($p = 0.21$)	BEV did not improve OS PFS was prolonged but did not reach the pre-specified target
RT + TMZ vs RT + TMZ + Bevacizumab [8]	6.2 vs 10.6 ($P < 0.001$)	16.7 vs 16.8 ($p = 0.10$)	BEV did not improve OS PFS was prolonged

17.9 Evolving Approaches in Adjuvant Radiotherapy

- Hypofractionation may be the next approach
  - Pattern of failure in GBM: 85–90% recurrence inside the radiation field or high dose area
  - Hypofractionation may result in higher BED and shorter OTT
- Phase II data [12] showed feasibility of this approach
- Additional benefit
  - No increase in toxicity
  - May reduce 33% machine time
  - Reduce cost of care
  - The following table summarizes various trials of hypofractionation in GBM

Author/Year	Schedule	Results (In months)	Toxicity	Comments
Omuro A et al./2014 [13] ($n = 40$)	HFSRT (6 × 6 Gy to contrast enhancement and 6 × 4 Gy to FLAIR hyperintensity with concomitant/adjuvant temozolomide and bevacizumab	Median OS—19 Median PFS—10	Acceptable	Safe and more convenient
Iuchi T et al./2014 [14] ($n = 46$)	PTV1 = surgical cavity and residual tumor on T1-weighted MR images with 5 mm margins PTV2 = PTV1 + 15 mm margins PTV3 = high intensity area on FLAIR 68, 40, and 32 Gy for PTV1, PTV2, and PTV3 in 8 fractions with concurrent and maintenance TMZ	Median OS—20.0	Radiation necrosis—20	Necrosis in the SVZ significantly correlated with prolonged survival
Reddy K et al./2013 [15] ($n = 24$)	PTV1 = surgical cavity and residual tumor on T1-weighted MR images with 5 mm margins 60 Gy in 10 fractions with TMZ PTV2 = T2 abnormality +5 mm margins	Median OS—16.6	No grade III or higher toxicity	Short treatment time with equal results
Cho et al./2010 [16]	A dose of 60 Gy and 50 Gy in 20 fractions to the periphery of the GTV and PTV, respectively with TMZ (20)	Median OS—14.8	Acceptable	Feasible and safe
Raymond et al./2009 [17]	A dose of 60 Gy and 40 Gy in 20 fractions to the periphery of the GTV and PTV, respectively	Median OS—14.4 Median PFS—7.7	Grade III/IV-1 case	Pattern of failure within 2 cm
Mallick et al./2018 [12] (phase II randomized trial)	A dose of 60 Gy and 50 Gy in 20 fractions to the periphery of the GTVhigh and GTVlow, respectively	Median OS:18.07 vs 25.18 Median PFS: 13.1 vs 14.1 (CRT vs HART)	One radionecrosis	NR

17.10 Recurrence

17.10.1 Diagnosis Is a Challenge as Progression and Radionecrosis Closely Resemble (Fig. 17.2)

Features	Radiation necrosis	Glioma recurrence
Magnetic resonance imaging—Suggestive features	Usually in the areas of high dose (>60 Gray) Contrast enhancement, edema, and mass effect may be seen in both Classical Swiss cheese appearance Involvement of corpus-callosum or peri-ventricular white matter	Can occur anywhere, but in-field recurrence is the most common Unlikely Less likely
Magnetic resonance spectroscopy	Low Cho/NAA <1.11 Low Ch/Cr ratio <1.17 Elevated lipid signals	High High Low
Diffusion weighted magnetic resonance imaging	Higher apparent diffusion coefficient (ADC)-maxima ADC in range of 2.3	Lower ADC values in range of 1.7
Perfusion magnetic resonance imaging	Cerebral blood volume (CBV) <0.6 Higher fractional anisotropy(FA) values—0.89 ± 0.15	CBV usually more than 2.6 Lower fractional anisotropy (FA) values—0.74 ± 0.14
Positron emission tomography	Low metabolic activity	High metabolic activity
Single photon emission computed tomography	Low metabolic activity	High metabolic activity

Taken from Mallick et al. [12]

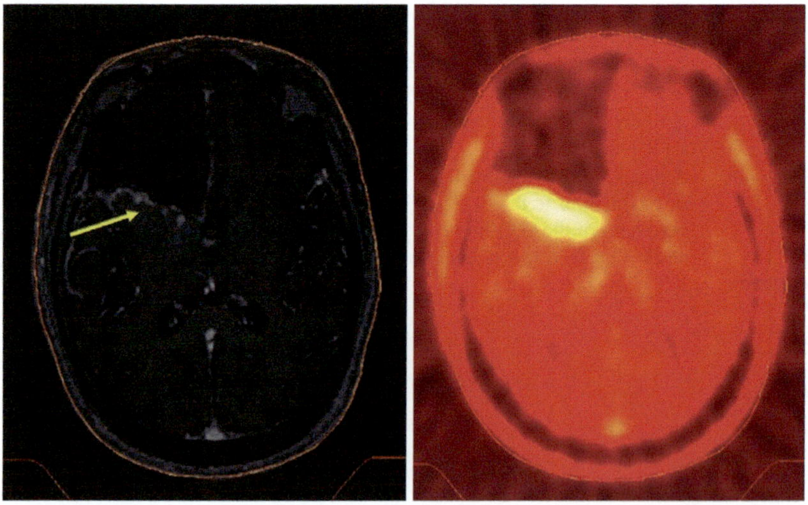

Fig. 17.2 Contrast enhanced MRI showing possible recurrence or radiation necrosis; Methionine PET suggesting high metabolic activity and recurrence

17.10.2 Management of Recurrent GBM [18]

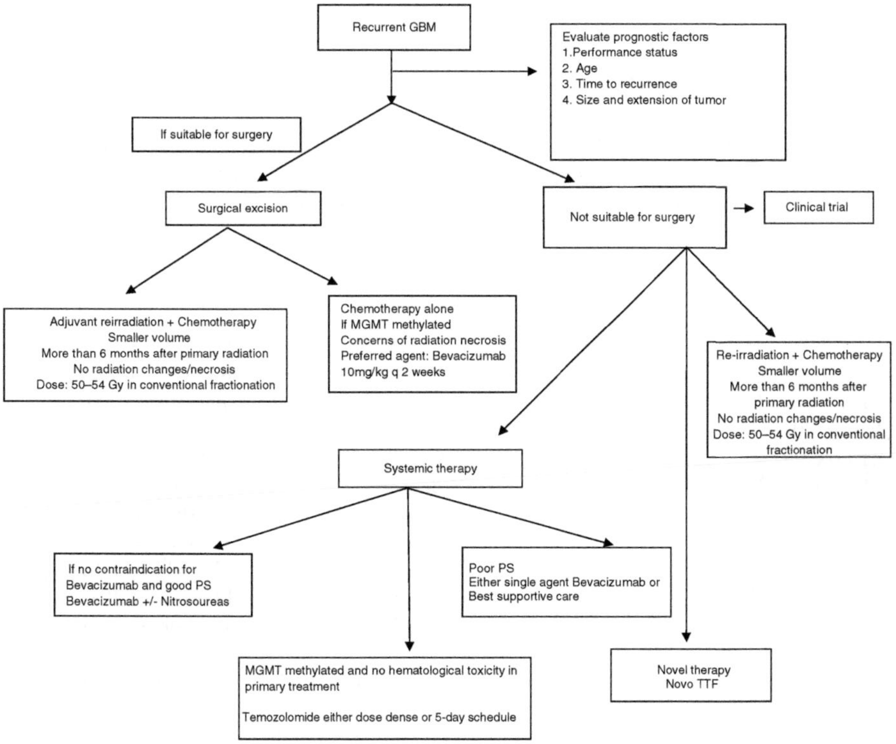

Prognostic factors of recurrence

1. Karnofsky performance status (KPS) > 60
2. Tumor size <4 cm
3. Progression after >6 months
4. Complete resection before ReRT

17.11 Special Variants

17.11.1 Pediatric GBM

- Pediatric GBM a distinct entity of GBM.
- Molecularly distinct from the adult counterpart.
- MGMT gene promoter methylation in 50% patients and p53 protein expression in 60% of cases.
- Available data shows adjuvant radiation with concurrent temozolomide followed by maintenance six cycles of temozolomide is standard.

- Target volume should be delineated to minimize the risk of radiation toxicity.
- Outcome is also not different.

17.11.2 Elderly GBM

- Elderly GBM is really challenging.
- Should evaluate patients for tolerance as outcome always remain guarded.
- Median survival 6–8 months after optimal treatment
- Treatment decision should be modified according to performance status, and age.

Treatment options	PFS (months)	OS (months)	Remark
BSC vs RT [19]	1.2 vs 3.7 ($P < 0.001$)	4.3 vs 7.3 ($p = 0.002$)	RT is better than BSC
RT 60 Gy vs RT 40 Gy [20]	NR	5.1 vs 5.6	SC not inferior to LC
RT 60 Gy vs TMZ alone [21]	3.3 vs 4.7 (EFS) p (non-inferiority) = 0.043	9.6 vs 8.6 Non-inferiority) = 0.033	TMZ not inferior to RT Subset: TMZ is better for MGMT methylated, and RT is better for MGMT unmethylated
RT 60 Gy vs RT 34 Gy vs TMZ alone [22]	NR	6 vs 7.5 vs 8.3 ($p = 0.01$); ($p = 0.24$)	LC is worse than SC TMZ alone is an option
RT 40 Gy vs RT 25 Gy [23]	4.2 vs 4.2 ($P = 0.716$)	6.4 vs 7.9 ($P = 0.988$)	USC not inferior to SC
RT 40 Gy vs RT 40 Gy + TMZ [24]	3.9 vs 5.3 ($P < 0.001$)	7.6 vs 9.3 ($P < 0.001$)	Addition of TMZ to short-course radiotherapy resulted in longer survival than short-course radiotherapy alone

17.11.3 Spinal GBM

- Very few case reports.
- Cervical tumor location most common, followed by thoracic location.
- Cervical tumors, in pediatric age has worse outcome.
- Addition of adjuvant RT and TMZ improves OS.
- Radiation dose 54–55 Gy may be tolerated well.

17.11.4 Gliosarcoma

1. Gliosarcoma, a rare biphasic tumor, two components: gliomatous and sarcomatous.
2. 1.8–2.8% of all cases of malignant gliomas.
3. More frequently arises in temporal lobes.
4. Lack of EGFR amplification in gliosarcoma.
5. A maximal safe surgical resection is considered the cornerstone of therapy.
6. Treatment strategies like GBM, adjuvant radiation with Temozolomide is the standard.
7. Extracranial metastases reported in 11%.
8. Most common site of metastasis: lungs (72%), liver (41%), and lymph nodes (18%).
9. Survival appears to be comparable with GBM.

17.11.5 GBM-PNET

1. GBM with areas of neuroectodermal differentiation (GBM-PNET).
2. Radiology: Usually present with a solid cystic lesion with heterogeneous contrast enhancement.
3. Cerebrospinal fluid spread in about 15–20% of the cases.
4. Adjuvant Local RT vs CSI. To be taken on a case to case basis.
5. The chemotherapy regimen also varies widely with some series using platinum-based chemotherapy as in PNET while others have used temozolomide-based chemotherapy alone as in GBM.

17.11.6 Other Variants of GBM

	Gliosarcoma (5%)	GBM-PNET (13%)	GBM-SC (9%)	GBM-GC (51%)	GBM-R (19%)	GBM-E (8%)
Age (median)	45	43.5 (17–82)	69.5 (23–92)	38 (1.9–76)	40 (12–79)	18 (3.5–69)
Gender (male: Female)	3:1		1:3.5	1:1	1:1.4	2:1
Radiology/ nature	Intense peripheral or irregular ring enhancement, calcification	Ring-enhancing reduced ADC	Multifocal, infiltrative	Demarcated	Well-circumscribed and enhancing mass	Sharply demarcated
Molecular changes	MGMT—31.25%, IDH1, p53, PTEN	N-myc, C-myc, IDH1 loss chromosome 10q	EGFR, EGFRvIII, PTEN	P53, PTEN, MDM2; mean survival: 18, 33, 79, 93, 105, ICD-09441/3 loss chromosome: 10	BRAF V600E mutation CDKN2A/2B deletion, and EGFR/MET copy number gain	P53, p21, EGFR BRAF: p.V600E mutation—50% H3F3A p.K27M mutation
Pattern of failure	Local		Local Leptomeningeal—5%	Local Leptomeningeal—5%	Local Leptomeningeal—31.3%	Local Leptomeningeal—37.8%
Survival Median OS	16.7 months	32 months	11 months	18.3 months	12 months	7.7 months

References

1. Kleihues P, Louis DN, Scheithauer BW, et al. The WHO classification of tumors of the nervous system. J Neuropathol Exp Neurol. 2002;61:215–25.
2. Yang M, Guo W, Yang C, et al. Mobile phone use and glioma risk: a systematic review and meta-analysis. PLoS One. 2017;12(5):e0175136.
3. Stupp R, Mason WP, van den Bent MJ, et al. Radiotherapy plus concomitant and adjuvant temozolomide for glioblastoma. N Engl J Med. 2005;352(10):987–96.
4. Stupp R, Hegi ME, Mason WP, et al. Effects of radiotherapy with concomitant and adjuvant temozolomide versus radiotherapy alone on survival in glioblastoma in a randomised phase III study: 5-year analysis of the EORTC-NCIC trial. Lancet Oncol. 2009;10(5):434–5.
5. Hegi ME, Liu L, Herman JG, et al. Correlation of O6-methylguanine methyltransferase (MGMT) promoter methylation with clinical outcomes in glioblastoma and clinical strategies to modulate MGMT activity. J Clin Oncol. 2008;26:4189–99.
6. Gilbert MR, Wang M, Aldape KD, et al. Dose-dense temozolomide for newly diagnosed glioblastoma: a randomized phase III clinical trial. J Clin Oncol. 2013;31(32):4085–91.
7. Gilbert MR, Dignam JJ, Armstrong TS, et al. A randomized trial of bevacizumab for newly diagnosed glioblastoma. N Engl J Med. 2014;370(8):699–708.
8. Chinot OL, Wick W, Mason W, et al. Bevacizumab plus radiotherapy-temozolomide for newly diagnosed glioblastoma. N Engl J Med. 2014;370(8):709–22.
9. Stupp R, Hegi ME, Gorlia T, et al. Cilengitide combined with standard treatment for patients with newly diagnosed glioblastoma with methylated MGMT promoter (CENTRIC EORTC 26071-22072 study): a multicentre, randomised, open-label, phase 3 trial. Lancet Oncol. 2014;15(10):1100–8.
10. Westphal M, Heese O, Steinbach JP, et al. A randomised, open label phase III trial with nimotuzumab, an anti-epidermal growth factor receptor monoclonal antibody in the treatment of newly diagnosed adult glioblastoma. Eur J Cancer. 2015;51(4):522–32.
11. Stupp R, Wong ET, Kanner AA, et al. NovoTTF-100A versus physician's choice chemotherapy in recurrent glioblastoma: a randomised phase III trial of a novel treatment modality. Eur J Cancer. 2012;48:2192–202.
12. Mallick S, Kunhiparambath H, Gupta S, et al. Hypofractionated accelerated radiotherapy (HART) with concurrent and adjuvant temozolomide in newly diagnosed glioblastoma: a phase II randomized trial (HART-GBM trial). J Neuro-Oncol. 2018;140(1):75–82.
13. Omuro A, Beal K, Gutin P, et al. Phase II study of bevacizumab, temozolomide, and hypofractionated stereotactic radiotherapy for newly diagnosed glioblastoma. Clin Cancer Res. 2014;20(19):502331.
14. Iuchi T, Hatano K, Kodama T, et al. Phase 2 trial of hypofractionated high dose intensity modulated radiation therapy with concurrent and adjuvant temozolomide for newly diagnosed glioblastoma. Int J Radiat Oncol Biol Phys. 2014;88(4):793800.
15. Reddy K, Gaspar LE, Kavanagh BD, et al. Prospective evaluation of health related quality of life in patients with glioblastoma multiforme treated on a phase II trial of hypofractionated IMRT with temozolomide. J Neuro-Oncol. 2013;114(1):1116.
16. Alessandro Bozzao Cho KH, Kim JY, Lee SH, et al. Simultaneous integrated boost intensity-modulated radiotherapy in patients with high-grade gliomas. Int J Radiat Oncol Biol Phys. 2010;78(2):390–7.
17. Panet-Raymond V, Souhami L, Roberge D, Kavan P, et al. Accelerated hypofractionated intensitymodulated radiotherapy with concurrent and adjuvanttemozolomide for patients with glioblastoma multiforme: a safety and efficacy analysis. Int J Radiat Oncol Biol Phys. 2009;73(2):473–8.
18. Keime-Guibert F, Chinot O, Taillandier L, Cartalat-Carel S, Frenay M, Kantor G, et al. Radiotherapy for glioblastoma in the elderly. N Engl J Med. 2007;356(15):1527–35. https://doi.org/10.1056/NEJMoa065901. PMID: 17429084.

19. Keime-Guibert F, Chinot O, Taillandier L, et al. Radiotherapy for glioblastoma in the elderly. N Engl J Med. 2007;356(15):1527–35.
20. Roa W, Brasher PM, Bauman G, et al. Abbreviated course of radiation therapy in older patients with glioblastoma multiforme: a prospective randomized clinical trial. J Clin Oncol. 2004;22(9):1583–8.
21. Wick W, Platten M, Meisner C, et al. Temozolomide chemotherapy alone versus radiotherapy alone for malignant astrocytoma in the elderly: the NOA-08 randomised, phase 3 trial. Lancet Oncol. 2012;13(7):707–15.
22. Malmström A, Grønberg BH, Marosi C, et al. Temozolomide versus standard 6-week radiotherapy versus hypofractionated radiotherapy in patients older than 60 years with glioblastoma: the Nordic randomised, phase 3 trial. Lancet Oncol. 2012;13(9):916–26.
23. Roa W, Kepka L, Kumar N, et al. International atomic energy agency randomized phase III study of radiation therapy in elderly and/or frail patients with newly diagnosed glioblastoma multiforme. J Clin Oncol. 2015;33(35):4145–50.
24. Perry JR, Laperriere N, O'Callaghan CJ, et al. Short-course radiation plus temozolomide in elderly patients with glioblastoma. N Engl J Med. 2017;376(11):1027–37.

Management of Gliomatosis Cerebri

18

Supriya Mallick and Prashanth Giridhar

18.1 Introduction

Gliomatosis cerebri (GC) is a rare form of diffuse glioma with growth pattern which involves at least three lobes of the cerebrum and frequently presents as bilateral growth and may have an infratentorial extension [1]. Gliomas of different grade and cell origin which includes astrocytes and oligodendrocytes, can grow in the pattern and this leads to difficulty in defining the molecular profile of the disease. The lesion has significant discordance in the pattern of presentation as it usually remains silent but the radiological imaging reveals extensive disease. Considered as a disease entity, the status has been removed from the 2016 WHO update on the classification of CNS Tumors [1]. The prognosis associated with GC is poor with around 25–50% of the patient population surviving less than a year from symptom onset [2, 3]. The peak incidence is seen between 20 and 50 years age with median age of 48 years (Range 46–53 years) [4] and equal gender involvement with slight male preponderance (1.5:1) [5]. The newer classification entitles the tumor to be classified under the subtype of glioma the disease resembles and they have been found to have poor prognosis compared to corresponding diffuse glioma of the similar molecular grade. The low incidence of the disease has led to poor understanding of the tumor biology and thereby lead to effective treatment options.

Classification GC can be classified into two types [6, 7]:

1. Primary GC—arises de novo; further classified as:
   (a) Type I—Classic type, when no obvious mass is present
   (b) Type II—Diffuse infiltrative pattern with associated tumor mass

S. Mallick · P. Giridhar (✉)
Department of Radiation Oncology, National Cancer Institute, AIIMS, Haryana, India

S. Mallick et al. (eds.), *Evidence based practice in Neuro-oncology*,
https://doi.org/10.1007/978-981-16-2659-3_18

2. Secondary GC—Infiltrative spread of tumor cells from previously diagnosed glioma and may be associated with prior radiation or antiangiogenic therapy.

Further, the tumors can be classified according to the histologic grade and molecular findings.

- GC—Diffuse Astrocytoma; IDH mutant or wild type
- GC—Anaplastic Astrocytoma; IDH mutant or wild type
- GC—Glioblastoma; IDH mutant or wild type
- GC—Oligodendroglioma; IDH mutant and 1p/19q co-deleted
- GC—Anaplastic Oligodendroglioma; IDH mutant and 1p/19q co-deleted

18.2 Symptoms and Signs

- Median age at diagnosis ranges from 46 to 53 years; Slight male preponderance.
- Clinical presentation is insidious and variable.
- Depend on the tumor location and patient's age.
- Include: Seizures, fatigue, mood changes, focal weakness, sensory loss, changes in thinking and memory with dementia-like features, and headaches.
- Signs: Corticospinal tract, spinocerebellar, sensory-motor and visual field deficits, cranial neuropathies, papilledema, myelopathy, hemiparesis, ataxia, and cranial neuropathies.

18.3 Diagnosis

A. *CT scan*:
   (a) Can be normal and lesion can present as isodense to normal brain parenchyma
   (b) Ill-defined asymmetry or subtle hypoattenuation of involved brain parenchyma
B. *MRI*: It forms the investigation of choice for establishing diagnosis [8, 9].
   (a) Loss of grey-white matter differentiation and diffuse gyral thickening.
   (b) Mass effect and enhancement are often minimal.
   (c) T1: iso- to hypoisotense to grey matter.
   (d) T2: hyperintense to grey matter.
   (e) T1 contrast: typically, no or minimal enhancement.
   (f) DWI: Usually no restriction.
C. *MR Spectroscopy*:
   (a) Elevated Choline: Creatine and Choline: NAA ratio
   (b) Marked elevation of myoinositol
   (c) Perfusion study: Low/normal rCBV—no vascular hyperplasia

D. *Positron emission Tomography (PET)*: [10, 11]
   (a) It is of not much use in establishing diagnosis; Usually shows hypometabolism
   (b) Can be useful for tumor extent or treatment response assessment
E. *Biopsy*: Histopathologically, the lesion needs to be confirmed with a biopsy. At least two different samples need to be taken from the same needle tract at different depths. This helps increase the sample amount without developing significant complications [12].

18.4 Histopathology and Molecular Classification: [13]

1. Gross: Firm, edematous areas with flatted gyri and loss of grey-white interphase.
2. Diffuse, irregular parenchymal infiltration of glial cells.
3. On microscopy, small, astrocytic cells with elongated fusiform nuclei which stain for glial fibrillary acidic protein (GFAP) are seen; however, neovascularization, high mitotic activity, and necrosis are uncommon.
4. Histological grading: GC is encompassed from Grade II-IV
   (a) *Grade II* gliomatosis cerebri are mid-grade tumors and they have a higher chance of recurrence.
   (b) *Grade III* and *IV*—They are fast-growing tumors and often become resistant to treatment.
5. Molecular Profiling: [14, 15]
   (a) There are no characteristic features or subgroups noted for GC.
   (b) Astrocytic variants are the most common with non-co-deleted 1p/19q variants.
   (c) Difference in the molecular profile of pediatric and adult GC is noted where IDH mutation and oligodendroglioma type is uncommon in the pediatric population.

18.5 Differential Diagnosis: [7, 16]

1. Progressive multifocal leukoencephalopathy
2. Multifocal/multicentric glioblastoma
3. Primary CNS lymphoma
4. Viral encephalitis
5. Behcet's disease
6. Vasculitis
7. Subacute sclerosing panencephalitis
8. Idiopathic intracranial hypertension
9. Tubercular encephalitis
10. Leukodystrophy
11. Primary progressive multiple sclerosis

Treatment There is no standard treatment defined for GC due to lack of clinical trials in view of the rare incidence. The disease has a poor prognosis with 25–50% patient population surviving for less than 1 year indicating towards the rapidly progressive nature of the disease.

A. *Surgery*:
   (a) Role of surgery lies in providing tissue diagnosis, as resection is not feasible considering the diffuse involvement of the cerebral regions by the tumor.
   (b) Partial resection can be considered when the patient presents with symptoms due to mass effect and T2-weighted or T1 contrast images can help in tumor debulking.
   (c) There is no significant benefit on the survival outcomes on the extent of resection (21 months with resection vs 18 months post biopsy; $p = 0.96$) [17].
B. *Radiation Therapy*: The role of radiation therapy presents a dilemma as the fields involved are extensive and the tumor appears radioresistant. The disease entity is treated similar to a high-grade glioma. Adult cases can be treated either with radiation alone or chemo-radiation; however, the role of chemotherapy is under evaluation in the children. The various trials conducted do not show any correlation between the radiation dose and treatment volumes with the survival outcomes and the dose delivered is controversial [18–21].
   (a) The radiation field involved and the doses delivered are:
   Whole Brain Radiation—20–60 Gy with 1.8–2Gy per fraction delivered. The total dose depends on the performance status of the patient and the toxicity assessment while on treatment [4, 5].
   Focal Radiation—54–66 Gy with 1.8 Gy per fraction delivered [17]. The volume included the visible disease on T2-weighted or T1 contrast images which constituted the gross tumor volume (GTV). A margin of 1–1.5 cm to this volume constituted the clinical target volume (CTV). An additional PTV was given with 0.5 cm to the CTV.
   (b) The survival outcomes differ in the literature with one study showing no difference in the overall survival (OS) with addition of radiation therapy and another contrasting study showed a significant OS benefit with addition of radiation therapy (27.5 vs 6.5 months; $p < 0.01$) [5, 7].
C. *Chemotherapy*:
   - Chemotherapy has been delivered either alone or with radiation therapy, but none of the studies in the literature available have shown a benefit in this disease entity.
   - In a prospective phase II single arm study, NOA-05, the role of procarbazine and lomustine was evaluated as upfront chemotherapy option in this disease [22]. It showed a progression-free survival benefit of 14 months and an overall survival of 30 months and they concluded, upfront addition of chemotherapy could have potential benefits in the management of GC.
   - Other chemotherapy options used include temozolomide (TMZ) and procarbazine, lomustine and vincristine (PCV) combination. On retrospective analysis of these chemotherapy options, the PFS benefit was of around

16 months and OS of 26 months for both TMZ and PCV. However, toxicities were found to be higher with PCV [6, 23].

- TMZ needs to specifically be considered for the cohort with oligodendroglial pathology with 1p/19q co-deletion as they have improved PFS and OS [24].

D. *Role of Immunotherapy*: Immunotherapy is an option being explored for high-grade gliomas; however, it has not been studied for GC. In view of IDH-mutations being common in GC, there are trials underway looking into the role of IDH inhibitors (NCT02746081).

18.6 Future Trials

- HIT-HGG-2013 trials is looking into the role of valproic acid and chloroquine along with radiation and TMZ in children (>3 years age) with high-grade glioma including GC.
- NCT02758366: A Phase II trial which is looking into the role of doxorubicin in combination with radiation and TMZ in high-grade glioma including GC.
- Another Phase I/II study is looking into the role of gefitinib with radiation therapy in pediatric patients with GC (NCT00042991).

18.7 Conclusion

Gliomatosis cerebri is a rare disease entity which has a poor prognosis and has not been completely understood. The treatment options continue to be a dilemma for the oncologists due to lack of clinical trials. Radiation with or without chemotherapy seems to be the current standard of care. Radiation therapy has shown significant improvement in the overall survival; however, the doses and treatment fields are still a controversy. The increasing role of molecular profiling will help better classify the disease and there is a need for prospective trials to comprehensively define the treatment options for this disease.

References

1. Louis DN, Perry A, Reifenberger G, von Deimling A, Figarella-Branger D, Cavenee WK, et al. The 2016 World Health Organization classification of tumors of the central nervous system: a summary. Acta Neuropathol. 2016;131:803–20.
2. Herrlinger U, Jones DT, Glas M, Hattingen E, Gramatzki D, Stuplich M, et al. Gliomatosis cerebri: no evidence for a separate brain tumor entity. Acta Neuropathol. 2015;131(2):309–19.
3. Jennings MT, Frenchman M, Shehab T, Johnson MD, Creasy J, LaPorte K, et al. Gliomatosis cerebri presenting as intractable epilepsy during early childhood. J Child Neurol. 1995;10(1):37–45.
4. Elshaikh MA, Stevens GH, Peereboom DM, Cohen BH, Prayson RA, Lee SY, et al. Gliomatosis cerebri. Cancer. 2002;95(9):2027–31.
5. Chen S, Tanaka S, Giannini C, et al. Gliomatosis cerebri: clinical characteristics, management, and outcomes. J Neurooncol. 2013;112(2):267–75.

6. Sanson M, Cartalat-Carel S, Taillibert S, Napolitano M, Djafari L, Cougnard J, et al. Initial chemotherapy in gliomatosis cerebri. Neurology. 2004;63(2):270–5.
7. Taillibert S, Chodkiewicz C, Laigle-Donadey F, Napolitano M, CartalatCarel S, Sanson M. Gliomatosis cerebri: a review of 296 cases from the ANOCEF database and the literature. J Neuro-Oncol. 2006;76(2):201–5.
8. Shin YM, Chang KH, Han MH, et al. Gliomatosis cerebri: comparison of MR and CT features. AJR Am J Roentgenol. 1993;161(4):859–62.
9. Bendszus M, Warmuth-metz M, Klein R, et al. MR spectroscopy in gliomatosis cerebri. AJNR Am J Neuroradiol. 2000;21(2):375–80.
10. Dexter MA, Parker GD, Besser M, Ell J, Fulham MJ. MR and positron emission tomography with fludeoxyglucose F 18 in gliomatosis cerebri. AJNR Am J Neuroradiol. 1995;16(7):1507–10.
11. Plowman PN, Saunders CA, Maisey MN. Gliomatosis cerebri: disconnection of the cortical grey matter, demonstrated on PET scan. Br J Neurosurg. 1998;12(3):240–4.
12. Greenfield JP, Castaneda Heredia A, George E, Kieran MW, Morales La Madrid A. Gliomatosis cerebri: a consensus summary report from the First International Gliomatosis cerebri Group Meeting, March 26-27, 2015, Paris, France. Pediatr Blood Cancer. 2016;63(12):2072–7.
13. Armstrong GT, Phillips PC, Rorke-Adams LB, Judkins AR, Localio AR, Fisher MJ. Gliomatosis cerebri: 20 years of experience at the Children's Hospital of Philadelphia. Cancer. 2006;107(7):1597–606.
14. Broniscer A, Chamdine O, Hwang S, Lin T, Pounds S, Onar-Thomas A, et al. Gliomatosis cerebri in children shares molecular characteristics with other pediatric gliomas. Acta Neuropathol. 2016;131(2):299–307.
15. Seiz M, Tuettenberg J, Meyer J, Essig M, Schmieder K, Mawrin C, et al. Detection of IDH1 mutations in gliomatosis cerebri, but only in tumors with additional solid component: evidence for molecular subtypes. Acta Neuropathol. 2010;120(2):261–7.
16. Rudà R, Bertero L, Sanson M. Gliomatosis cerebri: a review. Curr Treat Options Neurol. 2014;16(2):1–9.
17. Perkins GH, Schomer DF, Fuller GN, Allen PK, Maor MH. Gliomatosis cerebri: improved outcome with radiotherapy. Int J Radiat Oncol Biol Phys. 2003;56(4):1137–46.
18. Cozad SC, Townsend P, Morantz RA, Jenny AB, Kepes JJ, Smalley SR. Gliomatosis cerebri. Results with radiation therapy. Cancer. 1996;78(8):1789–93.
19. Artigas J, Cervos-Navarro J, Iglesias JR, Ebhardt G. Gliomatosis cerebri: clinical and histological findings. Clin Neuropathol. 1985;4(4):135–48.
20. Yanaka K, Kamezaki T, Kobayashi E, Matsueda K, Yoshii Y, Nose T. MR imaging of diffuse glioma. AJNR Am J Neuroradiol. 1992;13(1):349–51.
21. Fangusaro J, Warren KE. Unclear standard of care for pediatric high grade glioma patients. J Neuro-Oncol. 2013;113(2):341–2.
22. Glas M, Bahr O, Felsberg J, Rasch K, Wiewrodt D, Schabet M, et al. NOA-05 phase 2 trial of procarbazine and lomustine therapy in gliomatosis cerebri. Ann Neurol. 2011;70(3):445–53.
23. Soffietti R, Rudà R, Laguzzi E, Buttolo L, Pace A, Carapella C, et al. Treatment of gliomatosis cerebri with temozolomide: a retrospective study of the AINO. In: ASCO Annual Meeting Proceedings. Chicago: Italian Association for Neuro-Oncology; 2007.
24. Kaloshi G, Everhard S, Laigle-Donadey F, Marie Y, Navarro S, Mokhtari K, et al. Genetic markers predictive of chemosensitivity and outcome in gliomatosis cerebri. Neurology. 2008;70(8):590–5.

Pleomorphic Xanthoastrocytoma 19

Supriya Mallick and V. R. Anjali

19.1 Introduction [1–4]

- Pleomorphic xanthoastrocytoma (PXA) is a low-grade glial tumor constituting 1% brain tumors.
- Kepes et al. described it for the first time in 1993.
- Arises from subpial astrocytes.
- Most commonly affects second and third decade of life (10–30 years); Median age—33 years (Range 4.5–75).
- No gender predilection.
- Associated with seizure in young adults and considered a part of epileptogenic tumors spectrum.
- Nearly 50% of patients present with seizure and one-third present with headache.
- These tumors commonly arise from the temporal lobe (nearly 50%). Frontal, multi-lobar, and spinal location has also been described.
- Mostly localized but isolated reports of leptomeningeal dissemination.
- Most of the tumors are supratentorial, less than 10% are infratentorial.
- Tumor is superficial, proximal to dura but without involvement.

19.2 Classification

- Pleomorphic xanthoastrocytoma—WHO grade II tumors (80%)
- Anaplastic pleomorphic xanthoastrocytoma (APXA)—WHO grade III tumor (Anaplastic may be de novo or as progression from grade II)

S. Mallick (✉)
Department of Radiation Oncology, National Cancer Institute, AIIMS, Haryana, India

V. R. Anjali
Department of Radiation Oncology, All India Institute of Medical Sciences, New Delhi, India

S. Mallick et al. (eds.), *Evidence based practice in Neuro-oncology*,
https://doi.org/10.1007/978-981-16-2659-3_19

- Criteria
  - >5 mitoses per 10 high-power fields and increased cellularity
  - Presence of necrosis
  - MIB-1 labeling index >4%

19.3 Investigation

Anaplastic cases appear as high-grade glioma.

19.3.1 Appearance on CT Scan

- Solid and cystic component
- Appears hypodense or isodense
- Calcification rare
- With little or no surrounding edema
- Cause scalloping of the overlying bone

19.3.2 MRI

- Appears solid-cystic tumor with homogenous to heterogeneous contrast enhancement
- Peripheral eccentric cystic component (50–60%)
- Exhibit dural tail which is mostly reactive
- *T1*: Solid component iso to hypointense
- *T1 post contrast:* enhancement
- *T2*: Solid component iso- to hyperintense
- T2 FLAIR: Cystic areas show hyperintensity

19.4 Pathology

- Tumors are moderately cellular, predominantly pleomorphic and show foci of lymphoplasmacytic infiltration.
- Necrosis and mitoses rare.
- The tumor has been named pleomorphic as there is variable histological feature including spindle cells, polygonal cells, and multinucleated cells.
- Pathogenesis is driven by MAPK pathway.
- Nearly 60–70% patients harbor BRAF mutations; less common in anaplastic PXA.
- Mutation of BRAF confers better survival compared with wild variant.

- *Immunohistochemistry*
  - GFAP: Positive
  - S100: Positive
  - Vimentin: Positive
  - Reticulin: Positive
  - PAS: Positive
  - Synaptophysin, MAP2 and neurofilament: variable
  - Ki-67 proliferation index: <1%

19.5 Treatment

19.5.1 Surgical Excision [5]

- Surgical excision standard of care and a maximal safe resection is aim of surgery.
- Nearly 60% patients can undergo GTR.
- Patients with a GTR have good long-term survival.

	GTR (months)	STR (months)
PFS	48	14
OS	NR	62

19.5.2 Radiation [2, 3, 6–10]

- Data on adjuvant radiation is limited.
- Conflicting data about survival advantage in PXA.
- Radiation has been utilized more in salvage setting (76%).
- In APXA radiation may be more effective as these patients experience early failure.

19.5.2.1 Indications for Adjuvant RT

- PXA with GTR: Not required.
- PXA with STR or recurrence: RT should be considered.
- APXA: adjuvant radiation may be considered upfront even after GTR.

19.5.2.2 Target Volume

- In adjuvant or salvage setting:
  - Local radiation only (post-operative cavity plus any residual enhancing mass with 1 cm isotropic expansion as CTV and 3–5 mm additional as PTV)
- In CSF positive cases: CSI should be considered.

19.5.2.3 Dose

Local radiation: 54–60 Gy in conventional fractionation

CSI dose 30 Gy to the entire cranio-spinal axis followed by boost to the primary up to 56–60 Gy

19.5.3 Chemotherapy

- Role of chemotherapy controversial.
- V600E BRAF mutation in nearly 70% patients which constitutively activates RAS/RAF/MEK/ERK signaling pathway.
- BRAF inhibitor monotherapy or BRAF+MEK inhibitor are used in recurrent tumors.
- BRAF inhibitors, Vemurafenib and Dabrafenib showed promising result.

19.6 Results [2, 3]

Median OS of 183 months; PFS of 38 months.

	PFS (5 year)	OS (5 year)
PXA	Nearly 60%	Nearly 75%
APXA	Nearly 49%	Nearly 60%

19.6.1 Prognostic Factors (Favorable)

- Patients younger than 20 years
- GTR
- Grade II [PFS: (48 vs 11.7 months) and OS: (209 vs 68 months)]
- Adjuvant treatment

19.7 Follow-up

- Nearly 50% patients experience progression within 3 years.
- Local recurrence is most common pattern of recurrence. Leptomeningeal spread seen in 7% patients.
- Follow-up should be done with clinical examination and CEMRI of brain every 3 months for first 3 years and thereafter every 6 months.
- Re-surgery followed by radiation or chemotherapy may be used to salvage.

19.8 Treatment Algorithm

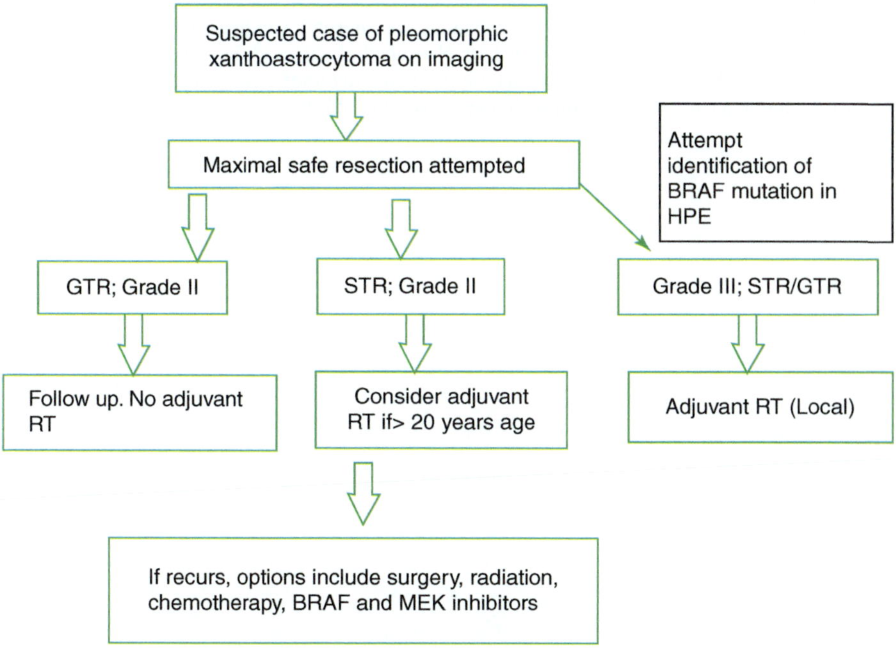

References

1. Patibandla MR, Nayak M, Purohit AK, Thotakura AK, Uppin M, Challa S. Pleomorphic xanthoastrocytoma with anaplastic features: a rare case report and review of literature with reference to current management. Asian J Neurosurg. 2016;11(3):319.
2. Mallick S, Benson R, Melgandi W, Giridhar P, Rath GK. Grade II pleomorphic Xanthoastrocytoma; a meta-analysis of data from previously reported 167 cases. J Clin Neurosci. 2018;54:57–62.
3. Mallick S, Giridhar P, Benson R, Melgandi W, Rath GK. Demography, Pattern of care, and survival in patients with xanthoastrocytoma: a systematic review and individual patient data analysis of 325 cases. J Neurosci Rural Pract. 2019;10(3):430–7.
4. Herpers MJ, Freling G, Beuls EA. Pleomorphic xanthoastrocytoma in the spinal cord. J Neurosurg. 1994;80(3):564–9.
5. Fouladi M, Jenkins J, Burger P, Langston J, Merchant T, Heideman R, et al. Pleomorphic xanthoastrocytoma: favorable outcome after complete surgical resection. Neuro-Oncology. 2001;3(3):184–92.
6. Nakamura M, Chiba K, Matsumoto M, Ikeda E, Toyama Y. Pleomorphic xanthoastrocytoma of the spinal cord. J Neurosurg Spine. 2006;5(1):72–5.
7. Gill M, Pathak HC, Madan R, Bhattacharya S, Choudhary GS. Primary spinal pleomorphic xanthoastrocytoma. Neurol India. 2010;58(5):771–3.

8. Simal-Julián JA, Sanchis-Martín R, Prat-Acín R, Miranda-Lloret P, Conde-Sardón R, Cárdenas-Ruiz-Valdepeñas E, et al. Spinal pleomorphic xantoastrocytoma. Neurocirugia (Astur). 2010;21(5):390–5.
9. Das S, Yip S, Hukin J, Cochrane D, Dunham C. Pleomorphic xanthoastrocytoma of the spinal cord: case report and literature review. Clin Neuropathol. 2014;33(3):190–6.
10. Jeong JY, Suh YL, Hong SW. Atypical teratoid/rhabdoid tumor arising in pleomorphic xanthoastrocytoma: a case report. Neuropathology. 2014;34(4):398–405.

Astroblastoma

20

Supriya Mallick

20.1 Introduction

Astroblastoma is a very uncommon tumor.

Bailey and Cushing in 1926 first described this tumor to arise from astroblasts.

These tumors are rare glial tumors which account for 0.45–2.8% of gliomas.

20.2 Clinical Presentation

1. Most commonly present in first two decades of life (median age 16 years).
2. Nearly one-third of cases are found in first decade and another one fifth in second decade of life.
3. Females are affected twice as frequently as males.
4. Most of the cases arise from frontal lobe (39%) and parietal lobe (26.7%).
5. Spine is an extremely rare subsite.
6. As these tumors arise in the cerebral hemisphere, most of the patients present with
   (a) Headache (nearly half)
   (b) Nausea and vomiting
   (c) Seizure (nearly 25%)

20.3 Radiology

- Mostly supratentorial, peripherally located.
- Tumor shows both solid and cystic components.
- Tumor has multiple intratumoral cysts showing "bubbly" appearance.

S. Mallick (✉)
Department of Radiation Oncology, National Cancer Institute, AIIMS, Haryana, India

S. Mallick et al. (eds.), *Evidence based practice in Neuro-oncology*,
https://doi.org/10.1007/978-981-16-2659-3_20

- Tumor is hyperintense to white matter on FLAIR and T2-weighted sequences and hypointense to isointense on T1-weighted images.
- It shows heterogeneous contrast enhancement and rim enhancement.
- Intraventricular location, intratumoral hemorrhage with a fluid-fluid level, and the presence of dural "tails" are less common but important findings.
- Calcification has also been reported in few cases.

20.4 Surgery

- Surgery is considered the cornerstone of therapy.
- Gross total resection (GTR) is possible in nearly 75% of cases. GTR confers 5-year OS of 73.1% compared to 15% after a STR.

20.5 Pathology

- These tumors are identified based on "astroblastic" pseudorosettes and perivascular hyalinization.
- World Health Organization recognized astroblastoma as a glial neoplasm of uncertain origin.
- Recent studies have found three variants: PXA-like AB (most common), HGG-like AB (poor outcome), and AB with MN1-BEND2 fusion (uncommon in adults).
- Grades of tumor: high and low grade. Patients with a low-grade tumor have significantly better OS compared to higher-grade tumors.

20.6 Adjuvant Therapy

20.6.1 Radiation

- Adjuvant radiation may be advocated in patents with high-grade tumors, subtotal resection or in patients with recurrent tumor. Mallick et al, in an IPD reported pooer survival outcome for higher grade patiets and in patients who undergo STR. Hence, the authors advocated intensification oftreatment in such cases with addition of radiation along with chemotherapy.
- Dose: radiation dose 54–56 Gy may be optimum.
- Technique: A conformal technique should be advised.
- Target should be single volume, CTV should include post-operative cavity, any residual enhancing volume, with 1 cm isotropic expansion. PTV should be 3–5 mm.

20.6.2 Chemotherapy

- Chemotherapy has been used in 60% of high-grade tumors.
- Regimen varies widely.

- Temozolomide is the most common drug used. It should be used in concurrent phase along with maintenance for a maximum of six cycles.
- A combination of cisplatin, etoposide with vincristine may be used.

20.6.3 Survival Outcome

- Median progression free survival (PFS) is 36 months and overall survival (OS) is 184 months.
- OS at 5 years was 79.5%.
- Nearly all patients had a local progression.

20.6.4 Prognostic Factors for Survival

- Extent of surgery (GTR vs STR)
- Grade of tumor (High vs Low)
- Size of tumor (>5 cm vs ≤5 cm)

20.7 Follow-up

- First follow-up after 1 month of surgery or radiation with clinical examination
- Follow-up every 3 months with contrast enhanced MRI brain for first 3 years
- Subsequently continue follow-up every 6 months with contrast enhanced MRI

20.8 Treatment Algorithm [69]

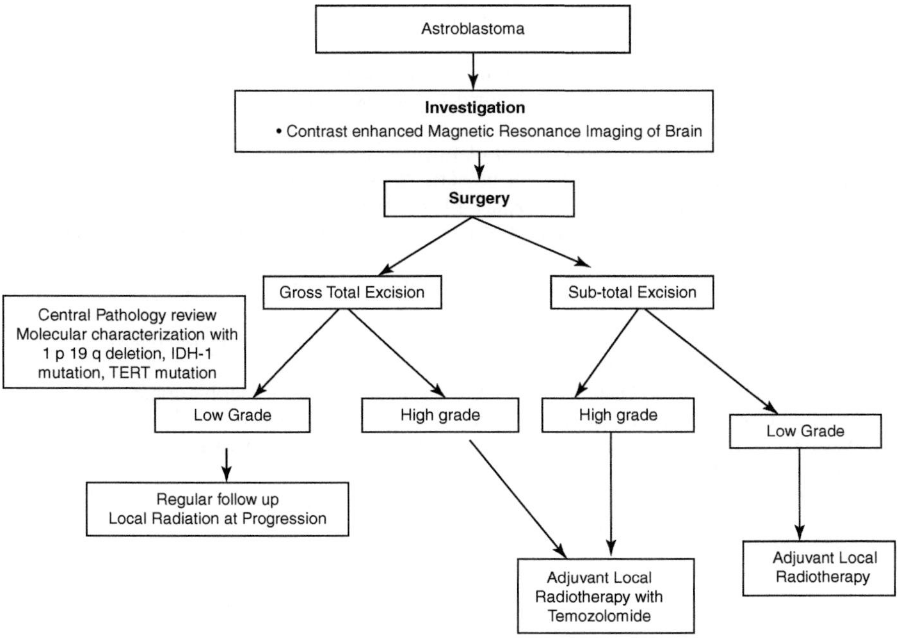

Suggested Reading

1. Bailey P, Cushing H. A classification of the tumors of the glioma group on a histogenetic basis with a correlated study of prognosis. Philadelphia: J.B. Lippincott Co.; 1926.
2. Sughrue ME, Choi J, Rutkowski MJ, Aranda D, Kane AJ, Barani IJ, et al. Clinical features and post-surgical outcome of patients with astroblastoma. J Clin Neurosci. 2011;18(6):750–4.
3. Ahmed KA, Allen PK, Mahajan A, Brown PD, Ghia AJ. Astroblastomas: a surveillance, epidemiology, and end results (SEER). World Neurosurg. 2014;82(1–2):e291–7.
4. Lehman NL, Hattab EM, Mobley BC, Usubalieva A, Schniederjan MJ, McLendon RE, et al. Morphological and molecular features of astroblastoma, including BRAFV600E mutations, suggest an ontological relationship to other cortical-based gliomas of children and young adults. Neuro-Oncology. 2016. pii: now118.
5. Brat DJ, Hirose Y, Cohen KJ, Feuerstein BG, Burger PC. Astroblastoma: clinicopathologic features and chromosomal abnormalities defined by comparative genomic hybridization. Brain Pathol. 2000;10(3):342–52.
6. Louis DN, Ohgaki H, Wiestler OD, et al. The 2007 WHO classification of tumours of the central nervous system. Acta Neuropathol. 2007;114:97–109.
7. Bale TA, Abedalthagafi M, Bi WL, Kang YJ, Merrill P, Dunn IF, et al. Genomic characterization of recurrent high-grade astroblastoma. Cancer Genet. 2016;209(7–8):321–30.
8. Yeo JJ, Low YY, Putti TC, Koh KM. Adult intraventricular astroblastoma. Singap Med J. 2016;57(1):53–4.
9. Yuzawa S, Nishihara H, Tanino M, Kimura T, Moriya J, Kamoshima Y, et al. A case of cerebral astroblastoma with rhabdoid features: a cytological, histological, and immunohistochemical study. Brain Tumor Pathol. 2016;33(1):63–70.
10. Singla N, Dhandapani SS, Kapoor A, Chatterjee D, Vashishta RK. Hemorrhage in astroblastoma: an unusual manifestation of an extremely rare entity. J Clin Neurosci. 2016;25:147–50.
11. Narayan S, Kapoor A, Singhal MK, Jakhar SL, Bagri PK, Rajput PS, Kumar HS. Astroblastoma of cerebrum: a rare case report and review of literature. J Cancer Res Ther. 2015;11(3):667.
12. Sabharwal P, Sadashiva N, Unchagi A, Mahadevan A, Pandey P. Intraventricular astroblastoma in an infant: a case report and review of the literature. Pediatr Neurosurg. 2015;50(6):325–9.
13. Yao K, Wu B, Xi M, Duan Z, Wang J, Qi X. Distant dissemination of mixed low-grade astroblastoma-arteriovenous malformation after initial operation: a case report. Int J Clin Exp Pathol. 2015;8(6):7450–6. eCollection 2015.
14. Samkari A, Hmoud M, Al-Mehdar A, Abdullah S. Well-differentiated and anaplastic astroblastoma in the same patient: a case report and review of the literature. Clin Neuropathol. 2015;34(6):350–8.
15. Mellai M, Piazzi A, Casalone C, Grifoni S, Melcarne A, Annovazzi L, et al. Astroblastoma: beside being a tumor entity, an occasional phenotype of astrocytic gliomas? Onco Targets Ther. 2015;8:451–60.
16. Janz C, Buhl R. Astroblastoma: report of two cases with unexpected clinical behavior and review of the literature. Clin Neurol Neurosurg. 2014;125:114–24.
17. de la Garma VH, Arcipreste AA, Vázquez FP, Aguilar RR, Castruita UO, Guerra RM. High-grade astroblastoma in a child: report of one case and review of literature. Surg Neurol Int. 2014;5:111.
18. Singh DK, Singh N, Singh R, Husain N. Cerebral astroblastoma: A radiopathological diagnosis. J Pediatr Neurosci. 2014;9(1):45–7.
19. Barakat MI, Ammar MG, Salama HM, Abuhashem S. Astroblastoma: case report and review of literature. Turk Neurosurg. 2016;26(5):790–4.
20. Yao K, Li TF, Zhu MW, Duan ZJ, Liu CQ, Wang JQ, et al. Mixed astroblastoma-arteriovenous malformation complex: a case report. Neurol India. 2013;61(4):439–42.
21. Nasit JG, Trivedi P. Recurrent low-grade astroblastoma with signet ring-like cells and high proliferative index. Fetal Pediatr Pathol. 2013;32(4):284–92.
22. Khosla D, Yadav BS, Kumar R, Agrawal P, Kakkar N, Patel FD, et al. Pediatric astroblastoma: a rare case with a review of the literature. Pediatr Neurosurg. 2012;48(2):122–5.

23. Agarwal V, Mally R, Palande DA, Velho V. Cerebral astroblastoma: a case report and review of literature. Asian J Neurosurg. 2012;7(2):98–100.
24. Bhattacharjee S, Pulligopu AK, Uppin MS, Sundaram C. Astroblastoma with bone invasion. Asian J Neurosurg. 2011;6(2):113–5.
25. Bergkåsa M, Sundstrøm S, Gulati S, Torp SH. Astroblastoma - a case report of a rare neuroepithelial tumor with complete remission after chemotherapy. Clin Neuropathol. 2011;30(6):301–6.
26. Johnson KA, Bonnin JM, Boaz JC, Douglas-Akinwande AC, Hattab EM. Anaplastic astroblastoma presenting as massive, sudden-onset, intraparenchymal hemorrhage. Pediatr Neurosurg. 2010;46(6):457–61.
27. Binesh F, Akhavan A, Navabii H, Mehrabaniyan M. Anaplastic astroblastoma: a rare glial tumour. BMJ Case Rep. 2011. pii: bcr0620114323.
28. Weintraub D, Monteith SJ, Yen CP, Schlesinger D, Rich T, Sheehan J. Recurrent astroblastoma treated with gamma knife radiosurgery. J Neuro-Oncol. 2011;103(3):751–4.
29. Mastrangelo S, Lauriola L, Coccia P, Puma N, Massimi L, Riccardi R. Two cases of pediatric high-grade astroblastoma with different clinical behavior. Tumori. 2010;96(1):160–3.
30. Kemerdere R, Dashti R, Ulu MO, Biçeroğlu H, Demiröz AS, Albayram S, et al. Supratentorial high grade astroblastoma: report of two cases and review of the literature. Turk Neurosurg. 2009;19(2):149–52.
31. Kantar M, Ertan Y, Turhan T, Kitis O, Anacak Y, Akalin T, et al. Anaplastic astroblastoma of childhood: aggressive behavior. Childs Nerv Syst. 2009;25(9):1125–9.
32. Salvati M, D'Elia A, Brogna C, Frati A, Antonelli M, Giangaspero F, et al. Cerebral astroblastoma: analysis of six cases and critical review of treatment options. J Neuro-Oncol. 2009;93(3):369–78.
33. Eom KS, Kim JM, Kim TY. A cerebral astroblastoma mimicking an extra-axial neoplasm. J Korean Neurosurg Soc. 2008;43(4):205–8.
34. Ganapathy S, Kleiner LI, Mirkin DL, Broxson E. Unusual manifestations of astroblastoma: a radiologic-pathologic analysis. Pediatr Radiol. 2009;39(2):168–71.
35. Denaro L, Gardiman M, Calderone M, Rossetto M, Ciccarino P, Giangaspero F, et al. Intraventricular astroblastoma. Case Rep. J Neurosurg Pediatr. 2008;1(2):152–5.
36. Fathi AR, Novoa E, El-Koussy M, Kappeler A, Mariani L, Vajtai I. Astroblastoma with rhabdoid features and favorable long-term outcome: report of a case with a 12-year follow-up. Pathol Res Pract. 2008;204(5):345–51.
37. Notarianni C, Akin M, Fowler M, Nanda A. Brainstem astroblastoma: a case report and review of the literature. Surg Neurol. 2008;69(2):201–5.
38. Unal E, Koksal Y, Vajtai I, Toy H, Kocaogullar Y, Paksoy Y. Astroblastoma in a child. Childs Nerv Syst. 2008;24(2):165–8.
39. Mangano FT, Bradford AC, Mittler MA, Valderrama E, Schneider SJ. Astroblastoma. Case report, review of the literature, and analysis of treatment strategies. J Neurosurg Sci. 2007;51(1):21–7. discussion 27.
40. Bannykh SI, Fan X, Black KL. Malignant astroblastoma with rhabdoid morphology. J Neuro-Oncol. 2007;83(3):277–8.
41. Tumialán LM, Brat DJ, Fountain AJ, Barrow DL. An astroblastoma mimicking a cavernous malformation: case report. Neurosurgery. 2007;60(3):E569–70. discussion E570.
42. Alaraj A, Chan M, Oh S, Michals E, Valyi-Nagy T, Hersonsky T. Astroblastoma presenting with intracerebral hemorrhage misdiagnosed as dural arteriovenous fistula: review of a rare entity. Surg Neurol. 2007;67(3):308–13.
43. Bell JW, Osborn AG, Salzman KL, Blaser SI, Jones BV, Chin SS. Neuroradiologic characteristics of astroblastoma. Neuroradiology. 2007;49(3):203–9.
44. Kaji M, Takeshima H, Nakazato Y, Kuratsu J. Low-grade astroblastoma recurring with extensive invasion. Neurol Med Chir (Tokyo). 2006;46(9):450–4.
45. Hata N, Shono T, Yoshimoto K, Mizoguchi M, Kawamura T, Nagata S, et al. An astroblastoma case associated with loss of heterozygosity on chromosome 9p. J Neuro-Oncol. 2006;80(1):69–73.

46. Miranda P, Lobato RD, Cabello A, Gómez PA, Martínez de Aragón A. Complete surgical resection of high-grade astroblastoma with long time survival: case report and review of the literature. Neurocirugia (Astur). 2006;17(1):60–3.
47. Kubota T, Sato K, Arishima H, Takeuchi H, Kitai R, Nakagawa T. Astroblastoma: immunohistochemical and ultrastructural study of distinctive epithelial and probable tanycytic differentiation. Neuropathology. 2006;26(1):72–81.
48. Lau PP, Thomas TM, Lui PC, Khin AT. 'Low-grade' astroblastoma with rapid recurrence: a case report. Pathology. 2006;38(1):78–80.
49. Huhn SL, Yung Y, Cheshier S, Harsh G, Ailles L, Weissman I, et al. Identification of phenotypic neural stem cells in a pediatric astroblastoma. J Neurosurg. 2005;103(5 Suppl):446–50.
50. Navarro R, Reitman AJ, de León GA, Goldman S, Marymont M, Tomita T. Astroblastoma in childhood: pathological and clinical analysis. Childs Nerv Syst. 2005;21(3):211–20.
51. Kim DS, Park SY, Lee SP. Astroblastoma: a case report. J Korean Med Sci. 2004;19(5):772–6.
52. Kim BS, Kothbauer K, Jallo G. Brainstem astroblastoma. Pediatr Neurosurg. 2004;40(3):145–6.
53. Caroli E, Salvati M, Esposito V, Orlando ER, Giangaspero F. Cerebral astroblastoma. Acta Neurochir. 2004;146(6):629–33.
54. Sugita Y, Terasaki M, Shigemori M, Morimatsu M, Honda E, Oshima Y. Astroblastoma with unusual signet-ring-like cell components: a case report and literature review. Neuropathology. 2002;22(3):200–5.
55. Port JD, Brat DJ, Burger PC, Pomper MG. Astroblastoma: radiologic-pathologic correlation and distinction from ependymoma. AJNR Am J Neuroradiol. 2002;23(2):243–7.
56. Shuangshoti S, Mitphraphan W, Kanvisetsri S, Griffiths L, Navalitloha Y, Pornthanakasem W, et al. Astroblastoma: report of a case with microsatellite analysis. Neuropathology. 2000;20(3):228–32.
57. Mierau GW, Tyson RW, McGavran L, Parker NB, Partington MD. Astroblastoma: ultrastructural observations on a case of high-grade type. Ultrastruct Pathol. 1999;23(5):325–32.
58. Thiessen B, Finlay J, Kulkarni R, Rosenblum MK. Astroblastoma: does histology predict biologic behavior? J Neuro-Oncol. 1998;40(1):59–65.
59. Yunten N, Ersahin Y, Demirtas E, Yalman O, Sener RN. Cerebral astroblastoma resembling an extra-axial neoplasm. J Neuroradiol. 1996;23(1):38–40.
60. Jay V, Edwards V, Squire J, Rutka J. Astroblastoma: report of a case with ultrastructural, cell kinetic, and cytogenetic analysis. Pediatr Pathol. 1993;13(3):323–32.
61. Cabello A, Madero S, Castresana A, Diaz-Lobato R. Astroblastoma: electron microscopy and immunohistochemical findings: case report. Surg Neurol. 1991;35(2):116–21.
62. Hoag G, Sima AA, Rozdilsky B. Astroblastoma revisited: a report of three cases. Acta Neuropathol. 1986;70(1):10–6.
63. Rubinstein LJ, Herman MM. The astroblastoma and its possible cytogenic relationship to the tanycyte. An electron microscopic, immunohistochemical, tissue- and organ-culture study. Acta Neuropathol. 1989;78(5):472–83.
64. Husain AN, Leestma JE. Cerebral astroblastoma: immunohistochemical and ultrastructural features. Case Rep J Neurosurg. 1986;64(4):657–61.
65. Yamashita J, Handa H, Yamagami T, Haebara H. Astroblastoma of pure type. Surg Neurol. 1985;24(2):218–22.
66. Pizer BL, Moss T, Oakhill A, Webb D, Coakham HB. Congenital astroblastoma: an immunohistochemical study. Case Rep J Neurosurg. 1995;83(3):550–5.
67. Sener RN. Astroblastoma: diffusion MRI, and proton MR spectroscopy. Comput Med Imaging Graph. 2002;26(3):187–91.
68. Cunningham DA, Lowe LH, Shao L, Acosta NR. Neuroradiologic characteristics of astroblastoma and systematic review of the literature: 2 new cases and 125 cases reported in 59 publications. Pediatr Radiol. 2016;46(9):1301–8.
69. Supriya Mallick, Rony Benson, Bhanuprasad Venkatesulu, Wineeta Melgandi, Goura K Rath. Patterns of care and survival outcomes in patients with astroblastoma: an individual patient data analysis of 152 cases. Child's Nervous System. 2017;33(8):1295–1302.

Nervous System Hemangioblastoma

21

Dodul Mondal

21.1 Introduction and Epidemiology

- Hemangioblastomas of CNS are classified as mesenchymal tumor of nonmeningoepithelial origin in 2016 WHO classification of CNS tumors [1].
- Comprises 1–2% of all intracranial tumors.
- Commonest cerebellar tumor in adult.
- Can be sporadic or part of genetic syndrome.
- Usual age at diagnosis is fourth decade in sporadic cases and third decade with VHL syndrome.
- Multiple intracranial hemangioblastomas (mostly cerebellar) are usually associated with von Hippel–Lindau (VHL) syndrome, which is autosomal dominant in nature, in approximately 10% cases.

21.2 Site of Origin

- Cerebellum—most common
- Brain stem
- Spinal cord
- Cerebrum (infrequent)
- Intramedullary (rare)

21.3 Clinical Manifestation

- Typically, dependent on site of origin.

D. Mondal (✉)
Department of Radiation Oncology, Indraprastha Apollo Hospitals, New Delhi, India

S. Mallick et al. (eds.), *Evidence based practice in Neuro-oncology*,
https://doi.org/10.1007/978-981-16-2659-3_21

- Cerebellar hemangioblastomas can produce hydrocephalus, ataxia, gait instability, or other cerebellar symptoms.
- Patient can also have symptoms associated with VHL.

21.4 Pathology

- Composed of endothelial lined capillary and sinusoid.
- Foamy cells with lipid laden polygonal cytoplasm.
- IHC shows positivity with neuron-specific enolase (NSE), vimentin, and S100.
- Epithelial membrane antigen (EMA) or glial fibrillary acidic protein (GFAP) negative [2].

21.5 Imaging

- Highly vascular tumor.
- Intense contrast enhancement on CT scan or MRI.
- Important to know resectability.
- Imaging should also include MRI orbit to rule out retinal pathology, CT scan of abdomen to rule out associated components of VHL like pheochromocytoma, renal tumor, etc.
- Most spinal cord hemangioblastomas are dorsally located [3].
- Fig. 21.1a–d shows MRI features of cerebellar and spinal hemangioblastoma.

Prognostic Factors Following factors are associated with better outcome after successful treatment:

- No preoperative neurologic deficits.
- Small volume tumor less than 0.5 cc.
- Dorsal location is better than ventral and easier to access during surgery.
- GTR is associated with better outcome.

21.6 Treatment

- Surgery is the primary treatment and usually curative.
- En-block removal (GTR) is the goal of surgery.
- Rapid neurological improvement achieved after surgical resection.
- Intralesional cysts are common and cyst decompression provides rapid relief.
- Radiation is useful in the following situations:
  - Incomplete resection.
  - Unresectable tumor.

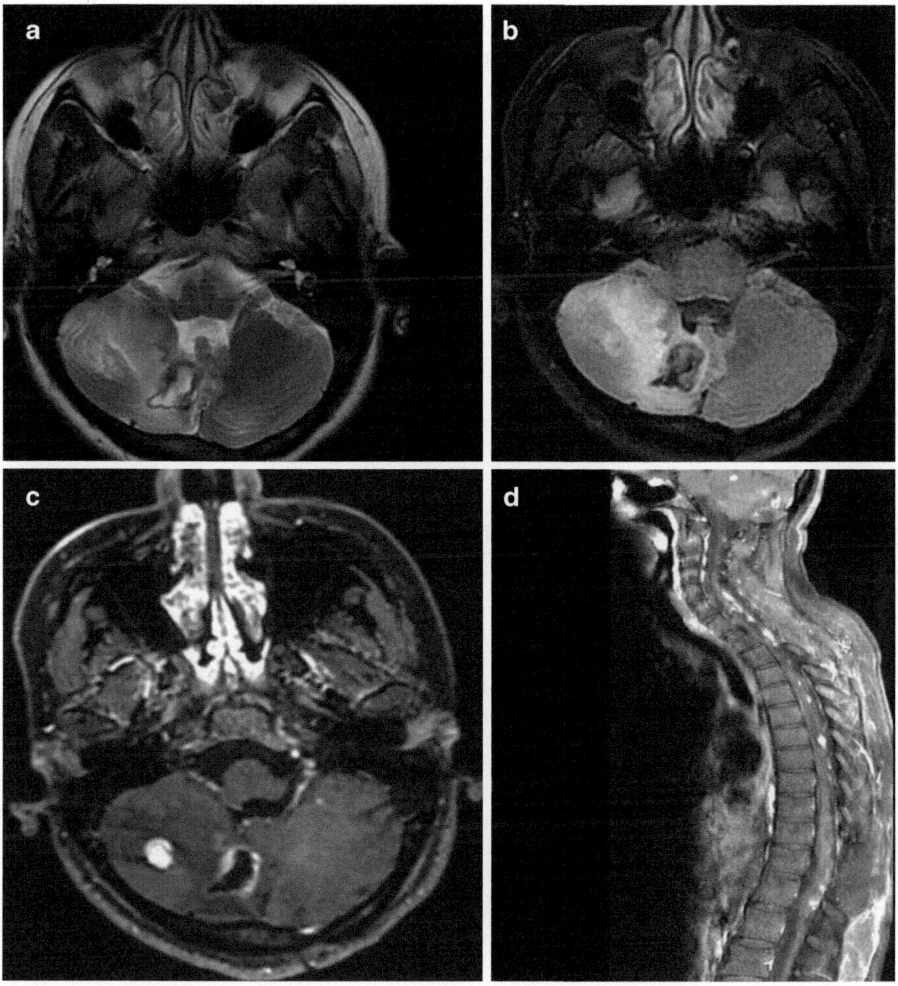

Fig. 21.1 (**a**–**d**) Hemangioblastoma involving cerebellum and cervicodorsal spine. The mass is heterogeneously hyperintense on T2WI (**a**) and FLAIR (**b**); with foci of hemorrhage (arrow in **b**). It shows a peripheral nodular area of strong contrast enhancement (arrow in **c**). Sagittal T1W contrast image reveals multiple other nodular enhancing foci on the dorsal aspect of cervicodorsal spine (arrows in **d**). (Courtesy: Dr. Leve Joseph)

- Recurrent tumor.
- Usual dose 54–60 Gy, conventional fractionation. Response is dose dependent with better outcome above 50 Gy.
- Small tumors or residual are suitable for radiosurgery as well.
- Reported series shows highly variable dose range with SRS.
- Range of median marginal dose with SRS 12–29.9 Gy and prescribed at 50–90% at the periphery [4].

Toxicity Toxicity after treatment of intracranial or spinal tumors are common and mostly related to their location, volume of resection, injury to vital structure during surgery, dose-volume-fractionation with radiation, etc. Acute side effects due to radiation can include but are not limited to dermatitis of scalp, alopecia, fatigue, acute rise if intracranial pressure due to cerebral edema, fits, neuritis, etc. Significant long-term side effects can include memory and cognitive impairment, myelitis, etc. It is important to select patient, surgical approach, post-surgical adjuvant treatment wisely. Wherever possible, radiation volume to be kept at minimum. Image guidance technique during daily radiation delivery can reduce PTV margin and consequential irradiated volume of brain.

Funding Nil

Conflict of Interest Nil

Acknowledgement Image courtesy and copyright: Dr. Leve Joseph, Associate Professor, Dept of Neuroradiology, AIIMS Delhi, India.

References

1. Louis DN, Perry A, Reifenberger G, et al. The 2016 World Health Organization classification of tumors of the central nervous system: a summary. Acta Neuropathol. 2016;131:803–20.
2. Asthagiri AR, Mehta GU, Zach L, et al. Prospective evaluation of radiosurgery for hemangioblastomas in von Hippel-Lindau disease. Neuro-Oncology. 2010;12(1):80–6.
3. Wanebo JE, Lonser RR, Glenn GM, et al. The natural history of hemangioblastomas of the central nervous system in patients with von Hippel-Lindau disease. J Neurosurg. 2003;98(1):82–94.
4. Pan J, Jabarkheel R, Huang Y, Ho A, et al. Stereotactic radiosurgery for central nervous system hemangioblastoma: systematic review and meta-analysis. J Neuro-Oncol. 2018;137(1):11–22. https://doi.org/10.1007/s11060-017-2697-0.

Rare/Uncommon Brain Tumors 22

Supriya Mallick

22.1 Ganglioglioma

22.1.1 Introduction [1, 2]

- Ganglioglioma (GG) was first described in 1926.
- Rare accounting for 0.4–1% of all brain tumors.
- Of all GG 1–6% are AGG.
- Tumor is composed of both ganglionic and glial component.
- Most commonly affect second decade of life (10–30 years); Median—6(Range 0.2–77 years).
- Slight male predilection.
- Associated with seizure in young adults and considered a part of epileptogenic tumors spectrum.
- Nearly 50% patient present with seizure and one-third present with headache.
- These tumors commonly arise from the temporal lobe (nearly 50%); However, tumor of frontal, multi-lobar, and spinal location (8%) has also been described.
- Mostly localized but isolated reports of leptomeningeal dissemination.
- Most of the tumors are supratentorial, less than 10% are infratentorial.
- Tumor is superficial, proximal to dura but without involvement.

22.1.2 Classification

1. Ganglioglioma WHO grade I are benign tumors typically associated with chronic drug-resistant epilepsy.

S. Mallick (✉)
Department of Radiation Oncology, National Cancer Institute, AIIMS, Haryana, India

S. Mallick et al. (eds.), *Evidence based practice in Neuro-oncology*,
https://doi.org/10.1007/978-981-16-2659-3_22

2. Anaplastic Ganglioglioma (AGG) WHO grade III (Anaplastic may be de novo or as progression from grade II—Criteria for grade II not clearly defined) [3].
3. Criteria:
   (a) >5 mitoses per 10 high-power fields and increased cellularity AND
   (b) Presence of necrosis OR
   (c) Angiogenesis

22.1.3 Investigation

- *CT*
  - Nearly 70% cases are solid cystic.
  - Appears iso- or hypodense mass.
  - Calcification in nearly one-third cases.
  - Post contrast enhancement in 50% cases.
- *MRI*
  - *T1: I*so- to hypointense
  - *T1 Post contrast: E*nhancement 40% cases
  - *T2: H*yperintense
  - T2 FLAIR: Peritumoral edema uncommon
  - DWI: Diffusion restriction 20%

22.1.4 Pathology

- Biphasic morphologic patterns.
- Proliferation of neuronal and glial cells (primarily astrocytic).
- The proportion of the glial cell determines the clinical behavior.
- Often progresses to higher grade tumors. When astrocytic component dedifferentiates it leads to GBM, and neuronal component leads to neuroblastoma.
- Nearly 25–50% cases show BRAF V600E mutations.
- AGG prognosis does not differ based on BRAF mutation status.
- These tumors are IDH negative.
- *Immunohistochemistry*
  - Synaptophysin: Positive
  - Neurofilament protein positive
  - MAP 2: positive
  - Chromogranin A: Positive
  - CD 34: Positive in 70–80%

22.1.5 Treatment

22.1.5.1 Surgical Excision

- Surgical excision standard of care and a maximal safe resection are aim of surgery.

- Complete seizure relief has been reported in 80% cases and > 75% relief in remaining cases.
- Nearly 60% of GG and 30% patients of AGG had undergone GTR in the literature.
- Patients with a GTR have long-term survival [4, 5].

	GTR (months)	STR (months)
PFS	42	12
OS	132	22

22.1.5.2 Radiation

- Data on adjuvant radiation is limited.
- In AGG radiation may be more effective as these patients experience early failure.

Indications for Adjuvant RT

- GG: Not required.
- AGG: adjuvant radiation may be considered upfront even after GTR.

Target Volume

- In adjuvant or salvage setting:
  - Local radiation only (postoperative cavity plus any residual enhancing mass with 1 cm isotropic expansion as CTV and 3–5 mm additional as PTV)

Dose

Local radiation: 54–60 Gy in conventional fractionation.

22.1.5.3 Chemotherapy

- Role of conventional chemotherapy limited.
- BRAF inhibitor promising.
- Resistance develops due to reactivation of the MAPK pathway through downstream molecules like the MEK1/2.
- Combination of BRAF and MEK1/2 inhibitors are being evaluated.

22.1.6 Results

Median OS of 29 months; PFS of 24 months.

	PFS (5 year) (%)	OS (5 year) (%)
GG	78	96
AGG	30	40

22.1.6.1 Prognostic Factors (Favorable for AGG)

- Pediatric patients
- GTR
- Cerebral lesion
- Single lobe involvement

22.1.7 Follow-up

- Follow-up should be done with clinical examination and CEMRI of brain every 3 months for first 3 years and thereafter every 6 months.
- CSF dissemination has been rarely reported.
- Extracranial spread is extremely rare.

22.1.8 Treatment Algorithm [1]

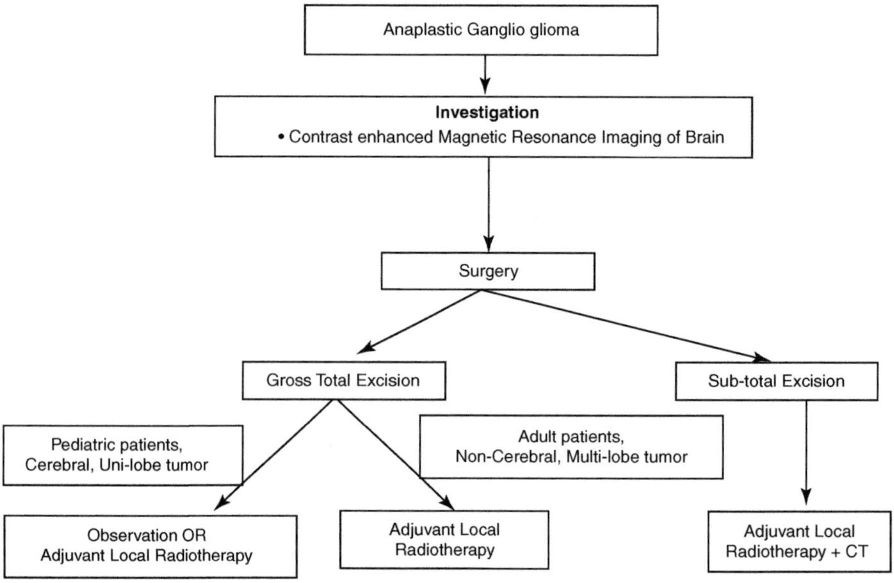

22.2 Plasmacytoma

22.2.1 Introduction

- Extremely rare.
- Only few cases have been published.
- Affects the skull, meninges.

- Median age nearly 50 years.
- Presentation: headaches, cranial nerve palsies, nausea, vomiting.

22.2.2 Investigation

Every effort should be made to rule out systemic myeloma.

The following investigation must be ordered:

- Bone marrow evaluation
- Serum protein electrophoresis
- Quantitative immunoglobulin analysis
- Urine protein electrophoresis
- Skeletal survey or while body PET CT
- *MRI*
  - *T1: I*so- to hyperintense
  - *T1 Post contrast: E*nhancement
  - *T2: Isointense*
- *Differential: Meningioma (one-third cases shoe dural tail sign) and metastasis*

22.2.3 Treatment

22.2.3.1 Surgical Excision

- Surgical excision standard of care and a maximal safe resection are aim of surgery
- Complete symptom relief possible with surgery

22.2.3.2 Radiation

- Data on adjuvant radiation is limited.

Indications for Adjuvant RT

- Adjuvant radiation may be considered upfront even after GTR.

Target Volume

- In adjuvant or salvage setting:
  - Local radiation only (Postoperative cavity plus any residual enhancing mass with 2 cm isotropic expansion as CTV and 3–5 mm additional as PTV)

Dose

Local radiation: 45–54 Gy in conventional fractionation

22.2.4 Chemotherapy

- Role of conventional chemotherapy limited

22.2.5 Follow-up

- Myeloma conversion happens.
- Follow-up should be done with clinical examination and CEMRI of brain every 3 months for first 3 years and thereafter every 6 months.

22.3 Rosette Forming Glioneuronal Tumors

Rosette forming glioneural tumors (RGNT) are a rare type of low-grade brain tumor. Because of benign nature of the disease, a complete surgical excision is optimum. Few aggressive cases have been reported and radiation has been applied.

- RGNT are rare tumors affecting young adults.
- Initially grouped as dysembryoblastic neuro-epithelial tumors (DNET) but were identified as a separate entity in 2007.
- Usually arise from the fourth ventricle but also from pineal region, midbrain, and pons.
- Most common symptom: headache, cerebellar ataxia.
- Imaging:
  - Hypointense on T1 and hyperintense on T2.
  - No post contrast enhancement seen.
  - Commonly shows a mass with cystic changes.
  - Micro-calcifications are commonly seen.
- Generally considered benign.
- Surgical excision has been the primary modality of treatment in RGNT.
- Good control rates.
- Immuno-positive for synaptophysin while negative for IDH 1, GFAP, P53, and NeuN.
- Radiation therapy alone has been used only in isolated cases.

22.4 Embryonal Tumors with Multilayered Rosettes (ETMR)

22.4.1 Introduction

- Eberhart et al. in the year 2000 first time described ETMR [6].
- Rare small round blue cell tumor of the central nervous system.
- One of the most aggressive brain tumors, WHO grade IV.
- Commonly affected children aged two to four.
- No gender predilection.
- Most are supratentorial, rarely found in infratentorial and in the spinal cord.
- Frontal lobe is affected most common.
- Presenting complaint: Raised intracranial pressure, seizures, neurological sign.
- Nearly 10–15% patients present with CSF dissemination.

22.4.2 Pathology [7, 8]

- Biphasic tumor
- Presence of multilayered true rosettes surrounded by primitive cells [Must for diagnosis]
- Characterized by undifferentiated neuroepithelial cells
- More than 90% shows amplification at 19q13.42
- High expression of LIN28A, an RNA binding protein
- IHC
- Vimentin: Positive
- Synaptophysin: Positive [neuropil]
- MIB-1 labeling index: very high

22.4.3 Radiographic Features

- CT
  - Large tumors.
  - Calcifications were present in 67%.
  - Contrast enhancement: weak and heterogeneous.
- MRI [6]
  - T1: hypointense
  - T1 post contrast: enhancement
  - T2: hyperintense
  - T2 FLAIR: hyperintense
  - DWI: Diffusion restricted

22.4.4 Treatment

Tri-modality approach has been used in nearly 30% cases.

22.4.4.1 Surgery

- Maximal safe surgical resection should be the aim.
- A GTR is achievable in most of the cases.

22.4.4.2 Radiation

- Data on adjuvant radiation is limited.
- All patients who are more than 2 years age should receive radiation.
- Entire cranio-spinal axis should be irradiated.
- Dose: CSI: 36 Gy and Boost to primary up to 56 Gy.
- But because of age criteria nearly 30–40% patients have received radiation.
- More than 90% of long-term survivors has received radiation (CSI).

22.4.4.3 Chemotherapy

- Chemotherapy has been used in nearly 75% cases.
- Combination Vincristine cyclophosphamide etoposide cisplatin.

22.4.5 Survival Outcome [7, 8]

- Prognosis is dismal.
- Median OS 18 months.
- For long-term survivor's median OS of 77.25 months.
- 1-year EFS 16–36% and 1-year OS 14–45%.

22.4.6 Follow-up

- Follow-up should be done with clinical examination and CEMRI of brain and spine every 3 months for first 3 years and thereafter every 6 months.
- Pituitary hormonal levels should be done every months and necessary medical management may be initiated.

22.5 Supratentorial PNET: WHO 2016 Has Dropped This Terminology

22.5.1 Introduction

- *Supratentorial PNET* was first described in 1973 by Hart et al.
- Rare poorly differentiated intracranial embryonal tumor.
- Of all pediatric brain tumors it accounts for 2.5%.
- Most commonly affect children and young adults; median age 11 years.
- There is slight male predilection.
- Symptoms depend on location; nearly 40% patients present with headache, vomiting.
- These tumors commonly arise from the cerebral cortex.
- Mostly localized but 15–20% present with leptomeningeal dissemination.

22.5.2 Investigation

- *CT*
  - Solid cystic mass
  - Appears iso- or hypodense mass
  - Calcification in nearly 70% cases
  - Heterogeneous contrast enhancement in 50% cases
- *MRI*
  - *Solid cystic mass*

- *T1: I*so- to hypointense
- *T1 Post contrast: E*nhancement heterogeneous
- *T2: H*yperintense
- DWI: Diffusion restriction in solid part

22.5.3 Pathology

- Morphology:
  - Highly cellular, heterogeneous
  - Poorly differentiated tumors comprising small round cells with high nucleo-cytoplasmic ratios
  - Variable amounts of glial, neuronal, and ependymal differentiation
- Highly proliferative tumor, MIB: median 35
- Loss of chromosome 14q is common
- Picard et al.

Group	Molecular Character	Survival
Group 1	Primitive neural	Females, younger, dismal prognosis
Group 2	Oligoneural	
Group 3	Mesenchymal	

- *Immunohistochemistry*
  - Synaptophysin: Positive
  - Neurofilament Protein positive
  - Neuron-specific enolase (NSE): Positive
  - Class III β-tubulin: Positive
  - GFAP, MIC 2-Negative

22.5.4 Treatment

22.5.4.1 Surgical Excision

- Few patients may require medium pressure ventriculo-peritoneal shunt.
- Surgical excision standard of care and a maximal safe resection are aim of surgery.
- Complete seizure relief has been reported in 40–50% cases and >75% relief in remaining cases.
- GTR not possible in: Large tumor size and extension to eloquent areas.
- Patients with a GTR have long-term survival.

22.5.4.2 Radiation

- Data on adjuvant radiation is limited.
- Entire cranio-spinal axis should be irradiated.
- Dose: CSI: 36 Gy and Boost to primary up to 56 Gy.

22.5.4.3 Chemotherapy

- Adjuvant chemotherapy with multidrug regimen is used.
- Combination of CCNU + VCR + prednisone q6wk × 8 has been used widely.
- Other regimen: 8 in 1 chemo with VCR, HU, BCNU, Procarbazine, CDDP, Ara C, prednisone, CTX.
- 6–12 cycles of an alternating regimen of VAC and IE is a reasonable approach.

22.5.5 Results

- Results variable
- 5-year PFS 30–60% and 5-year OS—nearly 15–55%

22.5.5.1 Prognostic Factors

1. Age (more than 3 years versus less than 3 years)
2. M stage (M0 versus M+ disease)
3. Extent of resection (complete versus incomplete resection)
4. CSI

22.5.6 Follow-up

- Follow-up should be done with clinical examination and CEMRI of brain and spine every 3 months for first 3 years and thereafter every 6 months.
- Pituitary hormonal level every 3 months.

22.6 ATRT

22.6.1 Introduction

- The tumor was described initially by Beckwith and Palmer in 1987.
- Rare small round blue cell tumor of the central nervous system.
- One of the most aggressive brain tumors, WHO grade IV.
- Commonly affected children aged less than 2 years.
- Most are infratentorial (50%) resembling medulloblastoma.
- Presenting complaint: raised intracranial pressure, vomiting.
- Incidence 1.38 per 100,000 person years [9].
- Nearly 15–30% patients present with CSF dissemination.

22.6.2 Pathology [10]

- Grade IV
- Presence of rhabdoid cell is the hallmark of AT/RT.

- Presence of neuroectodermal, epithelial, and mesenchymal elements.
- According to 2016 WHO classification, a diagnosis of AT/RT requires:
  - loss of INI1 tumor suppressor gene on chromosome 22
  - BRG1 gene
- IHC
  - Vimentin: Positive
  - EMA: Positive
  - SMA: Positive

22.6.3 Radiographic Features

- CT
  - Iso-hypodense mass
  - Hydrocephalus
  - Calcifications common
  - Contrast enhancement: heterogeneous
- MRI
  - T1: hypointense
  - T1 post contrast: enhancement
  - T2: hyperintense
  - DWI: Diffusion restricted

22.6.4 Treatment

22.6.4.1 Surgery

- MPVP shunt is required in all cases.
- Maximal safe surgical resection should be the aim.
- Surgical resection influenced by age, tumor location, and size.
- GTR is associated with better OS.

22.6.4.2 Radiation [11–13]

- Radiation is very crucial.
- As the tumor is seen in less than 3 years, a delicate balance is required to make between impact on neurodevelopment and tumor control.
- All patients who are more than 2 years age should receive radiation.
- Entire cranio-spinal axis should be irradiated.
- Dose: CSI: 36 Gy and Boost to primary up to 56 Gy.
- But because of age criteria nearly 30–40% patients have received radiation.
- Athale et al. showed significant benefit of addition of RT.

22.6.4.3 Chemotherapy

- Chemotherapy is crucial.
- Combination Vincristine cyclophosphamide etoposide cisplatin or vincristine, carboplatin, ifosfamide, and etoposide used.

- Overall response rate 42%.
- HDCT with autologous stem cell rescue has been used in younger children in order to defer radiotherapy.

22.6.5 Survival Outcome

- Prognosis is dismal.
- AT/RT typically recurs within 6 months.
- Cranio-spinal spread was found in 58% of patients at progression.

22.6.6 Follow-up

- Follow-up should be done with clinical examination and CEMRI of brain and spine every 3 months for first 3 years and thereafter every 6 months.
- Pituitary hormonal levels should be done every month and necessary medical management may be initiated.

22.7 Neurocytoma

22.7.1 Introduction [14]

- Rare benign neuronal tumor, accounting 0.1–0.5% of all brain tumors.
- Hassoun et al. in 1982 first described
- More common in young adults, in second and third decade of life.
- It arises from the neuronal cells of the septum pellucidum and the subependymal cells of the lateral ventricles.
- Two groups according to behavior
  - Typical neurocytomas, WHO grade II
  - Atypical neurocytomas; MIB-1 index >2%, necrosis, mitotic activity, and vascular proliferation
- Two groups according to location
  - Central
    Lateral ventricles around foramen of Monro: 50%
    Both lateral and third ventricles: 15%
    Bilateral: 15%
    Third ventricle in isolation: 5%
  - Extra-ventricular
    Cerebral parenchyma
    Thalamus
    Cerebellum
    Pons
    Pineal gland

Retina
Spinal cord
- Presentation
  - Insidious onset
  - Intra-ventricular tumors present with features of raised intracranial tension leading to obstructive hydrocephalus depending upon the proximity of the lesion to the foramen of Monroe
  - Common symptoms
    Headaches
    Seizures
    Nausea-vomiting
    Visual impairment
    Gait abnormality
    Memory disturbances

22.7.2 Pathology

- Morphology mimics oligodendrogliomas and ependymomas.
- Tumor cells are small and round with scant cytoplasm and stippled chromatin.
- A honeycomb pattern is created.
- Neuronal differentiation hallmark.
- Immunohistochemistry:
  - Synaptophysin: positive [In fibrillary and perivascular areas].
  - NeuN: Positive.
  - Neuron-specific enolase: Negative.
  - Vimentin: Negative.
  - EMA-Negative.
  - GFAP and IDH-1 R132H are negative.
  - IDH mutations and 1p19q co-deletion are absent.
- MIB labeling index: 10-year survival 90% vs 63% for MIB <2% vs >2%.
- Overexpression of N-Myc and under-expression of BIN-1 responsible for tumorigenesis.

22.7.3 Radiographic Features

- CT
  - Solid cystic tumors
  - Appears as dense mass
  - Calcifications in nearly 50% cases
  - Contrast enhancement: weak and heterogeneous
- MRI
  - T1: Hypointense to isointense
  - T1 post contrast: mild-moderate heterogeneous enhancement

- T2: Isointense to hyperintense
- T2 FLAIR: flow voids
- DWI: Diffusion restricted in solid component

22.7.4 Treatment

22.7.4.1 Surgery

- Mostly parenchyma is not involved.
- Maximal safe surgical resection should be the aim.
- Microsurgical complete resection is possible.
- Surgical route: transcortical-transventricular and interhemispheric-transcallosal-transventricular routes.
- A GTR is achievable in 30–50% of the cases.
- 5-year survival nearly 99% for GTR and 86% for STR.

22.7.4.2 Radiation [15–17]

- Indication in CN
  - Atypical Neurocytoma
  - Recurrent disease
  - STR [5-year PFS 67% vs53% favoring adjuvant radiation after STR]
  - Inoperable tumor
- Volume: Local radiation (CSI if there is CSF seeding)
- Dose: 50.4–56 Gy

22.7.4.3 Chemotherapy

- Limited data available.
- Used as salvage therapy.
- Drugs most used are carmustine, lomustine, prednisone, vincristine, and cisplatin.

22.7.5 Survival Outcome

22.7.5.1 Central Neurocytoma

	5-year Local Control (%)		5-year OS (%)	
Modality of treatment in CN	Typical	Atypical	Typical	Atypical
GTR	85	57	99	93
GTR + RT	89	53	95	90
STR	46	7	86	43
STR + RT	83	70	90	78

For extra-ventricular neurocytoma median progression free survival (PFS) 77 months (53.3–100.7) for the entire cohort. Median PFS for pediatric EVNs 180 months.

22.7.6 Follow-up

- Follow-up should be done with clinical examination and CEMRI of brain every 3 months for first 3 years and thereafter every 6 months.

22.7.7 Treatment Algorithm [15–17]

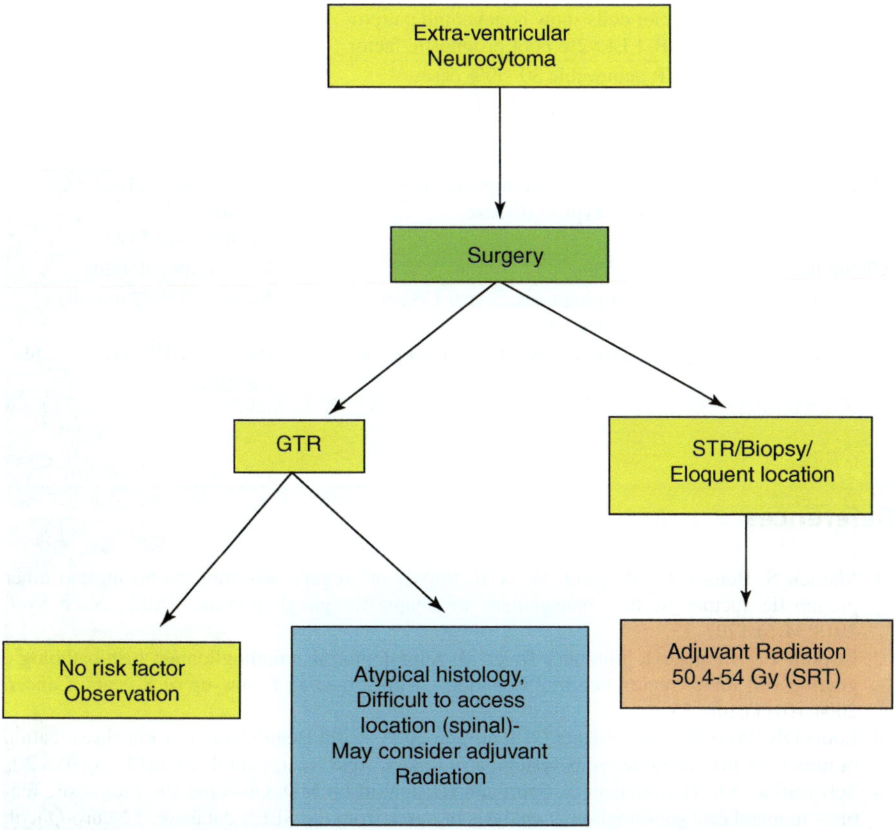

	CN	EVN
Incidence	0.1–0.5% of all brain neoplasms	15% of all neurocytoma
First described	1982	1989
Age group	Young adults Third decade	Median 30 years Range: 0.6–78 years Bimodal peak, second and fifth decade
Gender	Male predilection 1.2: 1	Male predilection 1.1: 1

	CN	EVN
Most common location	Lateral ventricle anterior horn Third ventricle	Frontal: 56% Temporal:17.5% Thoracic spine:8.5%
Presenting compliant	Raised ICT secondary to obstructive hydrocephalus	Seizure:32.3% Raised ICT:23.8%
Imaging	Mass attached to septum pellucidum Calcification in 50% cases	Calcification in 84% Heterogeneous enhancement 69.2%
Pathology	Tumor cells show honeycomb pattern MIB-1 LI >2% poor prognostic factor	Median MIB:5 75% grade II
Surgery	GTR achievable 30–50% cases 5-year OS 99% after GTR Risk of recurrence 3.8 times higher with STR than GTR	GTR:48.4% STR + RT:20.5%
Radiation	Adjuvant radiation improves outcome after STR and atypical disease	40% cases received radiation Median dose:54 Gy
Chemotherapy	Rarely used	Rare, Temozolomide
Survival	Excellent prognosis, 5-year OS >90% after GTR or STR + RT	Median PFS 77 months
Prognostic factors	GTR, MIB <4% has favorable prognosis	Pediatric EVN has favorable survival

References

1. Mallick S, Benson R, Melgandi W, et al. Impact of surgery, adjuvant treatment, and other prognostic factors in the management of anaplastic ganglioglioma. Childs Nerv Syst. 2018;34(6):1207–13.
2. Luyken C, Blümcke I, Fimmers R, et al. Supratentorial gangliogliomas: histopathologic grading and tumor recurrence in 184 patients with a median follow-up of 8 years. Cancer. 2004;101(1):146–55.
3. Louis DN, Perry A, Reifenberger G, et al. The 2016 World Health Organization classification of tumors of the central nervous system: a summary. Acta Neuropathol. 2016;131(6):803–20.
4. Selvanathan SK, Hammouche S, Salminen HJ, Jenkinson MD. Outcome and prognostic features in anaplastic ganglioglioma: analysis of cases from the SEER database. J Neuro-Oncol. 2011;105(3):539–45.
5. Terrier LM, Bauchet L, Rigau V, Amelot A, Zouaoui S, Filipiak I, et al. Natural course and prognosis of anaplastic gangliogliomas: a multicenter retrospective study of 43 cases from the French Brain Tumor Database. Neuro Oncol. 2017;19(5):678–88.
6. Dangouloff-Ros V, Tauziède-Espariat A, Roux C-J, et al. CT and multimodal MR imaging features of embryonal tumors with multilayered rosettes in children. AJNR Am J Neuroradiol. 2019;40(4):732–6.
7. Pei Y-C, Huang G-H, Yao X-H, et al. Embryonal tumor with multilayered rosettes, C19MC-altered (ETMR): a newly defined pediatric brain tumor. Int J Clin Exp Pathol. 2019;12(8):3156–63.
8. Tariq MU, Ahmad Z, Minhas MK, et al. Embryonal tumor with multilayered rosettes, C19MC-altered: report of an extremely rare malignant pediatric central nervous system neoplasm. SAGE Open Med Case Rep. 2017;5:2050313X17745208.

9. Woehrer A, Slavc I, Waldhoer T, et al. Incidence of atypical teratoid/rhabdoid tumors in children. Cancer. 2010;116(24):5725–32.
10. Dufour C, Beaugrand A, Le Deley M, et al. Clinicopathologic prognostic factors in childhood atypical teratoid and rhabdoid tumor of the central nervous system. Cancer. 2011;118(15):3812–21.
11. Buscariollo DL, Park HS, Roberts KB, Yu JB. Survival outcomes in atypical teratoid rhabdoid tumor for patients undergoing radiotherapy in a surveillance, epidemiology, and end results analysis. Cancer. 2011;118(17):4212–9.
12. Frühwald MC, Biegel JA, Bourdeaut F, et al. Atypical teratoid/rhabdoid tumors-current concepts, advances in biology, and potential future therapies. Neuro-Oncology. 2016;18(6):764–78.
13. Athale UH, Duckworth J, Odame I, Barr R. Childhood atypical teratoid rhabdoid tumor of the central nervous system. J Pediatr Hematol Oncol. 2009;31(9):651–63.
14. Sharma MC, Deb P, Sharma S, et al. Neurocytoma: a comprehensive review. Neurosurg Rev. 2006;29(4):270–85.
15. Mallick S, Benson R, Rath GK. Patterns of care and survival outcomes in patients with an extraventricular neurocytoma: an individual patient data analysis of 201 cases. Neurol India. 2018;66(2):362–7.
16. Mallick S, Roy S, Das S, et al. Role of adjuvant radiation in the management of central neurocytoma: experience from a tertiary cancer care center of India. Indian J Cancer. 2015;52:590–7.
17. Rades D, Schild SE. Value of postoperative stereotactic radiosurgery and conventional radiotherapy for incompletely resected typical neurocytomas. Cancer. 2006;106:1140–3.

Medulloblastoma and Other Embryonal Brain Tumors

23

Subhas Pandit and Simit Sapkota

23.1 Epidemiology

- Predominantly pediatric tumors with bimodal distribution, peaking at 3–4 years and again between 8 and 10 years.
- Overall, 70% of patients are diagnosed before the age of 10 years 20% in younger than 2 years and only 20% of patients are above 15 years.
- Male preponderance.
- Small numbers have inherited disorder with germline mutation. For example, Gorlin syndrome, Turcot syndrome, Li–Fraumeni syndrome.
- Advances in understanding of molecular biology of this molecular biology in the last decade have already changed our disease classification into a molecularly defined subgroups with distinct epidemiology, clinical course, and prognosis.
- Research is now focused on understanding aberrant pathways in these subgroups to develop appropriate treatment strategy and molecular therapy.

23.2 Pathology and Classification

The 2016 WHO classification of medulloblastoma reflects both traditional pathological system and recent understanding of molecular biology [1] (Table 23.1). Pathology and molecular biology of medulloblastoma has been discussed in a separate chapter.

S. Pandit (✉) · S. Sapkota
Kathmandu Cancer Center, Kathmandu, Nepal
e-mail: dr.subhas@kccrc.org

S. Mallick et al. (eds.), *Evidence based practice in Neuro-oncology*,
https://doi.org/10.1007/978-981-16-2659-3_23

Table 23.1 WHO classification of medulloblastoma

Medulloblastoma, histologically defined	Medulloblastoma, genetically defined
Medulloblastoma, classic	Medulloblastoma, WNT-activated
Medulloblastoma, desmoplastic/nodular	Medulloblastoma, SHH-activated, TP53 mutated
Medulloblastoma with extensive nodularity	Medulloblastoma, SHH-activated, TP53 wild-type
Medulloblastoma, large cell/anaplastic	Medulloblastoma, non-WNT/non-SHH Medulloblastoma, group 3 Medulloblastoma, group 4
Medulloblastoma NOS	

23.3 Clinical Presentation

- Increased intracranial pressure: Headache, especially early morning headache with vomiting is the most common sign, papilledema.
- Cerebellar signs due to mass effect in posterior fossa: ataxia, dysmetria, gait abnormalities.
- Infants: Non-specific symptoms like lethargy, irritability, feeding difficulty, enlarging head and bulging fontanelle, delayed developmental milestones, "Setting sun sign" with downward gaze.
- Cranial nerve palsy especially sixth nerve palsy. Infrequently facial weakness, hoarseness, and swallowing difficulty can occur due to lower cranial nerves palsy.
- Can present with acute alteration in consciousness due to intra-tumoral hemorrhage.
- Approximately 30–35% patients present with leptomeningeal metastasis. These are usually asymptomatic; however, some patients have neck and back pain.
- Extra-neural spread can be present in 1–5% patients. Most common extra-neural site is bone marrow and bone followed by peritoneum, lungs, and liver.

23.3.1 Workup

- Complete history and physical examination, neurological examination, fundus examination for papilledema
- Blood investigations: CBC, RFT, LFT
- Imaging studies: CT Head, MRI Brain and Spine
  - MRI spine if not done before surgery, should be done after 2 weeks of surgery.
  - Postoperative MRI brain is ideally done in 24–48 h postoperative.
- CSF Cytology
  - Avoided before surgery as ICP is high. Done 10–14 days after surgery to avoid false-positive.
- Endocrine evaluation, Audiometry, Neurocognitive evaluation

23.4 Imaging

Imaging characteristics are similar to common posterior fossa tumors like pilocytic astrocytoma, ependymoma, and AT/RT. MRI typically shows solid well-defined T1 hypointense and T2 iso- to hyperintense mass with heterogeneous contrast enhancement with restricted diffusion. Tumor edema is usually minimal; however, cystic and necrotic areas may be seen.

Prediction of molecular subtype based on imaging characteristics have been attempted to use MRI as surrogate for molecular subtype [2] (Table 23.2).

23.4.1 Staging and Risk Stratification

Modified Chang staging is currently used as only the "M" stage has prognostic value. M stage is based on subarachnoid metastasis, abnormal CSF cytology (M1), metastatic deposit in cranium (M2), spine (M3), and extra-neural metastasis (M4).

Traditionally, medulloblastoma has been divided into average-risk and high-risk groups based on age, M stage, extent of residual disease, and pathology.

	Average risk(all of following)	High risk (any of)
Age	>3 years	Less than 3 years
M stage	Mo in cranio-spinal MRI and CSF	M+
Residual tumor	Less than 1.5 cm^2	More than 1.5 cm^2
Histology	Non-neoplastic	Anaplastic histology

With the advent of molecular classification, new risk stratification was proposed based on molecular subtypes and key genetic abrasions. This allows evaluation of de-escalation strategy in low-risk groups and novel therapeutic agents in high-risk populations (Table 23.3).

23.4.2 Treatment Approach

Multimodal approach of maximal safe surgical resection, radiotherapy to craniospinal axis followed by boost and concurrent and adjuvant systemic chemotherapy are the current standard of care.

Table 23.2 Relation between imaging characteristics and molecular subtype

	Location	Enhancement	Metastasis
WNT	CP angle/CPA cistern		
SHH	Cerebellar hemisphere		Late/tumor bed
Group 3	Midline/fourth ventricle	Highly enhanced with ill-defined tumor margin.	Early/distant
Group 4	Midline/fourth ventricle	None/minimal	Early distant

Table 23.3 Prognostic classification for pediatric medulloblastomas according to molecular subgroups [3]

Risk group	5-year overall survival (%)	Molecular profile
Very high risk	<50	• SHH with TP53 mutation • Group 3: Metastatic and MYC amplified
High risk	50–75	• SHH: Localized with MYC amplification OR metastatic without TP53 mutation • Group 4: Metastatic
Standard risk	76–90	• SHH: Localized without TP53 mutation or MYCN amplification • Group 3 localized and without MYC amplification • Group 4 localized without loss of chromosome 11
Low risk	>90	• WNT group with age <16 years • Group 4 with loss of chromosome 11

Surgery Surgical resection is done after radiological diagnosis. Goal of surgery is to relieve pressure symptoms, provide local control, and obtain tissue for diagnostic evaluation. Gross total resection (GTR) has no or minimal survival benefit compared to near total resection in patients who receive adjuvant chemotherapy and radiotherapy. As the prognostic implication of the extent of resection is unclear in this era of molecular subtyping, maximal safe resection is favored over aggressive attempt for GTR with risk of increased neurological sequelae [4]. Routine ventriculoperitoneal shunting is not advised as it can lead to shunt dependency and complications. Medical decompression can be achieved using steroids, mannitol or furosemide. Dexamethasone is the preferred steroid, administered at a dose of 0.5–1 mg/kg IV loading dose, then 0.25–0.5 mg/kg/day IV or PO divided every 6 h. Surgery is performed as soon as possible. If CSF diversion is required before surgery, either an external ventricular drainage (EVD) or an endoscopic third ventriculostomy (ETV) is recommended.

Following radical resection, approximately 25% of patients experience postoperative cerebellar mutism, a symptom complex characterized by delayed onset mutism and irritability that may also be associated with truncal hypotonia, ataxia, and the inability to coordinate voluntary movements (posterior fossa syndrome).

Postoperative Treatment Is Stratified According to the Risk Groups Till date classical risk grouping is used; however, newer trials are allotting treatment according to biological group.

23.4.2.1 Average-Risk Medulloblastoma

Standard of care is maximal safe resection followed by reduced dose CSI of 23.4 Gy (in 1.8 Gy per fraction); and boost to tumor bed+1–1.5 cm margin upto 54–56 Gy along with concurrent weekly vincristine followed by 8 cycles of adjuvant chemotherapy (cisplatin, vincristine, CCNU or cyclophosphamide). With this strategy, 5- and 10-year event-free survivals were 81% and 75.8%; overall survivals were 87%

and 81.3% within the trial setting. Note also has to be made that with the above regimen, grade 3 or higher hematological toxicity occurred in >90% patients. So, cautious incorporation of this regimen is warranted in practice [5].

Attempts to reduce treatment intensity in this favorable subgroup have been done in various studies, some of which are discussed below:

1. Reduce radiotherapy volume:
   (a) French SFOP M4 trial: Avoided supratentorial volume. Only 18% EFS in 6 years. This dismal result showed that the entire neural axis should be irradiated [6].
2. Reduce CSI dose:
   (a) CCG 923/POG 8631: CSI with 23.4 Gy vs standard 36 Gy. Dose reduction in patients with average-risk disease yielded inferior event-free survival (EFS) when delivered without chemotherapy [7].
   (b) COG ACNS 0331 phase 3 study: CSI 23.4 Gy vs. reduced dose of 18 Gy in children 3–7 years age. Preliminary report suggests reduced dose is inferior to current standard of 23.4 Gy in terms of overall survival and leptomeningeal control.
3. Reduce boost volume.
   (a) St Jude MB 96, French M-SFOP 98 and MSFOP-2007 proved safety of reduced volume boost as relapse in posterior fossa were not increased. This strategy significantly reduces dose to temporal lobe, cochlea, and hypothalamus.
   (b) *COG ACNS 0031* randomized between focal boost vs posterior fossa: Survival rate following reduced volume boost to tumor bed with margin was comparable to treatment of the whole posterior fossa.
4. Altered fractionation: French M-SFOP 98 protocol and HIT-SIOP PNET 4 failed to show advantage of hyperfractionated radiotherapy (36 Gy CSI in 1.0 Gy fractions delivered twice daily) in terms of disease control or survival. Standard fractionation remains the current standard.

23.4.2.2 High-Risk Medulloblastoma

Apart from high-risk groups, incompletely staged patients and those unable to receive adjuvant chemotherapy should also be treated in this approach. High-risk patients are treated with higher dose CSI of 36 Gy with boost to the posterior fossa to 54–55.8 Gy with concurrent chemotherapy (Vincristine and/or carboplatin), followed by multi-agent chemotherapy. Metastatic deposits are boosted upto 45–50 Gy.

Current approach in this group is focused on intensification of chemotherapy with or without autologous stem cell support.

23.4.2.3 Medulloblastoma in Infants

- By definition, age <3 years are high risk and challenges are delay in diagnosis, increased surgical risk, under-treatment, and a potentially "more aggressive" biology.

- Radiotherapy is delayed or altogether avoided due to severe neurocognitive sequelae.
- Treatment strategy in recent trials have been to use intensive chemotherapy by using several approaches including systemic induction followed by high-dose chemotherapy with autologous stem cell support (Headstart, COG, SFOP), systemic chemotherapy with conformal local irradiation (UK, COG), and intraventricular methotrexate (HITSKK 92, HIT2000).
- SHH subgroup and nodular/desmoplastic histology have better outcome.
- Recognition and removal of poor prognostic groups like AT/RT which constituted 15–20% of infantile medulloblastoma has also apparently improved the outcome [8].

23.4.3 Radiotherapy Technique

- Traditionally cranio-spinal irradiation (CSI) is done using two lateral cranial fields and one or more direct posterior spinal fields with the patient in a prone position.
- Field border for spinal fields: Superior—C4–C7, Inferior 1–2 cm below termination of thecal sac (usually S2), laterally 1 cm lateral to vertebral body.
- Field border for cranial field: Superiorly/posteriorly flash; Inferior—Matching with spinal field, 1 cm margin to cribriform plate, middle cranial fossa, and vertebral body.
- Supine position is more stable and comfortable for patients especially when anesthesia is required. Advantage of a prone position is direct visualization and monitoring of junctions and skin-gaps; which has become less important in the era of image guidance.
- General anesthesia may be needed in young children during simulation and treatment.
- Whole body immobilization using a vacuum bag with a five-point head mask is ideal. CT scan is obtained from vertex to mid-thigh with proper immobilization and laser markings.
- Thin slice CT simulation is preferred with 1–3 mm thickness, especially in skull base area.
- MRI-CT fusion is recommended for better target and OAR delineation.
- Proton RT gives better dosimetric results and early clinical outcomes show it gives equivalent tumor control with low dose organs like lung, heart, face, abdomen, and pelvis [9].

23.4.4 Target Volumes in CSI Treatment

High-precision conformal radiotherapy is the current standard of care. Modern techniques like 3D-CRT, IMRT, VMAT, tomotherapy, or proton RT with image guidance using portal imaging or cone-beam CT may be used according to availability. However, principle of volume delineation remains the same irrespective of treatment technique.

GTV Any residual enhancing or non-enhancing tumor as well as wall of resection cavity.

CTV—CSI Includes entire subarachnoid space in cranium and spine.

Cranial CTV includes the whole brain within the inner table of skull. Care must be taken to contour cribriform fossa, space between orbits, inferior part of temporal lobes, pituitary fossa and ventricular cisterns. Switching between brain and bone window helps in contouring (Figs. 23.1, 23.2, and 23.3a, b). Attempts to spare the cochlea by excluding CSF within the internal auditory canal should be avoided.

Following skull base foramina are delineated to cover the dural sheath of cranial nerves.

- Superior orbital fissures
- Optic canals
- Foramen rotundum
- Foramen ovale
- Internal acoustic meatus
- Jugular foramen
- Hypoglossal canal

Spinal CTV should include spinal subarachnoid space encompassing spinal nerve roots laterally (Fig. 23.2). Sacral nerve roots need not be covered. Lower extent of the spinal field should be at the end of thecal sac best seen in the MRI. In the past lower border of S2 was taken. This has implication in gonadal dosage.

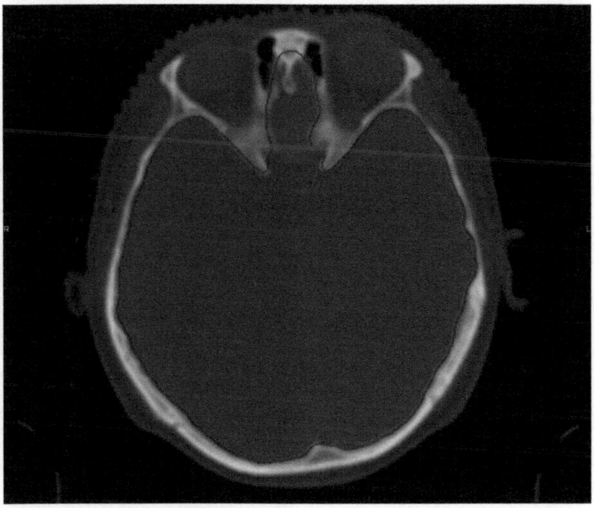

Fig. 23.1 Cribriform plates should be adequately covered

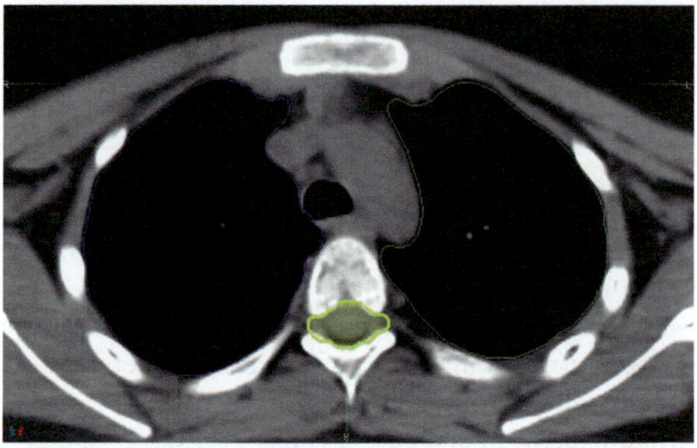

Fig. 23.2 Lateral extension to cover spinal nerve roots

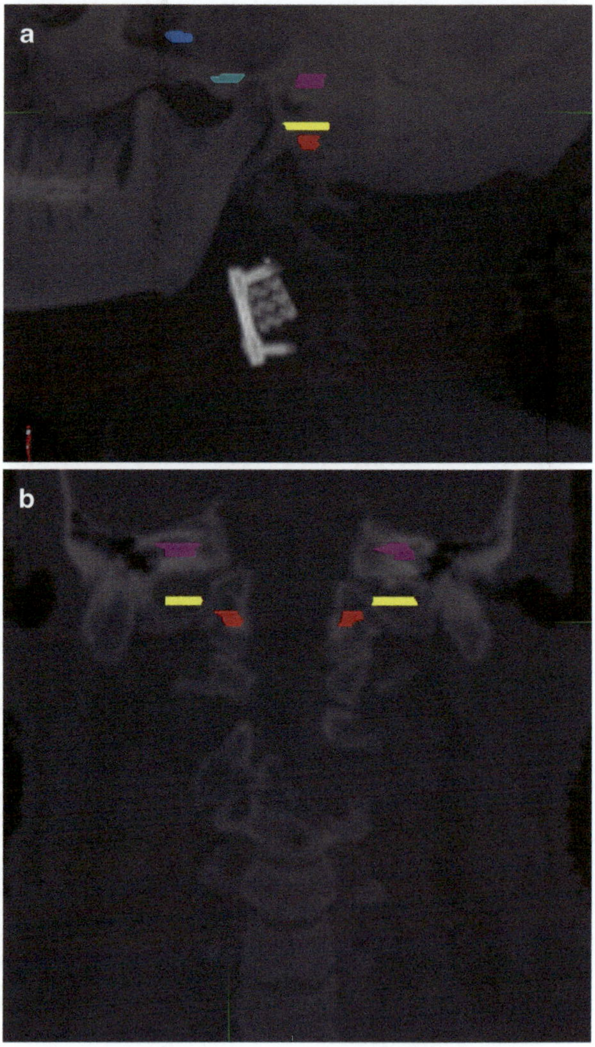

Fig. 23.3 (**a**, **b**) Sagittal and coronal DRR demonstration skull base foramina. (blue—foramen rotundum, yellow—jugular foramen, pink—internal acoustic meatus, green—foramen ovale)

Entire vertebral body should be uniformly irradiated in growing children to avoid asymmetrical spinal growth.

Boost CTV Traditionally, the entire posterior fossa was boosted; however, recent trials have shown that tumor bed boost is equally effective and less toxic, at least in the average-risk group.

In tumor bed boost technique, CTV is GTV + 1 to 1.5 cm margin. CTV is not extended beyond the tentorium or bones. For tumor that was "touching" brainstem in preoperative image, margin of 2–3 mm is taken inside the brainstem, otherwise brainstem is not included in boost CTV.

In posterior fossa boost technique, anterior border is taken at posterior clinoid and the entire brainstem is included.

PTV Geometrical expansion of 3–10 mm depending on reproducibility and departmental preference. With proper image guidance, tighter PTV of 3–5 mm for cranial field and a bit wider margin of 5–10 mm for spinal field can be done. Image guided techniques are valuable to reduce PTV margins.

CSI treatment volume is irregular and usually >40 cm in length. So, it can't be encompassed in a conventional field. Multiple fields are used to cover it, creating multiple field junctions. Avoiding field matching errors in these junctions is the real challenge in CSI.

1. "**Feathering**" refers to moving field junctions regularly during treatment. Field is moved cephalad or caudally around every 9 Gy by 1–2 cm to smoothen out areas of overdose or underdose in cord.
2. Collimator rotation and couch rotation technique to match divergence of cranial and spinal fields.
3. Extended SSD technique to cover entire spine in one field. Trade-off of extended SSD will be higher exit dose.
4. Using half beam blocks.
5. Entirely avoiding junctions by treating the whole neuraxis in a single setup using newer techniques like IG-IMRT with helical tomotherapy or VMAT.

23.4.5 Monitoring During Radiotherapy

Blood Counts Neutropenia and thrombocytopenia are common. When necessary, G-CSF and platelet transfusion may be employed during CSI. Undue prolongation of overall treatment time is detrimental. If CSI can't be continued, boost treatment can be given rather than treatment gap.

Nausea/Vomiting More common in adolescent and adult. 5-HT antagonists are commonly used. Low dose steroids may be needed if refractory.

Vincristine Vincristine causes poly-neuropathy and is frequently observed in adult patients.

23.4.6 Chemotherapy

Medulloblastoma is a chemo-sensitive tumor and has shown to be sensitive to alkylating agents (cyclophosphamide, CCNU [lomustine]), platinum salts (cisplatin, carboplatin), etoposide, antimetabolite (methotrexate), and camptothecins (topotecan).

Chemotherapy may be used in following settings:

1. Average-risk group to reduce CSI dose.
2. High-risk group to improve outcome.
3. Infantile medulloblastoma.
4. As part of myeloablative regimen with autologous stem cell rescue.
5. In a recurrent setting.

Original "Packer regimen" consisted of weekly vincristine with radiotherapy, followed by a combination of eight 6 weekly cycles of cisplatin 75 mg/m^2, lomustine 75 mg/m^2, and vincristine 1·5 mg/m^2 .

COG A9961 study showed that results are equivalent if cyclophosphamide is used instead of lomustine. Recent COG trial used both regimes alternatively. Cisplatin ototoxicity is a major toxicity in these regimens and requires close monitoring by audiometry.

In infants, intensive regimens including carboplatin/cisplatin, etoposide, cyclophosphamide, vincristine with or without the use of systemic high-dose methotrexate and/or intrathecal methotrexate are used.

23.4.7 Side Effects and Long-Term Issues

As survival has improved, long-term side effects and quality of life are major issues.

- Neurocognitive effects: Serial decrease in intelligence coefficient, more in children younger than 8 years. Reduction of boost volume and use of proton therapy seem to preserve cognition [9].
- Neuroendocrine effects are seen in upto 80% of patients. Growth hormone deficiency and primary hypothyroidism are common. Radiation dose >42 Gy to hypothalamic-pituitary axis was associated with increased risk of TSH deficiency [10].
- Second malignancy: In a trial involving 379 patients, incidence of second malignancy was 4.2% in 10 years [5]. Use of radiotherapy and alkylating agents are implied.

- Skeletal problems: Growth retardation, skeletal deformity, e.g., kyphosis, scoliosis.
- Survivors are more likely to have academic underachievement, unemployment, not being married or having children, inability to live independently, and social isolation.
- Long-term follow-up to address issues regarding education, employment, and family life.

23.5 Ongoing Trials and Future Direction

- The use of histone deacetylases (HDAC) inhibitors, which disrupts chromatin remodeling, has been suggested in WNT MB patients as it can disrupt chromatin remodeling.
- Vismodegib, arsenic trioxide, bromodomain inhibitors, and aurora kinase inhibitors are being tested in pre-clinical and early phase clinical trials for SHH medulloblastoma.
- Bromodomain inhibitors, HDAC inhibitors, and PI3K inhibitors are potential therapeutic options being tested for Group 3 tumors.
- Newer trials are molecularly informed and are enrolling patients based on biology of tumors. Currently, two large studies, PNET5 in Europe and SJMB12 in America are enrolling patients based on their molecular biology.
- Both studies are evaluating a reduction in the intensity of therapy for average-risk WNT patients with a reduction of CSI to 18 Gy in PNET5 and 15 Gy in SJMB12.
- Moreover, SJMB12 is evaluating the use of Vismodegib (SHH pathway inhibitor) as maintenance therapy in SHH patients, and the addition of Pemetrexed and Gemcitabine for select Group 3 and 4 patients.

23.6 Other Embryonal Tumors

Traditionally non-cerebellar embryonal tumors were referred to as CNS-PNET. However, new genetic and epigenetic information has completely changed our understanding of these tumors and the term CNS-PNET itself was removed from the 2016 WHO classification of CNS tumors. Instead, the classification refers to the following diagnoses within the overarching embryonal group: Embryonal tumor with multilayered rosettes (ETMR) C19MC-altered, ETMR NOS, medulloepithelioma, CNS neuroblastoma, CNS ganglioneuroblastoma, atypical teratoid/rhabdoid tumors, CNS embryonal tumor NOS and CNS embryonal tumor with rhabdoid features. These heterogeneous groups of tumor share many features with medulloblastoma like small, blue cells in histological appearance, early age of onset, tendency of dissemination into neuronal-axis and they are treated similar to high-risk medulloblastoma.

References

1. Louis DN, Ohgaki H, Wiestler OD, Cavenee WK, editors. WHO classification of tumours of the central nervous system. 4th ed. Lyon: IARC; 2016.
2. Perreault S, Ramaswamy V, Achrol AS, et al. MRI surrogates for molecular subgroups of medulloblastoma. AJNR Am J Neuroradiol. 2014;35(7):1263–9.
3. Ramaswamy V, Remke M, Bouffet E, et al. Risk stratification of childhood medulloblastoma in the molecular era: the current consensus. Acta Neuropathol. 2016;131(6):821–31.
4. Thompson EM, Hielscher T, Bouffet E, et al. Prognostic value of medulloblastoma extent of resection after accounting for molecular subgroup: a retrospective integrated clinical and molecular analysis. Lancet Oncol. 2016;17(4):484–95.
5. Packer RJ, Zhou T, Holmes E, et al. Survival and secondary tumors in children with medulloblastoma receiving radiotherapy and adjuvant chemotherapy: results of Children's Oncology Group trial A9961. Neuro-Oncology. 2013;15(1):97–103.
6. Bouffet E, Bernard JL, Frappaz D, et al. M4 protocol for cerebellar medulloblastoma: supratentorial radiotherapy may not be avoided. Int J Radiat Oncol Biol Phys. 1992;24(1):79–85.
7. Thomas PR, Deutsch M, Kepner JL, et al. Low-stage medulloblastoma: final analysis of trial comparing standard-dose with reduced-dose neuraxis irradiation. J Clin Oncol. 2000;18(16):3004–11.
8. Rorke LB, Packer RJ, Biegel JA. Central nervous system atypical teratoid/rhabdoid tumors of infancy and childhood: definition of an entity. J Neurosurg. 1996;85(1):56–65.
9. Yock TI, Yeap BY, Ebb DH, et al. Long-term toxic effects of proton radiotherapy for paediatric medulloblastoma: a phase 2 single-arm study [published correction appears in Lancet Oncol. 2020 Mar;21(3):e132]. Lancet Oncol. 2016;17(3):287–98.
10. Laughton SJ, Merchant TE, Sklar CA, et al. Endocrine outcomes for children with embryonal brain tumors after risk-adapted craniospinal and conformal primary-site irradiation and high-dose chemotherapy with stem-cell rescue on the SJMB-96 trial. J Clin Oncol. 2008;26(7):1112–8.

Ependymoma

24

Prashanth Giridhar, Aruna Nambirajan, and Rony Benson

24.1 Introduction

Ependymomas are a heterogeneous group of glial neoplasms that morphologically and ultrastructurally recapitulate the ependymal cells lining the ventricles and central canal of the spinal cord. They account for ~2% of all brain tumors and 6–7% of gliomas [1]. They occur across all age groups; median age at diagnosis irrespective of location and subtype is 43 years [1]. They are more commonly diagnosed in males as compared to females (male: female ratio of 1.32:1). The chapter will describe in brief the workup and management of ependymoma.

The World Health Organization classification (revised fourth edition) classifies ependymal tumors [2] as follows:

- Subependymoma, WHO grade 1
- Myxopapillary ependymoma, WHO grade 1
- Ependymoma, WHO grade 2
- Anaplastic ependymoma, WHO grade 3
- Ependymoma, *RELA* fusion-positive

P. Giridhar (✉)
Department of Radiation Oncology, National Cancer Institute, AIIMS, Haryana, India

A. Nambirajan
Department of Pathology, AIIMS, New Delhi, India

R. Benson
Department of Medical Oncology, Mar Sleeva Medicity Palai, Cherpunkal, India

S. Mallick et al. (eds.), *Evidence based practice in Neuro-oncology*,
https://doi.org/10.1007/978-981-16-2659-3_24

24.2 Workup

History and physical examination
- MRI brain and spine with a gadolinium (Gd)-containing contrast agent
- CSF cytology
- Biopsy
- Complete blood count, liver and kidney function tests

24.3 Management of Ependymoma

Ependymomas are managed primarily with maximal safe resection followed by radiotherapy. There is no role of chemotherapy in adjuvant setting except in young children to help delay radiotherapy.

Histology	Surgical extent	Adjuvant therapy
Myxopapillary ependymoma (spine)	Gross total resection if feasible	Adjuvant RT to local site to 50.4 Gy
Supratentorial RELA fusion-positive ependymoma	Gross total resection if feasible	Adjuvant RT to local site to 59.4 Gy
Anaplastic ependymoma	Gross total resection if feasible	Adjuvant RT to local site to 59.4 Gy
Posterior fossa ependymoma with balanced 1q	Gross total resection if feasible	Adjuvant RT to local site to 54 Gy
Posterior fossa ependymoma with 1q gain	Gross total resection if feasible	CSI to 36 Gy followed by local boost till 54 Gy
Age <3 years	Gross total resection if feasible	Adjuvant chemotherapy f/b RT after 3 years
CSF positive or disseminated disease	Gross total resection if feasible	CSI to 36 Gy followed by local boost till 54 Gy in brain and 50.4 Gy in spine

24.4 Volumes to Be Included for Post-operative Radiotherapy

GTV	Tumor bed + residual disease
CTV	GTV + 1.0–1.5 cm margin
PTV	CTV+ 0.5 cm margin

24.5 Future Directions

The last decade has witnessed an improved understanding of the molecular underpinnings of the biological heterogeneity of ependymal tumors that extends beyond histological grade. The lack of objective grading criteria, marked inter-observer variability in grading, and the morphological heterogeneity of these tumors limit the usefulness of WHO grade in accurate prognostication. Treatment guidelines based on molecular subgroups are evolving and grade-based treatment decisions are likely to become obsolete in the near future. A global ependymoma consensus conference in 2015 recommended that in patients >12 months of age, maximal safe resection, and focal radiotherapy should be standard of care in PF-A posterior fossa ependymomas; for the PF-B posterior fossa ependymomas, radiotherapy may be withheld in view of their good prognosis [3] heralding the era of molecular subgroup-based treatment decisions in ependymoma management. There have been contradictory reports as well [4, 5] and the prognostic impact of molecular subgrouping needs to be validated in independent cohorts.

The Consortium to Inform Molecular and Practical Approaches to CNS Tumor Taxonomy – Not Official WHO update 7 [6] has recommended updating the classification of ependymal tumors as follows:

1. Subependymoma, WHO grade 1
2. Myxopapillary ependymoma, WHO grade 2
3. Supratentorial ependymoma, *C11orf95* fusion-positive
4. Supratentorial ependymoma, *YAP1* fusion-positive
5. Supratentorial ependymoma, WHO grade 2/3
6. Posterior fossa ependymoma, PFA subgroup
7. Posterior fossa ependymoma, PFB subgroup
8. Posterior fossa ependymoma, WHO grade 2/3
9. Spinal ependymoma, *MYCN* amplified
10. Spinal ependymoma, WHO grade 2/3

It is to be noted that 7 of the 10 proposed subtypes are defined (epi)genetically and do not have an assigned grade. The upcoming WHO classification of CNS tumors is likely to incorporate these recommendations. Routine identification of these molecular subtypes will enable inclusion in clinical trials, confirmation of the prognostic significance, and development of better therapies.

References

1. Ostrom QT, Gittleman H, Fulop J, et al. CBTRUS statistical report: primary brain and central nervous system tumors diagnosed in the United States in 2008–2012. Neuro-Oncology. 2015;17(Suppl 4):iv1–iv62.
2. Ellison DW, McLendon R, Wiestler OD, et al. Ependymoma. In: Louis DN, Ohgaki H, Wiestler OD, Cavenee WK, editors. WHO classification of tumours of the central nervous system, Revised. 4th ed. Lyon: IARC; 2016. p. 106–12.

3. Pajtler KW, Witt H, Sill M, et al. Molecular classification of ependymal tumors across all CNS compartments, histopathological grades, and age groups. Cancer Cell. 2015;27:728–43.
4. Merchant TE, Bendel AE, Sabin ND, et al. Conformal radiation therapy for pediatric ependymoma, chemotherapy for incompletely resected ependymoma, and observation for completely resected. Supratentorial Ependymoma J Clin Oncol. 2019;37(12):974–83.
5. Tanrıkulu B, Danyeli AE, Özek MM. Is H3K27me3 status really a strong prognostic indicator for pediatric posterior fossa ependymomas? A single surgeon, single center experience. Childs Nerv Syst. 2020;36:941–9.
6. Ellison DW, Aldape KD, Capper D, et al. cIMPACT-NOW update 7: advancing the molecular classification of ependymal tumors. Brain Pathol. 2020;30(5):863–6. https://doi.org/10.1111/bpa.12866.

Intracranial Germ Cell Tumour

25

Prashanth Giridhar and Rony Benson

25.1 Introduction

Intracranial germ cell tumours are rare. Incidence rates are higher in both males and females of Asian/Pacific Islander descent [1]. Germinomas were the most common histologic subtype in males overall and in adolescent females, while germinoma and teratoma were equally common in females aged 0–9 years [1]. They arise most commonly from the pineal gland [2]. Germinoma is primarily treated (and usually cured) by radiotherapy. Aim of surgery is diagnostic only—there is no need to debulk these tumours. Germinomas sometimes secrete Placental Alkaline Phosphatase (PLAP) but are usually non-secreting. On the other hand, non-germinomatous germ cell tumours (with the exception of mature teratoma) usually secrete tumour markers. Therefore if a germ cell tumour is suspected serum and CSF beta-HcG, AFP and PLAP should be obtained pre-operatively. If these are positive, no biopsy is required. If a biopsy or during a surgical procedure a smear result confirms a germinoma, then the operation should be stopped and no attempt at debulking should be made.

25.2 Histological Subtypes

1. Germinoma
2. Non-germinomatous germ cell tumour
   Subtype:
   (a) Embryonal carcinoma
   (b) Yolk sac tumour

P. Giridhar (✉)
Department of Radiation Oncology, National Cancer Institute, AIIMS, Haryana, India

R. Benson
Department of Medical Oncology, Marsleeva Medicity Palai, Cherpunkal, India

S. Mallick et al. (eds.), *Evidence based practice in Neuro-oncology*,
https://doi.org/10.1007/978-981-16-2659-3_25

(c) Choriocarcinoma
(d) Teratoma
(e) Mixed germ cell tumour

25.3 Workup: (Both Germinoma and NGGCT)

Normal AFP (>2 years): <15 ng/mL
Normal AFP (Adults): <10 ng/mL
Normal HCG (Non-pregnant): <5 mU/mL

CBC/LFT/KFT

Serum AFP, Beta-HCG

CE-MRI Brain and spine

CSF cytology (Perform after MRI. 2 mL volume; To reach lab within 2 h; DO NOT REFRIGERATE)

CSF AFP, Beta-HCG

Tissue Biopsy (*If not feasible, rely on imaging and markers*)

Pituitary function tests (S. GH, S. TSH, S. ACTH, S. Prolactin, S. LH, S. FSH)

Ophthalmology review

Pulmonary function tests (If bleomycin given)

Thyroid function tests (S.T3, S.T4) (Based on volume of radiation)

2-D Echocardiography (Based on volume of radiation)

DTPA renal function scan (If suspicion of kidney disorder) (Based on volume of radiation)

Neurocognitive function assessment at baseline (Based on volume of radiation)

Rule out diabetes insipidus—Serum sodium, serum osmolality, and urine osmolality (*Endocrinology referral needed*)

Rule out gonadal dysfunction—Testosterone level in males; prolactin level in females

25.4 Management

Germinoma (CR after NACT; uni- or bifocal)	3 cycles BEP or 4 cycle EP f/b Whole ventricular RT
Germinoma (CR after NACT; multifocal)	3 cycles BEP or 4 cycle EP f/b CSI
Germinoma (PR after NACT; uni- or bifocal)	3 cycle BEP or 4 cycle EP f/b Whole ventricular RT + Boost
Germinoma (PR after NACT; multifocal)	3 cycles BEP or 4 cycle EP f/b CSI + Boost
Mature teratoma	Surgery

Non-germinomatous GCT (localized—CR after NACT)	3 cycles BEP or 4 cycle EP f/b Whole ventricular RT
Non-germinomatous GCT (localized—PR/SD after NACT)	3 cycle BEP or 4 cycle EP f/b Whole ventricular RT + Boost
Non-germinomatous GCT (CSF +; spinal mets+; multifocal)	3 cycles BEP or 4 cycle EP f/b CSI + Boost if PR/SD
Growing teratoma syndrome	On suspicion, attempt surgery

25.5 Radiotherapy Dose Prescription

Germinoma (CR after NACT; uni- or bifocal)	Whole ventricular RT including pre-chemo volume + 2 cm (23.4 Gy/13#/2.5 weeks)
Germinoma (CR after NACT; multifocal)	CSI (23.4 Gy/13#/2.5 weeks)
Germinoma (PR after NACT; uni- or bifocal)	Whole ventricular RT including pre-chemo volume + 2 cm (23.4 Gy/13#/2.5 weeks) + Boost (Till 45 Gy)
Germinoma (PR after NACT; multifocal)	CSI (23.4 Gy/13#/2.5 weeks) + Boost (Till 45 Gy)
Mature teratoma	
Non-germinomatous GCT (localized—CR after NACT)	Whole ventricular RT including pre-chemo volume +2 cm (36 Gy/20#/4 weeks) + Boost (Till 54 Gy)
Non-germinomatous GCT (localized—PR/SD after NACT)	Whole ventricular RT (36 Gy/20#/4 weeks) including pre-chemo volume + 2 cm + Boost (Till 59 Gy if feasible)
Non-germinomatous GCT (CSF +; spinal mets+; multifocal)	CSI (36 Gy/20#/4 weeks) including pre-chemo volume + 2 cm + Boost if PR/SD (Till 54 Gy)
Growing teratoma syndrome	

25.6 Volumes to Be Included in Craniospinal Irradiation

CSI volumes	Volumes	Comments
Brain CTV CSI	Entire brain with meninges up to foramen magnum Carefully include cribriform plate, optic nerves, temporal lobes	
Spine CTV CSI	Entire spinal canal with exiting nerve roots Lower end of thecal sac to be identified on MRI and contoured	Identify thecal sac ending on sagittal cut of T2 weighted MRI Seen as grey structure surrounded by white Including entire vertebral body in CTV for homogenous dose distribution is faculty call
Boost CTV	Pre-chemotherapy volume + 2 cm	Brainstem is a natural barrier; if involved then to be included
PTV	CTV + 5 mm	

25.7 Volumes to Be Included in Whole Ventricular Radiotherapy

WVRT volumes	Volumes	Comments
Whole ventricles	The WVV should encompass the lateral, third, and fourth ventricles Care should be taken to ensure the suprasellar, pineal and pre-pontine cisterns are included in the WVV	Planning CT should be fused to the most recent T2 MRI sequence for delineation of the whole ventricular volume (WVV) The third ventricle protrudes anteriorly to form the supraoptic and infundibular recesses (suprasellar cistern) The third ventricle protrudes posteriorly to form the suprapineal and pineal recesses (pineal cistern)
Boost CTV	Pre-chemotherapy volume + 2 cm	Brainstem is a natural barrier; if involved then to be included
PTV	CTV + 5 mm	

References

1. Poynter JN, Fonstad R, Tolar J, Spector LG, Ross JA. Incidence of intracranial germ cell tumors by race in the United States, 1992-2010. J Neuro-Oncol. 2014;120(2):381–8. https://doi.org/10.1007/s11060-014-1562-7.
2. Arora RS, Alston RD, Eden TO, Geraci M, Birch JM. Comparative incidence patterns and trends of gonadal and extragonadal germ cell tumors in England, 1979 to 2003. Cancer. 2012;118(17):4290–7.

Management of Pineal Region Tumors

26

Supriya Mallick

26.1 Introduction

Pineal region tumors are rare tumors that are deep-seated in the brain parenchyma and the incidence is 0.5–1% of the total intracranial neoplasm in adults [1] and is ten times more common in children with an incidence of 5–10% of all intracranial neoplasms in the pediatric population [2]. The tumors arise from diverse cell origins and most common types are the germ cell variants followed by pineal parenchymal tumors (PPTs) [3, 4]. These tumors arise from the pineal region and represent a heterogeneous group of lesions with the spectrum ranging from the benign to malignant potentials. The role of imaging and diagnostic molecular markers has simplified the diagnosis of the different types of pineal region tumors. Due to the rare occurrence of the disease, treatment options have not been clearly defined in the different types of tumors associated with the pineal region. Surgical excision is the mainstay in the majority of the subtypes followed by adjuvant radiation therapy or chemotherapy depending on the type of disease and the resection undertaken [5]. This chapter aims at providing a comprehensive review defining the various types of pineal region tumors, the investigations associated and the clinical presentations, and clinical management of various tumors associated in the region.

26.2 Pathological Classification of Pineal Region Tumors: [6]

1. *Germ cell Tumors* (50–75% of pineal region tumors)
   (a) Germinoma (65%)
   (b) Non-germinomatous GCT (17%)
      - Yolk sac/Endodermal sinus tumor

S. Mallick (✉)
Department of Radiation Oncology, National Cancer Institute, AIIMS, Haryana, India

S. Mallick et al. (eds.), *Evidence based practice in Neuro-oncology*,
https://doi.org/10.1007/978-981-16-2659-3_26

- Embryonal carcinoma
- Choriocarcinoma
- Mixed malignant GCT

(c) Teratoma (18%)
- Mature
- Immature
- Teratoma with malignant transformation

2. *Pineal Parenchymal tumors* (15–30%)
   (a) Pineocytoma (WHO Grade I)
   (b) Pineal parenchymal tumor of intermediate differentiation (WHO Grade II/III)
   (c) Pineoblastoma (WHO Grade IV)
   (d) Papillary tumor of the pineal region
3. *Uncommon:*
   (a) Astrocytoma
   (b) Ependymoma
   (c) Meningioma
   (d) Hemangioblastoma
   (e) Lymphoma
   (f) ATRT
   (g) Trilateral Retinoblastoma
   (h) Metastatic deposit

26.3 Epidemiology

- Accounts for less than 1% of all intracranial neoplasms.
- They are ten times more common in the pediatric population as compared to the adults.
- Germ cell tumors constitute close to 50% of all tumor types associated with the pineal region.
- Male predominance has been noted for germ cell tumors; However, PPT do not show gender predisposition and the median age of presentation is 12 years [7].

26.4 Clinical Presentation: [8, 9]

- Symptoms related to hydrocephalus are the most common presenting findings and include headache, nausea and vomiting.
- Other features are seen depending on the area of compression: visual field defects (sellar region), various endocrinopathies (suprasellar region), and Parinaud's syndrome.
- Parinaud's syndrome includes pseudo-Argyll-Robertson pupil, upward gaze palsy, convergence-retraction nystagmus, Collier's sign.

- Compressive hypothalamus syndrome including the classic triad: diabetes insipidus, precocious puberty, and visual symptoms.
- Drop metastases can present as radiculopathy or myelopathy.

Role of Markers Specific proteins are secreted by the NGGCT and immature teratomas and can be detected in serum or the cerebrospinal fluid (CSF) [10]. These markers, if proven to be positive for NGGCT, can help avoid surgery and also reflect not only the prognosis but also the treatment response and help monitor disease recurrence [11].

- Alpha-Fetoprotein (AFP): Positive in yolk sac tumors, embryonal carcinoma, and immature teratoma.
- β-hCG: Choriocarcinoma and some embryonal tumors; some GCT with syncytiotrophoblastic component (poor prognosis as compared to other GCTs).
- Placental Alkaline Phosphatase: Positive in germ cell tumors.
- AFP >10 ng/ml and/or β-hCG >50 mIU/ml is considered positive [5].
- AFP in the serum and CSF correlates however, the β-hCG does not and hence, if serum levels are not diagnostic, CSF sample needs to be taken.
- AFP levels >1000 ng/ml have a poorer prognosis [11].

26.5 Endoscopy

- In non-secreting pineal region tumors, tissue diagnosis is required for further management decisions and endoscopic guided biopsy can be taken.
- Endoscopic third ventriculostomy (ETV) can help in providing the biopsy sample and also provide therapeutic options in case of obstructive hydrocephalus, screening of the ventricular system for CSF dissemination and obtain CSF sampling for cytology and markers [12].

26.6 Imaging: [13]

- Computed tomography and MRI constitute the imaging modalities which aid the diagnosis of the pineal region tumors.
- Normal appearance of the pineal gland
  - On CT: soft-tissue attenuation with calcification, however, it is rare in children under 5 years of age.
  - On MRI: heterogeneous intermediate signal intensity due to the presence of calcification and cystic spaces.
- The pineal region is identified with these four important landmarks: internal cerebral veins, tectum of the midbrain, superior aspect of the cerebellar vermis, posterior aspect of the tentorial incisura.

- Germ cell tumors:
  - CT scan: Mature teratoma show fat density; however, it cannot differentiate between the other variants; germinomas appear more homogenous than NGGCTs; calcification is noted along with hyperdense lesion [14].
  - MRI: *T1*: isointense to grey matter; *T2*: isointense to grey matter; ***T1 contrast: may be homogenous or show vivid enhancement; DWI shows restriction in germinoma*** [15].
- Pineocytoma: They are well demarcated on CT scan and are iso- to hyperattenuating; on MRI, T1—hypo- to isointense and T2—hyperintense; On T1 contrast—homogenous enhancement; tumor size is usually less than 3 cm.
- PPTID: No specific difference in imaging findings from pineoblastoma and pineocytoma; high signal intensity on T2 and T1 contrast show enhancement.
- Pineoblastoma: On CT—Large ($\geq$ 3 cm) lobulated hyperattenuating mass with exploding calcifications at the periphery; almost all patients present with obstructed hydrocephalus; on MRI—T1 appears hypo- to isointense and T2 iso- to mildly hyperintense; T1 contrast shows heterogeneous enhancement.
- Entire neuraxial screening with MRI is important to detect disseminated metastasis in the spine.

26.7 Overview of the Management of Various Pineal Region Tumors

1. Pineoblastoma: [7, 16, 17]
   (a) Maximal safe resection
   (b) Adjuvant radiation therapy: CSI with a dose of 36 Gy with a local boost to a total dose of 54–55.8 Gy
   (c) If residual disease—Radiosurgery
   (d) Adjuvant chemotherapy
2. Pineocytoma: [18]
   (a) Maximal safe resection
   (b) If gross total resection—Observation
   (c) Incomplete resection—Adjuvant radiation therapy needs to be considered and dose as per the Mayo data is 50–55 Gy
3. Germinoma: [19, 20]
   (a) If localized disease, consider induction chemotherapy with carboplatin/etoposide
      - If complete response: 23.4 Gy whole ventricular irradiation
      - If a partial response or stable disease: Second-look surgery followed by evaluation for the role of radiation therapy (if residual disease)
      - If a partial response or stable disease with limited residual disease: Radiation therapy with 24 Gy whole ventricular irradiation followed by 12–21 Gy boost [21]
   (b) If metastatic disease: Radiation therapy with 24 Gy craniospinal irradiation followed by a 16–21 Gy boost

4. NGGT:
   (a) If localized disease, consider induction chemotherapy with carboplatin/etoposide or ifosfamide/etoposide)
      - Complete or partial response: Radiation therapy with 30.6 Gy whole ventricular irradiation and 23.4 Gy boost
      - If partial response, stable or progressive disease:
        If molecular markers normal: Consider second-look surgery
        If elevated: Surgery followed by chemotherapy or radiotherapy after multidisciplinary discussion
   (b) Metastatic disease:
      - Consider induction chemotherapy with carboplatin/etoposide or ifosfamide/etoposide.
      - Can be switched to thiotepa and etoposide or consider autologous stem cell transplant.
      - Radiation therapy: CSI to a dose of 36 Gy with 18 Gy local boost and 9 Gy boost can be delivered to the metastatic sites.
5. Mature Teratoma: Maximal safe resection followed by observation

26.8 Future Prospects

- In the pathogenesis of GGCTs, activation of the MAPK pathway by KIT or RAS mutation has led to the exploration of the aspect of molecular targeting agents, for example, imatinib, in the management of GGCTs with the KIT mutation. Further genetic studies will help establish the role of molecular oncology in the management of GGCTs in the future [22, 23].
- ACNS 1123: A phase II trial which is evaluating the response assessment with radiation in localized intracerebral germ cell tumors. The results will evaluate the survival outcomes and the toxicity profile associated with reduced radiation doses and the effects of the reduced fields.

26.9 Conclusion

Pineal region tumors are a heterogeneous group of tumors with the rare occurrence and varying prognosis and no well-established treatment options. The diagnosis can be made by imaging and serum and CSF markers which enable the treatment decisions. Advances in the techniques of radiation therapy delivery have led to improved outcomes and reduced toxicities associated. In this chapter, we presented the management options available for pineal-based tumors. The subsequent management depends upon the treatment response based on the disease etiology. Molecular Oncology and future trials in these groups of tumors would help establish evidence-based management which would help further improve the outcomes.

References

1. Cushing H. Intracranial tumours. Notes upon a series of two thousand verified cases with surgical-mortality percentages pertaining thereto. Springfield: Charles C Thomas; 1932. p. 150.
2. Fuller BG, Kapp DS, Cox R. Radiation therapy of pineal region tumors: 25 new cases and a review of 208 previously reported cases. Int Radiat Oncol Biol Phys. 1994;28:229–45.
3. Hirato J, Nakazato Y. Pathology of pineal region tumors. J Neuro-Oncol. 2001;54:239–49.
4. Louis DN, Ohgaki H, Wiestler OD, Cavenee WK, Burger PC, Jouvet A, Scheithauer BW, Kleihues P. The 2007 WHO classification of tumours of the central nervous system. Acta Neuropathol. 2007;114:97–109.
5. Fontana EJ, Garvin J, Feldstein N, Anderson RC. Pediatric considerations for pineal tumor management. Neurosurg Clin N Am. 2011;22:395–402. ix.
6. Louis DN, Ohgaki H, Wiestler OD, Cavenee WK WHO classification of tumours of the central nervous system. 4th ed Revised.
7. Mandera M, Marcol W, Kotulska K, et al. Childhood pineal parenchymal tumors: clinical and therapeutic aspects. Neurosurg Rev. 2010;34(2):191–6.
8. Sugiyama K, Uozumi T, Kiya K, et al. Intracranial germ-cell tumor with synchronous lesions in the pineal and suprasellar regions: report of six cases and review of the literature. Surg Neurol. 1992;38(2):114–20.
9. Nagasawa DT, Lagman C, Sun M, et al. Pineal germ cell tumors: two cases with review of histopathologies and biomarkers. J Clin Neurosci. 2017;38:23–31.
10. Qaddoumi I, Sane M, Li S, Kocak M, Pai-Panandiker A, Harreld J, Klimo P, Wright K, Broniscer A, Gajjar A. Diagnostic utility and correlation of tumor markers in the serum and cerebrospinal fluid of children with intracranial germ cell tumors. Childs Nerv Syst. 2012;28:1017–24.
11. Lee D, Suh YL. Histologically confirmed intracranial germ cell tumors; an analysis of 62 patients in a single institute. Virchows Arch. 2010;457:347–57.
12. Morgenstern PF, Souweidane MM. Pineal region tumors: simultaneous endoscopic third ventriculostomy and tumor biopsy. World Neurosurg. 2013;S18:e19.
13. Dumrongpisutikul N, Intrapiromkul J, Yousem DM. Distinguishing between germinomas and pineal cell tumors on MR imaging. AJNR Am J Neuroradiol. 2012;33(3):550–5.
14. Chang T, Teng MM, Guo WY, et al. CT of pineal tumors and intracranial germ-cell tumors. AJNR Am J Neuroradiol. 1989;10(5):1039–44.
15. Borja MJ, Plaza MJ, Altman N, et al. Conventional and advanced MRI features of pediatric intracranial tumors: supratentorial tumors. AJR Am J Roentgenol. 2013;200(5):W483–503.
16. Tate MC, Rutkowski MJ, Parsa AT. Contemporary management of pineoblastoma. Neurosurg Clin N Am. 2011;22:409–12. Ix.
17. Gilheeney SW, Saad A, Chi S, Turner C, Ullrich NJ, Goumnerova L, Scott RM, Marcus K, Lehman L, De Girolami U, Kieran MW. Outcome of pediatric pineoblastoma after surgery, radiation and chemotherapy. J Neuro-Oncol. 2008;89:89–95.
18. Clark AJ, Sughrue ME, Aranda D, Parsa AT. Contemporary management of pineocytoma. Neurosurg Clin N Am. 2011;22:403–7. Ix.
19. Khatua S, Dhall G, O'Neil S, Jubran R, Villablanca JG, Marachelian A, Nastia A, Lavey R, Olch AJ, Gonzalez I, Gilles F, Nelson M, Panigrahy A, McComb G, Krieger M, Fan J, Sposto R, Finlay JL. Treatment of primary CNS germinomatous germ cell tumors with chemotherapy prior to reduced dose whole ventricular and local boost irradiation. Pediatr Blood Cancer. 2010;55:42–6.
20. Rogers SJ, Mosleh-Shirazi MA, Saran FH. Radiotherapy of localised intracranial germinoma: time to sever historical ties? Lancet Oncol. 2005;6:509–19.
21. Dhall G, Khatua S, Finlay JL. Pineal region tumors in children. Curr Opin Neurol. 2010;23:576–82.

22. Fukushima S, Otsuka A, Suzuki T, Yanagisawa T, Mishima K, Mukasa A, Saito N, Kumabe T, Kanamori M, Tominaga T, Narita Y, Shibui S, Kato M, Shibata T, Matsutani M, Nishikawa R, Ichimura K. Mutually exclusive mutations of KIT and RAS are associated with KIT mRNA expression and chromosomal instability in primary intracranial pure germinomas. Acta Neuropathol. 2014;127:911–25.
23. Sakuma Y, Sakurai S, Oguni S, Satoh M, Hironaka M, Saito K. c-kit gene mutations in intracranial germinomas. Cancer Sci. 2004;95:716–20.

27 Skull Base Chordoma and Chondrosarcoma

Soumyajit Roy

27.1 Introduction

The skull base is the anatomic junction of the neural and facial viscero-cranium. It holds a clinically unique importance because it supports the brain and contains the neurovascular structures entering or exiting the skull [1]. Because of a conglomeration of a number of important neurovascular structures, the regional anatomy is highly complex [2]. Skull base malignancies may originate within the skull base or involve it by growth from either the intracranial dura or extracranial structures [3]. These malignancies belong to different histological subtypes. We would review skull base chordoma and chondrosarcoma here. Although different with respect to tissue of origin and immunohistochemical phenotypes, these two malignant entities are often clubbed together due to similarity in their clinical presentations and radiological findings [4–6].

27.2 Origin and Histology of Chordoma and Chondrosarcoma

Chordomas are rare and considered to be slow growing yet locally aggressive neoplasms of the axial skeleton; about one-third occurs in the clivus [7–9]. Notochordal rests are believed to be the origin for these malignancies. This potentially explains the similarity of chordomas with fetal notochord with respect to light and electron microscopic findings including immunohistochemical and ultrastructural features. Microscopic examination shows the tumors to be pseudo-encapsulated by fibrous strands forming thick hyalinized septa or thin septa

S. Roy (✉)
Radiation Oncology Branch, National Cancer Institute, Bethesda, MD, USA

Department of Radiation Oncology, Rush University Medical Center, Chicago, IL, USA

S. Mallick et al. (eds.), *Evidence based practice in Neuro-oncology*,
https://doi.org/10.1007/978-981-16-2659-3_27

creating lobules. These lobules appear as sheets of vacuolated physaliferous cells or pools of mucin. The sheets of cells contain varying amounts of intracytoplasmic mucin, ranging from hardly visible to quantities causing the cells to rupture. Immunohistochemistry is an important tool in the diagnosis of chordoma. Chordomas are positive for cytokeratin, EMA, S-100, and vimentin [4, 10]. The staining for cytokeratin is diffuse and strong while EMA staining is mostly focal and noted in about 80% of cases. Expression of brachyury, a key transcription factor in notochord development, appears to be a sensitive and specific biomarker of chordoma, and immunohistochemical staining for brachyury is routinely performed to help distinguish chordoma from cartilaginous tumors [11]. Other immunostains are variable depending upon the type of chordoma. In one study, all 16 classic chordomas stained for keratin, while only 32% of 25 chondroid chordomas did. In the same study, 44% of the classic and 85% of the chondroid chordomas were positive for S-100 protein [12, 13].

Chordomas are divided into three overlapping and sometimes coexisting histopathologic types: conventional (typical), chondroid, and dedifferentiated (atypical). Typical chordomas are the most common; chondroid chordomas appear to confer a better prognosis, although many pathologists believe this variant is a low-grade chondrosarcoma. The aggressive "dedifferentiated" variety accounts for only 5% of cases and portends poor outcome [14]. Most of the dedifferentiated chordomas are aneuploid on DNA analysis compared to only one-fourth of conventional chordomas.

Chondrosarcomas originate from cartilage. They mostly involve long bones and pelvis; only about 2% of all chondrosarcoma occur within the skull base. They account for 0.1–0.2% of all intracranial tumors [15]. Most skull base chondrosarcomas arise in the middle fossa, followed by the posterior or anterior fossae. Four different histological subtypes are recognized: conventional, mesenchymal, clear-cell, or dedifferentiated type. However, there is no consistent correlation with outcome. In contrast, prognosis is determined largely by grade, which is determined depending on mitotic rate, cellularity, nuclear size of the cells, and chondroid matrix [4]. Osteoid metaplasia may be present within a chondrosarcoma and is distinguished from the chondroblastic variant of osteosarcoma by the absence of osteoid that is lined by cytologically atypical osteoblasts [10]. In a number of large retrospective series, grade I and II chondrosarcomas comprised majority of those from skull base origin. Collectively, the preponderance of low-grade tumors in this location leads to its overall good prognosis [15–17]. Other rare types of chondrosarcoma that may occur at the skull base include high-grade mesenchymal chondrosarcoma and the so-called extra-skeletal myxoid chondrosarcoma.

In one retrospective study of 109 patients with chordoma, chondroid chordoma, and chondrosarcoma who were treated by a single surgeon with maximum resection followed by proton beam radiotherapy in majority of the cases, patients with chondrosarcoma had a significantly better recurrence-free survival compared to those with chordoma, whereas patients with chondroid chordoma had a poor recurrnce-free survival similar to chordoma patients [4].

27.3 Clinical Presentation

The clinical presentation of a skull base tumor depends in part upon the site of origin of the tumor and its direction of growth. Majority of the patients complain of headache and intermittent diplopia. Progressive worsening of headache is common with upper clivus tumors while neck pain develops with lower clivus tumors. Invasion of the cavernous sinus often results in diplopia and facial numbness. Lower clivus tumors initially compress the lower cranial nerves; brainstem compression may follow. The diagnosis of a clival chordoma is often delayed because of inconspicuous and non-specific symptoms.

The clinical features of chordoma and chondrosarcoma are mostly similar [18]. In one series of 48 patients with chordoma and 49 with low-grade chondrosarcoma, headache and diplopia from abducens nerve palsy was common in both groups [18, 19]. Normal neurologic examinations were more common in patients with chordoma, while visual loss, facial numbness, and multiple cranial neuropathies were more common in patients with chondrosarcoma. These differences probably reflect the tendency of chordomas to originate from the clivus and chondrosarcomas from the petro-clival fissure [18].

27.4 Imaging Findings

Most chondrosarcomas of the skull base occur in a para-median location, unlike chordomas, which tend to be more midline. On non-enhanced CT scans, these tumors appear as isodense to hyperdense soft tissue density masses. Evidence of erosion of adjacent bone or heterogeneous enhancement after contrast administration are common findings of chondrosarcoma. Ring or arc calcification is found in about half of the chondrosarcomas [15, 20]. On MRI, they show a low signal intensity on T1-weighted images with good heterogeneous enhancement. Some describe "whorls" or a "honeycomb" pattern of enhancement. Not all tumors however will show enhancement. The low-grade lesions can have a poor blood supply and will only enhance mildly. T2-weighted scans show high signal intensity of the lesion and this can be further delineated on T2-weighted fluid attenuated inversion recovery imaging [10].

The classic appearance of clival chordoma with high-resolution CT is that of a centrally located, well-circumscribed, expansile soft tissue mass that arises from the clivus with associated extensive lytic bone destruction. Intra-tumoral calcifications usually represent bone destruction rather than dystrophic calcifications in the tumor itself. However, the chondroid variant is more likely to demonstrate true intra-tumoral calcifications. There is moderate to marked enhancement following administration of iodinated contrast material [16]. On conventional spin-echo T1-weighted MR images, chordoma has intermediate to low signal intensity and is easily recognized within the high signal intensity of the fat of the clivus. On T2-weighted images, chordoma has high signal intensity, a finding that likely reflects the high fluid content. The majority of chordomas demonstrate moderate to marked enhancement following contrast material injection. The enhancement pattern of the tumor may have a "honeycomb" appearance [14]. Fat suppression is useful for identifying enhanced tumor margins. Even though multiple synchronous tumors or drop metastases from an intracranial chordoma are very rare, screening of the entire spine at the initial evaluation of a skull base chordoma patient should be considered [21].

Ecchordosis physaliphora (EP), a benign notochordal remnant, is described in approximately 2% of autopsy specimens and can mimic clival chordoma radiographically. EP usually appears as a small (<2 cm), well-circumscribed focus of T2 hyperintensity on the dorsal surface of the clivus in the midline. Unlike chordoma, EP is non-enhancing and rarely shows evidence of clival erosion on CT. A characteristic bony stalk or pedicle at the base of the mass is present in approximately two-thirds of cases and is a useful diagnostic clue [22].

27.5 Management

Although these tumors are slow growing, they are locally invasive tumors and there is a high incidence of local recurrence that can result in death due to uncontrolled local disease. For patients with chondrosarcoma, metastases occur in approximately 10% of cases. For patients with chordoma, dedifferentiation or sarcomatous transformation occurs in 2–8% of chordomas [10, 13, 23].

There exist no randomized controlled trials or large population-based or multicentric retrospective studies which could define the optimal management for skull base chordomas or chondrosarcomas. The treatment recommendations are generally based on evidence gleaned from small retrospective series and expert consensus [24]. Majority of such evidence support combined modality management with maximally safe surgical extirpation followed by local radiotherapy.

27.6 Surgery

Surgery offers two-pronged benefit in the management of these skull base tumors [25]. It reduces the tumor bulk and yields tissue for histological diagnosis. In 2014, Jones et al. reviewed 226 patients of skull base chondrosarcoma in the Surveillance, Epidemiology and End Results (SEER) database treated between 1973 and 2009. They showed that the 10-year survival for patients treated with surgery was significantly superior compared to those treated conservatively (69.3% versus 53.9%). Surgery predicted better overall survival by univariate analysis (hazard ratio [HR] 0.420, $p = 0.03$). This reaffirms the results of smaller case series [8].

Although the goal is to attain complete resection, it is often not feasible due to proximity with critical neural or vascular structures. Both open (transcranial and transfacial) and endoscopic endonasal surgical approaches are successfully used for these tumors [14]. There is evidence from observational studies that gross total resection of chordomas is better facilitated with a midline endoscopic endonasal approach [26].

27.7 Radiation Therapy

Adjuvant radiotherapy is widely accepted as a treatment option for skull base chordomas and chondrosarcomas. It has been shown in different series that addition of adjuvant radiotherapy improves both progression-free and overall survival. For skull base chordomas, adjuvant radiotherapy is indicated even after macroscopically complete surgery. For inoperable patients, definitive radiotherapy after biopsy is the treatment of choice [24].

Preoperative and postoperative CT scan and MRI (at least T2-weighted sequence) is essential for radiotherapy target delineation. Additionally, a baseline examination including cranial nerves, visual acuity, visual field assessment, audiometry and pituitary gland function, should be undertaken to assess baseline reserve and foresee potential radiotherapy-induced toxicities. The primary clinical target volume (CTV1) should encompass all volumes at risk for microscopic disease, including areas of preoperative tumor extension. An additional target volume (CTV2) receiving an additional boost of radiation should include any residual microscopic disease in the tumor bed after surgery. In patients with macroscopically incomplete resection, CTV3 should be delineated considering all gross imaging-detectable residual disease [24, 27]. Tumor seeding and recurrence along the surgical pathway for skull base chordoma is reported in less than 5% of cases. However, to include the entire surgical access path in CTV1 remains controversial [28]. The inclusion of the retropharyngeal space in the target volume definition as a possible site of recurrence remains controversial too. Because chordomas are radioresistant, a dose of at least 74 Gy should be delivered to CTV2 and CTV3, using conventional fractionation (1.8–2 Gy) [24, 25, 29]. Moderate hypofractionation with 16–22 fractions of 3–4.2 GyE per fraction to CTV2 and CTV3 (66–67.6 GyE total dose) and at least 36 GyE to CTV1 is feasible using carbon ions [30, 31]. Daily image guidance is strongly recommended. In a series of 42 patients with skull base chondroma ($n = 24$) and chondrosarcoma ($n = 18$), image-guided intensity modulated radiotherapy, the 5-year overall survival and local control rates were 85.6% and 65.3%, and 87.8% and 88.1%, respectively. In total, 10 patients progressed locally: 8 were chordoma patients and 2 chondrosarcoma patients (both high-grade). Median radiotherapy was 70 Gy and 76 Gy at 2 Gy/fraction for chondrosarcoma and chordoma, respectively [32].

Overall, surgery with adjuvant radiotherapy clearly has a survival benefit over surgery alone in chondrosarcoma. However, radiotherapy is associated with a risk of long-term complications including cognitive decline and risk of secondary malignancy. On these grounds, radiotherapy is reserved for grade I chondrosarcomas only at recurrence [15].

Stereotactic radiotherapy has been used in skull base chordomas and chondrosarcomas. A review of the literature identified five published series of SRS for chordoma [33, 34]. In total, 168 patients were treated with a median follow-up of 34–60 months. Progression-free survival at 5 years ranged from 21% to 72% . However, SRS is limited to tumors with a relatively small tumor volume [33]. Fractionated stereotactic radiotherapy has been practiced for relatively larger tumors. In a small series of 12 patients, median dose of 66.6 Gy (range, 48.6–68.4 Gy) was delivered in 180 cGy fractions prescribed to the 90% isodose line that covered the target volume to achieve a median isocenter dose of 74 Gy (range, 54–76 Gy). Overall survival was 76.4% at 5 years, and 46.9% and 37.5% of patients were free of progression at 24 and 60 months, respectively. Six patients had disease progression after radiation with a median time to progression of 17.3 months. One patient was salvaged with radiosurgery and surgical resection, with stable disease almost 7 years since diagnosis [34].

Charged-particle radiotherapy offers theoretical advantage of predictable dose deposition at a given depth, essentially allowing for dose escalation within the tumor and superior sparing of adjacent critical regions of interest. This technically results in superior therapeutic ratio compared to photon-based radiotherapy. However, in absence of any randomized controlled trial or other prospective evidence, comparison of photon to charged-particle therapy has been mostly limited to retrospective series [29]. Some series employed combined proton or photon therapy techniques owing to logistical issues with proton therapy availability [35]. Hybrid proton or photon therapy plans are not mandatory for robust proton therapy plan generation, but may be helpful in certain cases, such as improving prescription dose coverage uniformity and reducing range-out errors in cases of implanted hardware [29]. Ongoing randomized trials at Heidelberg University comparing proton therapy to carbon therapy in low and intermediate grade chondrosarcomas and photon to proton in chordomas would add further clarity to this ongoing controversy.

27.8 Role of Systemic Treatment

Historically, chondrosarcomas have been considered resistant to chemotherapy. The highest benefit being observed in mesenchymal and dedifferentiated chondrosarcoma. There have been isolated reports of successful treatment with ifosfamide plus doxorubicin or single-agent methotrexate [36]. Small series claim potential benefit of anti-angiogenic therapy with ramucirumab or multi-kinase inhibitors like pazopanib [37]. A variety of molecular targets may be therapeutically relevant in chordoma, including platelet-derived growth factor receptor (PDGFR), epidermal growth factor receptor (EGFR), vascular endothelial growth factor receptor (VEGFR), mammalian target of rapamycin (mTOR), and INI1 [38, 39]. In some cases, participation in clinical trials investigating immunotherapy and other novel approaches may be considered for the treatment of residual, recurrent, or metastatic disease [40].

Relevant Conflicts of Interest None.

Support or Funding None.

References

1. Patel CR, Fernandez-Miranda JC, Wang WH, Wang EW. Skull base anatomy. Otolaryngol Clin N Am. 2016;49(1):9–20. https://doi.org/10.1016/j.otc.2015.09.001.
2. Rhoton AL Jr. The anterior and middle cranial base. Neurosurgery. 2002;51:273–302. Available from: http://www.embase.com/search/results?subaction=viewrecord&from=export&id=L36324028.
3. Kunimatsu A, Kunimatsu N. Skull base tumors and tumor-like lesions: a pictorial review. Polish J Radiol. 2017;82:398–407.
4. Almefty K, Pravdenkova S, Colli BO, Al-Mefty O, Gokden M. Chordoma and chondrosarcoma: similar, but quite different, skull base tumors. Cancer. 2007;110(11):2457–67.
5. Heffelfinger MJ, Dahlin DC, MacCarty CS, Beabout JW. Chordomas and cartilaginous tumors at the skull base. Cancer. 1973;32(2):410–20. Available from: http://www.ncbi.nlm.nih.gov/pubmed/4722921.
6. Rosenberg AE, Brown GA, Bhan AK, Lee JM. Chondroid chordoma - a variant of chordoma: a morphologic and immunohistochemical study. Am J Clin Pathol. 1994;101(1):36–41.
7. Al-Mefty O, Colli BO, Angtuaco EJ, Hug EB. Chordomas and chondrosarcomas of the cranial base. In: Textbook of neuro-oncology. Elsevier; 2005. p. 357–65.
8. Jones PS, Aghi MK, Muzikansky A, Shih HA, Barker FG, Curry WT. Outcomes and patterns of care in adult skull base chordomas from the Surveillance, Epidemiology, and End Results (SEER) database. J Clin Neurosci. 2014;21(9):1490–6.
9. Chambers KJ, Lin DT, Meier J, Remenschneider A, Herr M, Gray ST. Incidence and survival patterns of cranial chordoma in the United States. Laryngoscope. 2014;124(5):1097–102.
10. Chordoma and chondrosarcoma of the skull base - UpToDate. [cited 2020 Feb 10]. Available from: https://www.uptodate.com/contents/chordoma-and-chondrosarcoma-of-the-skull-base/print.
11. Vujovic S, Henderson S, Presneau N, Odell E, Jacques TS, Tirabosco R, et al. Brachyury, a crucial regulator of notochordal development, is a novel biomarker for chordomas. J Pathol. 2006;209(2):157–65. Available from: http://www.ncbi.nlm.nih.gov/pubmed/16538613.
12. Mitchell A, Scheithauer BW, Krishnan Unni K, Forsyth PJ, Wold LE, McGivney DJ. Chordoma and chondroid neoplasms of the spheno-occiput: an immunohistochemical study of 41 cases with prognostic and nosologic implications. Cancer. 1993;72(10):2943–9.
13. Meis JM, Giraldo AA. Chordoma. An immunohistochemical study of 20 cases. Arch Pathol Lab Med. 1988;112(5):553–6. Available from: http://www.ncbi.nlm.nih.gov/pubmed/2451900.
14. Koutourousiou M, Snyderman CH, Fernandez-Miranda J, Gardner PA. Skull base chordomas, vol. 44. Otolaryngolgic Clinics of North America; 2011. p. 1155–71.
15. Awad M, Gogos AJ, Kaye AH. Skull base chondrosarcoma. J Clin Neurosci. 2016;24:1–5.
16. Weber AL, Brown EW, Hug EB, Liebsch NJ. Cartilaginous tumors and chordomas of the cranial base. Otolaryngol Clin N Am. 1995;28(3):453–71. Available from: http://www.ncbi.nlm.nih.gov/pubmed/7675464.
17. Gay E, Sekhar LN, Rubinstein E, Wright DC, Sen C, Janecka IP, et al. Chordomas and chondrosarcomas of the cranial base: results and follow-up of 60 patients. Neurosurgery. 1995;36(5):887–97.
18. Korten AGGC. Intracranial chondrosarcoma: review of the literature and report of 15 cases. J Neurol Neurosurg Psychiatry. 1998;65(1):88–92. Available from: http://www.ncbi.nlm.nih.gov/pubmed/9667567.
19. Volpe NJ, Liebsch NJ, Munzenrider JE, Lessell S. Neuro-ophthalmologic findings in chordoma and chondrosarcoma of the skull base. Am J Ophthalmol. 1993;115(1):97–104. Available from: http://www.ncbi.nlm.nih.gov/pubmed/8420385.
20. Murphey MD, Walker EA, Wilson AJ, Kransdorf MJ, Temple HT, Gannon FH. Imaging of primary chondrosarcoma: radiologic-pathologic correlation. Radiographics. 2003;23:1245–78.
21. McMaster ML, Goldstein AM, Bromley CM, Ishibe N, Parry DM. Chordoma: incidence and survival patterns in the United States, 1973–1995. Cancer Causes Control. 2001;12(1):1–11.

22. Park HH, Lee K-S, Ahn SJ, Suh SH, Hong C-K. Ecchordosis physaliphora: typical and atypical radiologic features. Neurosurg Rev. 2017;40(1):87–94. Available from: http://www.ncbi.nlm.nih.gov/pubmed/27206421.
23. Hruban RH, Traganos F, Reuter VE, Huvos AG. Chordomas with malignant spindle cell components. A DNA flow cytometric and immunohistochemical study with histogenetic implications. Am J Pathol. 1990;137(2):435–47. Available from: http://www.ncbi.nlm.nih.gov/pubmed/1696786.
24. Sommer J. Building a global consensus approach to chordoma: a position paper from the medical and patient community. Lancet Oncol. 2015;16(2):e71–83.
25. Fernandez-Miranda JC, Gardner PA, Snyderman CH, Devaney KO, Mendenhall WM, Suárez C, et al. Clival chordomas: a pathological, surgical, and radiotherapeutic review. Head Neck. 2014;36:892–906. Available from: http://www.ncbi.nlm.nih.gov/pubmed/23804541.
26. Koutourousiou M, Gardner PA, Tormenti MJ, Henry SL, Stefko ST, Kassam AB, et al. Endoscopic endonasal approach for resection of cranial base chordomas: outcomes and learning curve. Neurosurgery. 2012;71(3):614–24.
27. Noël G, Feuvret L, Ferrand R, Boisserie G, Mazeron JJ, Habrand JL. Radiotherapeutic factors in the management of cervical-basal chordomas and chondrosarcomas. Neurosurgery. 2004;55(6):1252–60.
28. Arnautović KI, Al-Mefty O. Surgical seeding of chordomas. J Neurosurg. 2001;95(5):798–803.
29. Ahmed SK, Brown PD, Foote RL. Protons vs photons for brain and skull base tumors. [cited 2020 Feb 11]; Available from: https://doi.org/10.1016/j.semradonc.2017.11.001.
30. Mizoe J-E, Hasegawa A, Takagi R, Bessho H, Onda T, Tsujii H. Carbon ion radiotherapy for skull base chordoma. Skull Base. 2009;19(3):219–24. Available from: http://www.ncbi.nlm.nih.gov/pubmed/19881902
31. Schulz-Ertner D, Karger CP, Feuerhake A, Nikoghosyan A, Combs SE, Jäkel O, et al. Effectiveness of carbon ion radiotherapy in the treatment of skull-base chordomas. Int J Radiat Oncol Biol Phys. 2007;68(2):449–57.
32. Sahgal A, Chan MW, Atenafu EG, Masson-Cote L, Bahl G, Yu E, et al. Image-guided, intensity-modulated radiation therapy (IG-IMRT) for skull base chordoma and chondrosarcoma: preliminary outcomes. Neuro-Oncology. 2015;17(6):889–94.
33. Kano H, Niranjan A, Dade Lunsford L. Radiosurgery for chordoma and chondrosarcoma. Prog Neurol Surg. 2019;34:207–14. Available from: http://www.ncbi.nlm.nih.gov/pubmed/31096236.
34. Bugoci DM, Girvigian MR, Chen JCT, Miller MM, Rahimian J. Photon-based fractionated stereotactic radiotherapy for postoperative treatment of skull base chordomas. Am J Clin Oncol. 2013;36(4):404–10. Available from: http://www.ncbi.nlm.nih.gov/pubmed/22772429.
35. Noël G, Habrand JL, Mammar H, Pontvert D, Haie-Méder C, Hasboun D, et al. Combination of photon and proton radiation therapy for chordomas and chondrosarcomas of the skull base: the Centre de Protonthérapie D'Orsay experience. Int J Radiat Oncol Biol Phys. 2001;51(2):392–8.
36. Italiano A, Mir O, Cioffi A, Palmerini E, Piperno-Neumann S, Perrin C, et al. Advanced chondrosarcomas: role of chemotherapy and survival. Ann Oncol. 2013;24:2916–22.
37. Jones RL, Katz D, Loggers ET, Davidson D, Rodler ET, Pollack SM. Clinical benefit of antiangiogenic therapy in advanced and metastatic chondrosarcoma. Med Oncol. 2032;34:167.
38. Colia V, Stacchiotti S. Medical treatment of advanced chordomas. Eur J Cancer. 2017;83:220–8. Available from: http://www.ncbi.nlm.nih.gov/pubmed/28750274.
39. Lebellec L, Chauffert B, Blay J-Y, Le Cesne A, Chevreau C, Bompas E, et al. Advanced chordoma treated by first-line molecular targeted therapies: outcomes and prognostic factors. A retrospective study of the French Sarcoma Group (GSF/GETO) and the Association des Neuro-Oncologues d'Expression Française (ANOCEF). Eur J Cancer. 2017;79:119–28. Available from: http://www.ncbi.nlm.nih.gov/pubmed/28478340.
40. Frezza AM, Botta L, Trama A, Dei Tos AP, Stacchiotti S. Chordoma: update on disease, epidemiology, biology and medical therapies. Curr Opin Oncol. 2019;31(2):114–20. Available from: http://www.ncbi.nlm.nih.gov/pubmed/30585858.

28 Intracranial Hemangiopericytoma

Dodul Mondal, V. R. Anjali, and Supriya Mallick

28.1 Introduction

- Intracranial hemangiopericytoma (I-HPC) rare tumour, accounts for <1% of intracranial tumours.
- These are mesenchymal tumours arising from the Zimmermann pericytes.
- Zimmermann pericytes are contractile cells that surround the endothelial lining of capillaries and post capillary venules.
- HPC was first described by Stout and Murray in 1942.
- Begg and Garret reported first primary I-HPC originating from meninges in 1954.
- I-HPC usually seen to arise from meninges of brain and spinal cord.
- Median age of presentation is 52 years [1].
- No gender predilection, but in some series slightly higher male preponderance is reported.
- Mostly supratentorial location, closely related to meninges.

28.2 Pathology

- Initially it was considered as a subtype of meningioma known as angioblastic meningioma.
  - In 2016 WHO classification HPC/solitary fibrous tumour (SFT) was put together as a single entity and adapted grading system. CNS tumours which

D. Mondal (✉)
Department of Radiation Oncology, Apollo Hospitals, New Delhi, India

V. R. Anjali
Department of Radiation Oncology, All India Institute of Medical Sciences, New Delhi, India

S. Mallick
Department of Radiation Oncology, National Cancer Institute, AIIMS, Haryana, India

S. Mallick et al. (eds.), *Evidence based practice in Neuro-oncology*,
https://doi.org/10.1007/978-981-16-2659-3_28

share inversion 12q13, fusing the NAB2 and STAT6 genes, resulting in STAT6 nuclear expression were grouped under SFT/HPC.
  - SFT/HPC was divided three grades.
    Grade I: Tumours highly collagenous, low cellularity and spindle cells [Previously: solitary fibrous tumour].
    Grade II: Typically, more cellular, less collagenous with plump cells and staghorn vasculature [Previously: hemangiopericytoma].
    Grade III: Most aggressive with 5 or more mitoses per 10 high power field and/or necrosis plus moderate to high nuclear atypia or cellularity corresponds to anaplastic HPC [2].
- Immunohistochemistry.
  - Dense intercellular reticulin deposition
    Positive for
      STAT 6-Specific, sensitivity of 90-98% [3, 4].
      Vimentin
      CD 34 (>90%).
      CD 99
    Negative
      EMA
      SMA
      S 100
- Nuclear expression of STAT6 is a specific marker for SFT/HPC. It has high sensitivity of 90–98% [3, 4]. Inversion at chromosome 12q13 produces NGFI-A binding protein 2 (NAB2)- signal transducer and activator of transcription 6 (STAT6) fusion, results in a chimeric protein in which repressor domain of NAB2 is replaced by carboxyl-terminal transactivation domain from STAT6. This fusion product induces EGR gene expression which promotes cell proliferation and plays important role in tumorigenesis. Most frequent fusion breakpoint seen in intracranial SFT/HPC is NAB2 exon6-STAT6 exon16/17/18 [5].

28.3 MGS Grading System for I-HPC

Predictors for aggressive behaviour

- Mitotic activity
- Nuclear atypia
- Cellular density
- Haemorrhage
- Necrosis

Marseille Grading System (MGS) used histological variables like mitotic activity, hypercellularity and necrosis for risk stratification in I-HPC/SFT.

- Recently MGS system has been modified by Macagno et al. in 2018, summarised in Table 28.1.

Table 28.1 Marseille Grading System (MGS) of hemangiopericytoma

MGS grade	MGS 2012	MGS 2018(updated)	Survival outcome
Grade I	Mitotic activity ≤5/10HPF No necrosis No hypercellularity	Mitotic activity <5/10HPF (independent of necrosis)	
Grade II	II a • Mitotic activity ≤5/10HPF • No necrosis • Hypercellularity II b • Mitotic activity >5/10HPF • No necrosis	Mitotic activity ≥5/10HPF No necrosis	The 5, 10 year OS of MGS grade II is 89% and 72%
Grade III	Mitotic activity >5/10HPF and necrosis and hypercellularity	Mitotic activity ≥5/10HPF and necrosis	The 5, 10 year OS of MGS grade III is 68% and 25%

28.4 Imaging

28.4.1 CT Scan

- Enhancement
- Erosion of adjacent bone
- Hyperostosis-Absent
- Calcification-Absent

28.4.2 MRI

Features on various sequences include

- *T1:* isointense
- *T1 C+ (Gd)*
  - Heterogeneous enhancement
  - Isointense in low grade (Gr I–II) and heterogenous in high grade (Anaplastic) with T1/T2 (Fig. 28.1)
  - dural tail sign is seen (30% cases)-narrow
- Serpentine flow void due to high vascularity*T2.*
  - isointense
  - multiple flow voids on MRI
- *MR spectroscopy.*

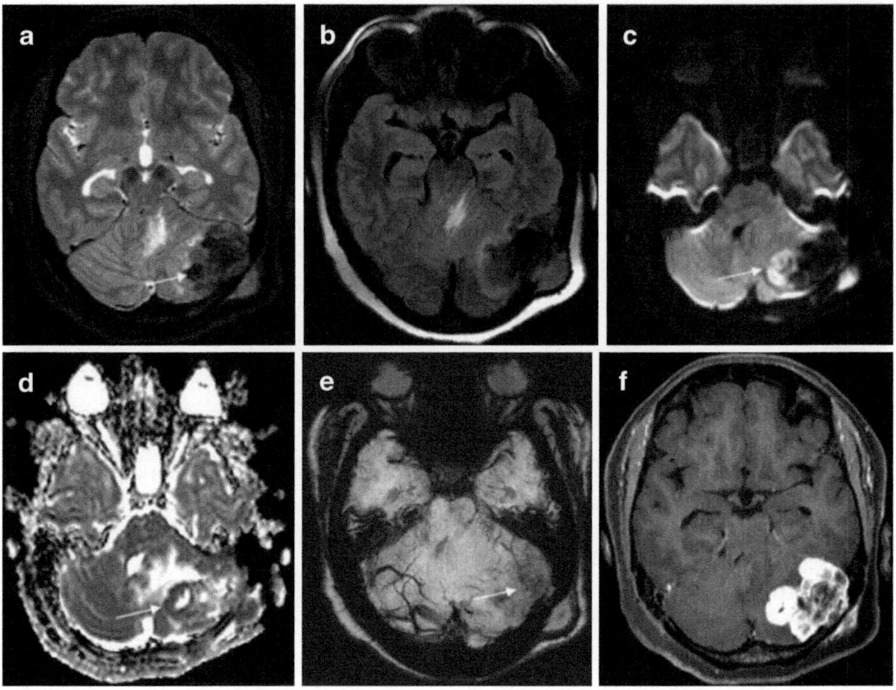

Fig. 28.1 (**a**–**f**) Hemangiopericytoma of left cerebellar hemisphere. The mass is heterogeneous on T2WI (**a**) and FLAIR (**b**); with foci of hypointense haemorrhage on lateral aspect (arrow in **a**). The medial part of the mass shows diffusion restriction (arrows in **c**, **d**). Note the serpentine hypointense flow voids within the mass (arrow in **e**) and strong contrast enhancement (**f**). The mass erodes the calvarium and has an extracranial extension. (Courtesy: Dr. Leve Joseph)

  - alanine peak absent (present in meningiomas)
  - high myoinositol
- *DWI.*
  - intermediate restricted diffusion (less than meningioma)

28.5 Surgery

- The primary treatment of HPC is surgical resection.
- Gross total resection (GTR) is the cornerstone.
- Median survival (13 years vs 9.75 years) in GTR vs STR [6].
- After surgery average time for first recurrence 47 months.
- Favourable Prognostic factors
  - GTR
  - Adjuvant RT
  - Good KPS

- Unfavourable Prognostic factors
  - Infratentorial tumour
  - Advancing age
  - Multifocal/disseminated disease

28.6 Radiation

- High predilection for local recurrence necessitates the need for adjuvant radiotherapy.
- Adjuvant radiotherapy increases average time of recurrence from 34–75 months to 62–92 months [7].
- High grade tumours recur early at 3.3 years, compared to low grade tumours at 10 years ($p = 0.004$) [8].
- 10 year OS with grade II and grade III tumours were 84.6% and 35.7%, respectively [9].

28.6.1 Indications for Adjuvant RT

- HPC grade II with GTR: Not required
- HPC with STR or Recurrence: RT should be considered
- Grade III HPC: adjuvant radiation may be considered upfront even after GTR

28.6.2 Target Volume

- In adjuvant or salvage setting:
  - Local radiation only
  - CTV:
    Grade II HPC: Post-operative cavity plus any residual enhancing mass visible on T2 weighted MRI image with 1.5 cm isotropic expansion.
    Grade III HPC: Post-operative cavity plus any residual enhancing mass visible on T2 weighted MRI image with 2 cm isotropic expansion, If any bone erosion should be included.
  - PTV: 3–5 mm additional margin as PTV

28.6.3 Dose

Local radiation: 54–60 Gy in conventional fractionation.

28.7 Chemotherapy

- Data on chemotherapy is sparse
- Mostly used in in metastatic and recurrent setting
- Effective chemotherapeutic regimen is still not established
- Most effective drugs are
  - Vincristine, Adriamycin, Cyclophosphamide (VAC)
  - Ifosfamide, Cisplatin, Etoposide (ICE)
  - Temozolomide
- In AHPC concurrent and adjuvant Temozolomide is prescribed, however, data is limited

AHS or Adaptive Hybrid Surgery A relatively new concept in neurosurgery for critically located tumours, it involves planned microsurgical maximum safe resection to avoid serious surgical complication with postop SRS to remaining tumour. Surgical resection reduces gross tumour volume to facilitate SRS. With large tumour volume, toxicity of SRS becomes higher compared to its efficacy in tumour control owing to sensitive organ at risks. With AHS, microsurgery and SRS complement each other. A special software (Adaptive Hybrid Surgery Software, Brainlab Inc., Germany) is used before proceeding to simulate the entire process and define "SRS-ideal" tumour volume.

28.8 Follow-up

- I-HPC is distinct from other CNS tumours by its potential to metastasize outside the brain.
- Overall rate of extracranial metastasis was 28%.
- Most common sites for metastasis are bone, lung, and liver.
- AHPC is associated with 1.88 ($p = 0.016$) times increased risk of extracranial metastasis [10].
- When compared with grade II tumours, grade III HPC were 2.53 times likely to metastasize.
- Infiltration of brain parenchyma/invasiveness has been reported to be significantly associated with metastasis and decreased survival (mena [11], kinslow [12]).
- Bone invasion and soft tissue invasion reported to be associated with decreased metastasis free survival.
- CEMRI brain every 3 months for 2 years and thereafter 6 months should be advised during follow up.

References

1. Trifiletti DM, Mehta GU, Grover S, Sheehan JP. Clinical management and survival of patients with central nervous system hemangiopericytoma in the National Cancer Database. J Clin Neurosci. 2017;44:169–74.
2. Louis DN, Ohgaki H, Wiestler OD, Cavenee WK, Ellison DW, Figarella-Branger D, Perry A, Reifenberger G, von Deimling A. WHO classification of tumours of the central nervous system. International Agency for Research on Cancer (IARC); 2016.
3. Mohajeri A, Tayebwa J, Collin A, et al. Comprehensive genetic analysis identifies a pathognomonic NAB2/ STAT6 fusion gene, non-random secondary genomic imbalances, and a characteristic gene expression profile in solitary fibrous tumor. Genes Chromosomes Cancer. 2013;52:873–86.
4. Doyle LA, Vivero M, Fletcher CD, Mertens F, Hornick JL. Nuclear expression of STAT6 distinguishes solitary fibrous tumor from histologic mimics. Mod Pathol. 2014;27(3):390–5.
5. Nakada S, Minato H, Nojima T. Clinicopathological differences between variants of the NAB2-STAT6 fusion gene in solitary fibrous tumors of the meninges and extra-central nervous system. Brain Tumor Pathol. 2016;33(3):169–74.
6. Rutkowski MJ, Sughrue ME, Kane AJ, Aranda D, Mills SA, Barani IJ, Parsa AT. Predictors of mortality following treatment of intracranial hemangiopericytoma. J Neurosurg. 2010;113:333–9.
7. Guthrie BL, Ebersold MJ, Scheithauer BW, Shaw EG. Meningeal hemangiopericytoma: histopathological features, treatment, and long-term follow-up of 44 cases. Neurosurgery. 1989;25:514–22.
8. Ecker RD, Marsh WR, Pollock BE, Kurtkaya-Yapicier O, McClelland R, Scheithauer BW, et al. Hemangiopericytoma in the central nervous system: treatment, pathological features, and long-term follow up in 38 patients. J Neurosurg. 2003;98:1182–7.
9. Combs SE, Thilmann C, Debus J, et al. Precision radiotherapy for hemangiopericytomas of the central nervous system. Cancer. 2005;104:2457–65.
10. Ratneswaren T, Hogg FA, Gallagher MJ, Ashkan K. Surveillance for metastatic hemangiopericytoma-solitary fibrous tumors-systematic literature review on incidence, predictors and diagnosis of extra-cranial disease. J Neuro Oncol. 2018; https://doi.org/10.1007/s11060-018-2836-2.
11. Mena H, Ribas JL, Pezeshkpour GH, et al. Hemangiopericytoma of the central nervous system: a review of 94 cases. Hum Pathol. 1991;22:84–91.
12. Kinslow CJ, Rajpara RS, Wu C, et al. Invasiveness is associated with metastasis and decreased survival in hemangiopericytoma of the central nervous system. J Neuro-Oncol. 2017;133:409–17. https://doi.org/10.1007/s11060017-2450-8.

29 Management of Choroid Plexus Carcinoma

Supriya Mallick

29.1 Introduction

Choroid plexus tumors (CPTs) are primary brain tumors of neuro-ectodermal origin which have been derived from choroid plexus epithelium and have shown vascularization from choroidal arteries. These are rare intraventricular tumors accounting for about 1–4% of pediatric brain tumors. CPTs are classified as *choroid plexus papilloma (Grade 1), atypical CPT (Grade 2) and choroid plexus carcinoma.*

As per the WHO classification, Choroid plexus carcinoma (CPC) has been defined as Grade III tumors and is associated with poor prognosis. The first case of CPC was first described by Rokitansky in the year 1844 [1]. They occur in very young children and are very aggressive and appear disseminated at presentation in 20% of the cases [2]. The median age of incidence is 3 years and is a very rare in adults [3]. Surgery forms the mainstay of treatment and requires gross total resection as the treatment option. *However, the role of adjuvant therapy has not been well defined and the use of chemotherapy or radiation therapy remains a controversy.* The role of molecular testing has opened up avenues for future trials and interventions to improve the outcomes in a disease entity with poor prognosis.

29.2 Epidemiology

- CPC occur predominantly in children and typically occur in the first 5 years of life.
- The incidence is rare and constitutes about 1–4% of all pediatric brain tumors [4, 5].

S. Mallick (✉)
Department of Radiation Oncology, National Cancer Institute, AIIMS, Haryana, India

S. Mallick et al. (eds.), *Evidence based practice in Neuro-oncology*,
https://doi.org/10.1007/978-981-16-2659-3_29

- The median age at diagnosis is 3.5 years with slight male preponderance (1.2:1) [6].
- They represent a neurological manifestation of Li-Fraumeni syndrome [7] and various mutations are commonly encountered including somatic and germline p53 mutations [8].
- Sites of origin correlate with the presence of choroid plexus and include each lateral ventricle, third ventricle, and fourth ventricle. Ectopic locations include the suprasellar and pineal regions.
- Cases have been reported wherein there has been a transformation from Grade I CPT to CPCs [9].

29.3 Clinical Presentation

- Symptoms commonly are due to the presenting hydrocephalus and they include increased head circumference, progressive headache, nausea, vomiting, ataxia, and double or blurred vision. Younger children present most commonly with irritability, lethargy, increased head circumference with protruding scalp veins, and sutural diastasis. Seizures are a rare presentation [10].
- Tumors located in the third ventricle can present with *bobble-head doll syndrome* due to *bilateral thalamic compression*.
- Due to the increased propensity of the tumors to bleed, deteriorating level of consciousness is a worrying symptom and requires urgent management.
- Papilledema is the most common clinical finding and is visible on fundoscopy.
- Focal neurological dysfunction can be noted when the tumor invades the adjacent brain parenchyma.

29.4 Genetic Mutations and Pathogenesis in CPC

- CPC occur mostly sporadically, however rarely, syndromic associations have been noted [11].
- A gain of chromosome 9 and loss of chromosome 10 have been seen to be associated with improved survival in CPCs [12].
- The most recognized syndrome associated with CPC is Li-Fraumeni syndrome caused by TP53 germline mutation [12].
- The disease variant is very rare in adults and if present, viral infections with polyoma viruses SV40, JC, and BK could be involved in the development of CPC [8].

29.5 Imaging

A. *Computed Tomography*: [13]

- The tumor appears heterogeneous and iso to hyperdense to gray matter and calcification may be seen in around 20% cases.
- Areas of necrosis and cystic regions can be noted in contrast-enhanced CT scans. There may be areas of hemorrhage and perilesional edema.
- Intraventricular tumor noted with a lobulated 'cauliflower-like' edge with homogenous contrast enhancement.
- Hydrocephalus is a very common feature.

B. *Magnetic Resonance Imaging (MRI)*:
- MRI helps in the preoperative stage by providing details regarding the tumor extent and vascularization and post-operatively to determine the extent of resection and follow-up tool for picking up recurrence early.
- T1-weighted—iso to hypointense.
- T2-weighted—iso to hypointense with hyperintense areas of necrosis.
- T2 GRE—blooming may be noted in areas of calcification or hemorrhage.
- T1 contrast—shows heterogeneous enhancement.
- CPCs have the propensity of tumor seeding and hence, the entire neural axis needs to be screened.
- Presence of curvilinear signal void which represents enlarged intratumoral blood vessels [14].
- Leptomeningeal enhancement if present indicates dissemination.

C. *MR Spectroscopy*: [15]
- It helps differentiate the high-grade and low-grade tumors.
- In CPCs: Choline and lactate peak and decrease of N-acetyl aspartate levels.

29.6 Histopathology: [16]

- On gross appearance, the tumors appear lobulated masses with necrotic and cystic areas.
- To make the diagnosis of CPCs, 4 out of the 5 factors need to be present:
  - mitotic figures (≥5 per 10 high power fields)
  - nuclear atypia
  - increased nucleus to cytoplasm ratio
  - necrosis
  - distorted papillary structure
- Microcalcification, hemorrhage, and presence of invasion of adjacent brain parenchyma are other characteristic features of CPCs that may be present.
- Immunohistochemistry:
  - Cytokeratin, S100, and GFAP positive; p53 protein-positive in 50% of patients who have TP53 mutation (Cytokeratin positivity rules out the possibility of ependymoma and meningioma)
  - KIR7.1 is found to be positive in around 50% cases
  - S11, transthyretin, and EMA are negative
  - MiB index reported with CPC is around 13% [17]

29.7 Differential Diagnosis

- Choroid plexus papilloma and atypical choroid plexus papilloma
- Central neurocytoma
- Intraventricular meningioma
- Choroid plexus metastases
- Rarely, PNET, ATRT, and astrocytoma

29.8 Treatment of CPCs

29.8.1 Surgery

Surgery forms the mainstay of treatment and gross total resection (GTR) needs to be achieved for better outcomes. However, GTR can be achieved in less than 50% of patients. Even after GTR, these tumors have a high propensity for local recurrence and metastases and are associated with poor prognosis with 5-year survival ranging from 11% to 86% [18, 19]. Surgical resection in itself is a challenge due to the deep-seated nature of the tumor and proximity to critical and vascular structures. Hemorrhage is a common complication associated with surgical procedures, especially in young children. If the tumor appears large and unresectable, some studies have advocated biopsy followed by neo-adjuvant chemotherapy and then surgical resection [20].

29.8.2 Adjuvant Therapy

CPCs have a high propensity for local recurrence and Leptomeningeal metastases even after GTR. Adjuvant therapy in the form of radiotherapy or chemotherapy has been shown in a recent systematic review and meta-analysis to improve overall survival. We summarize the role of radiotherapy and chemotherapy in the adjuvant setting followed by a management algorithm for CPC.

29.8.3 Radiation Therapy

The role of radiation therapy is not well defined in CPCs, however, it has shown to have a significant impact on survival outcomes in CPCs especially in incomplete resections [21]. There is also no consensus on the radiation target volumes and the doses to be prescribed. Based on available data, the recommended doses for treatment are:

- *Maximal safe resection and no drop metastases*: Radiation to post-operative tumor bed with a dose of 54 Gy.
- *Spinal metastasis present*: CSI 23.4-36 Gy with boost 54 Gy and a boost to the metastases to a dose of 45–54 Gy.

- The volume includes the visible disease on T1 contrast images which constituted the Gross tumor volume (GTV). A margin of 1.5–2 cm to this volume constituted the Clinical Target Volume (CTV). An additional PTV was given with 0.5 cm to the CTV.

29.8.4 Chemotherapy

The addition of chemotherapy as an adjuvant treatment has been found to reduce the recurrence risk and also improve the survival outcomes when combined with surgery, especially after GTR [20]. Chemotherapy can help buy time for infants and young children not suitable for radiation therapy and neo-adjuvant chemotherapy has been shown to increase resectability and reduce the risk of bleeding [22]. The most effective regimen used in CPC has been the combination of etoposide and platinum compounds [23]. The role of chemotherapy in relapsed disease is insufficient and also, the prognosis in CPC after relapse is dismal as it presents with distant dissemination. The role of high-dose chemotherapy with an autologous stem cell transplant is also controversial.

29.9 Conclusion

CPCs are a challenge to the treating oncologist and the treatment decisions require a multi-disciplinary approach. Due to the rarity of the disease, the designing of clinical trials has been difficult and this is reflected in the lack of consensus on the role of adjuvant treatment. The genomic studies done in the future will help guide the establishment of newer therapies for improving the outcomes in the disease with poor prognosis.

29.10 Management Protocol for Choroid Plexus Carcinoma

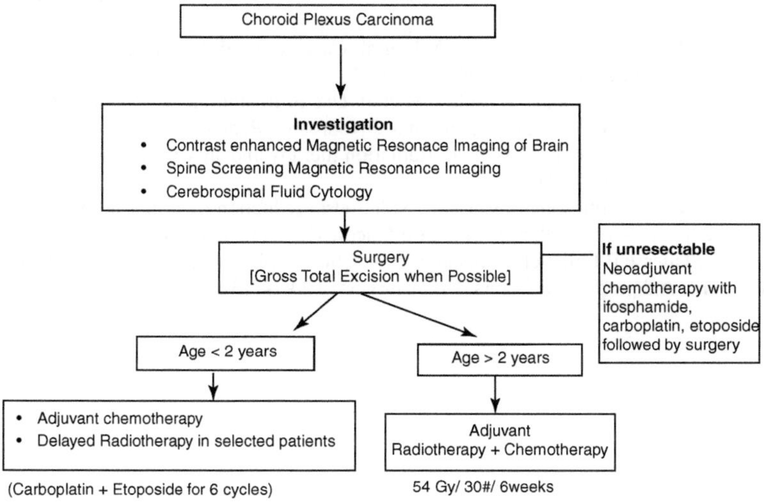

References

1. Dhormann GJ, Collias JC. Choroid plexus carcinoma. J Neurosurg. 1975;43:225–31.
2. Wrede B, Hasselblatt M, Peters O, Thall PF, Kutluk T, Moghrabi A, Mahajan A, Rutkowski S, Diez B, Wang X, Pietsch T, Kortmann RD, Paulus W, Jeibmann A, Wolff JE. Atypical choroid plexus papilloma: clinical experience in the CPT-SIOP-2000 study. J Neuro-Oncol. 2009;95:383–92.
3. Kim T, Park MR, Hong EK, et al. Choroid plexus carcinoma in adults: two case reports. Brain Tumor Res Treat. 2019;7(1):48–52.
4. Custodio G, Taques GR, Figueiredo BC, et al. Increased incidence of choroid plexus carcinoma due to the germline TP53 R337H mutation in southern Brazil. PLoS One. 2011;6(3):e18015.
5. Grygotis LA, Chew FS. Choroid plexus carcinoma of the lateral ventricle. AJR Am J Roentgenol. 1997;169(5):1400.
6. Wolff JE, Sajedi M, Brant R, et al. Choroid plexus tumors. Br J Cancer. 2002;87:1086–191.
7. Hinduja S, DeLozier CD, Knezevich SR, et al. A rare case of choroid plexus carcinoma that led to the diagnosis of Lynch syndrome (hereditary nonpolyposis colorectal cancer). Clin Neurol Neurosurg. 2017;158:46–8.
8. Kamaly-Asl ID, Shams N, Taylor MD. Genetics of choroid plexus tumors. Neurosurg Focus. 2006;20(1):E10.
9. Jeibmann A, Wrede B, Peters O, et al. Malignant progression in choroid plexus papillomas. J Neurosurg. 2007;107:199–202.
10. Ellenbogen RG, Winston KR, Kuspsky WJ. Tumors of the choroid plexus in children. Neurosurgery. 1989;25:327–35.
11. Gupta N. Choroid plexus tumors in children. Neurosurg Clin N Am. 2003;4:621–31.
12. Rickert CH, Wiestler OD, Paulus W. Chromosomal imbalances in choroid plexus tumors. Am J Pathol. 2002;160:1105–13.
13. Hopper KD, Foley LC, Nieves NL, Smirniotopoulos JG. The intraventricular extension of choroid plexus papillomas. AJNR Am J Neuroradiol. 1987;8:469–72.
14. Han SR, Yee GT, Joo M, et al. Choroid plexus carcinoma in an adult. J Korean Neurosurg Soc. 2006;40:122–4.
15. Krieger MD, Panigrahy A, McComb JG, et al. Differentiation of choroid plexus tumors by advanced magnetic resonance spectroscopy. Neurosurg Focus. 2005;18(6A):E4.
16. Louis DN, Ohgaki H, Wiestler OD, Cavenee WK. WHO classification of tumours of the central nervous system. 4th ed. Revised.
17. Misaki K, Nakada M, Mohri M, et al. MGMT promoter methylation and temozolomide response in choroid plexus carcinoma. Brain Tumor Pathol. 2011;28:259–63.
18. Packer RJ, Perilongo G, Johnson D, et al. Choroid plexus carcinoma of childhood. Cancer. 1992;69(2):580–5.
19. Pierga JY, Kalifa C, Terrir-Lacombe MJ, et al. Carcinoma of the choroid plexus: a pediatric experience. Med Pediatr Oncol. 1993;21:480–7.
20. Lafay-Cousin L, Mabbot DJ, Halliday W, et al. Use of ifosfamide, carboplatin, and etoposide chemotherapy in choroid plexus carcinoma. J Neurosurg Pediatr. 2010;5(6):615–21.
21. Wolff JE, Sajedi M, Coppes MJ, et al. Radiation therapy and survival in choroid plexus carcinoma. Lancet. 1999;353:2126.
22. Souweidane MM, Johnson JH Jr, Lis E. Volumetric reduction of a choroid plexus carcinoma using preoperative chemotherapy. J Neuro-Oncol. 1999;43(2):167–71.
23. Berger C, Thiesse P, Lellouch-Tubiana A, et al. Choroid plexus carcinomas in childhood: clinical features and prognostic factors. Neurosurgery. 1998;42(3):470–5.

Meningioma

30

Yousra Izzuddeen

30.1 Introduction and Epidemiology

- Meningioma is a common tumor of the central nervous system [1]. It is usually slow growing but sometimes can be aggressive [2].
- It is a dural-based tumor arising from the arachnoid cap.
- According to the Central Brain Tumor Registry in the United States (CBTRUS), there were 145,916 new cases of meningioma diagnosed from 2011 to 2015 [3].
- The median age of presentation is 65 years and it is rare in children, except in those with hereditary syndromes such as neurofibromatosis type 2 (NF2) or previous history of radiotherapy [4].
- It is more common in females with a female is to male ratio of approximately 2 to 3 is to 1 [5].

30.2 Etiology and Risk Factors

- Radiation: History of radiation in the form of previous therapeutic cranial irradiation, treatment of head and neck cancers [6] or incidental radiation exposures like dental radiographs, diagnostic imaging, or radiation accidents [7, 8].
- Genetic Predisposition: NF-2, an autosomal dominant disorder caused by mutations of a tumor suppressor gene on chromosome 22, is prone for multiple intracranial meningiomas and sometimes spinal meningiomas and is often anaplastic or other aggressive variants of meningiomas [9, 10]. It has also been seen in patients with Shwannomatosis, caused by mutation in the tumor suppressor gene, SMARCB1 gene.

Y. Izzuddeen (✉)
Department of Radiation Oncology, AIIMS, New Delhi, India

S. Mallick et al. (eds.), *Evidence based practice in Neuro-oncology*,
https://doi.org/10.1007/978-981-16-2659-3_30

- Hormonal factors: Progesterone and androgen receptors positivity seen in about 60% patients, while estrogen receptors in 10% of cases [11] although the inhibition of these have not shown any clinical response.
- Others: A small increase in incidence of meningiomas have been associated with breast cancer and obesity, although the underlying cause remains unclear [12, 13].

30.3 Pathology

30.3.1 WHO Classification

This is based on morphological characteristics [14] *and divided into 3 groups*:

WHO Grade I: Benign meningioma, with no characteristics of aggressive subtypes.

WHO Grade II: They include atypical meningiomas, with increased mitotic activity (>4 mitoses per 10 hpf), brain invasion or 3 or more of the following features—increased cellularity, small cells with high nuclear to cytoplasmic ratio, prominent nucleoli, uninterrupted patternless or sheet like growth or foci of spontaneous or geographic necrosis. They *include atypical, clear cell, and choroidal meningiomas.*

WHO Grade III: They include malignant meningioma >20 mitosis per 10 hpf, loss of meningioma growth pattern, abundant atypical forms, and multifocal areas of necrosis. They include *anaplastic, papillary, and rhabdoid meningiomas.*

30.4 Molecular Pathology

- An increasing number of genetic alterations have been identified [15] and may improve further biological classification based on prognosis of the same.
- Some alterations associated with aggressive histological subtypes including *telomerase reverse transcriptase (TERT) promoter region and inactivation of the dystrophin (DMD) gene* [16].
- The incidence of 22q LOH increases with WHO grade, is present in WHO grade I tumors and in around 80% of WHO grades II and III tumors.
- Non-NF2 mutations that may have therapeutic implications in future and include mutation in Smoothened (SMO), an activator of Hedgehog pathway (3–6%), mutations in v-akt murine thymoma viral oncogene homolog 1 (AKT 1) in around 10% and activators of phosphatidylinositol 3-kinase (PI3K) pathways in around 4–7% of meningiomas [17].

30.5 Clinical Features and Evaluation

- Symptoms depends mainly on location of tumor, main symptoms include headache, seizures, vomiting, weakness of extremities, change in behavior or symptoms of obstructive hydrocephalous.

- Skull base tumors may present with cranial nerve abnormalities while sphenoid wing meningiomas may present with seizures.
- Spinal meningioma may present with lower limb weakness.
- Evaluation includes history, a proper physical examination and *a contrast enhanced MRI* (magnetic resonance imaging). With increase in the use of neuro-imaging, there has been steady increase in the reporting of incidental meningiomas, and the management of the same is being extensively reviewed.

Most common sites of meningioma

- Falx and para-saggital—25%
- Convexity—20%
- Sphenoid ridge—20%
- Posterior fossa—10%
- Supra-sellar—10%
- Intra-ventricular— <2%
- Spinal— <2%

30.6 Neuro-Radiology

- On a contrast enhanced MRI a typical meningioma is an *extra-axial, dural based* mass that is iso or hypo intense on T1 and iso- or hyper intense on T2 weighted images with strong homogenous contrast enhancement with a marginal dural thickening called the 'dural tail'.
- A classic sign of its extra-axial location is the presence of CSF cleft, between the tumor and underlying brain cortex which contain CSF or vessels trapped between tumor and underlying cortex.
- The calcification inside the tumor is better appreciated in CT.

30.6.1 High Grade Features Include

- Intra-tumoral cystic change
- Peri-tumoral brain edema
- Extension of tumor to skull base
- Low ADC values

30.6.2 Radiological Differentials

- Lymphoma
- Plasmacytoma
- Gliosarcoma
- Hemangiopericytoma—erosion of adjacent bone (hyperostosis of adjacent bone in meningioma) with no intra-tumoral calcification
- Benign conditions like tuberculosis

30.7 Management

30.7.1 Surgery

- Surgery remains the corner stone of treatment in any type of meningioma and the extent of resection has been correlated with risk of recurrences [18].
- The resection for meningioma has been described by Simpson and commonly known as Simpson Grading System [19].

Grade 1: Complete resection, including dural attachment, and any abnormal bone
Grade 2: Complete resection, with coagulation of dural attachment
Grade 3: Complete resection, without resection or coagulation of dural attachment
Grade 4: Subtotal resection
Grade 5: Tumor biopsy only

- Although most convexity meningiomas can be completely removed by surgery, those surrounding or invading major draining veins or venous sinuses, skull base meningiomas, cavernous sinus meningiomas and optic nerve meningiomas can be difficult to resect by surgery.
- The major factors that determine the risk or recurrence after surgery include the extent of resection [19], tumor grade [20] and previous recurrence [21] and these in turn become important factors in determining adjuvant treatment.

30.8 Post-operative Radiotherapy

30.8.1 Grade 1 Meningioma

- After gross total resection, most patients can be kept on observation.
- Radiotherapy in grade 1 meningiomas is usually reserved for those patients with a subtotal resection (STR) or recurrent meningiomas. There remains controversy if a patient can be kept on active surveillance even after a STR [22]. There are reports on increased progression free survival even after a Simpson 4 resection [23].
- Definitive radiotherapy, in the form of conventional radiotherapy or stereotactic radiotherapy, is used where surgical resection is not possible, as in the case of meningiomas of the optic nerve sheath, cavernous sinus and base of skull.
- In such a setting, fractionated radiotherapy is preferred when the lesion is near critical organs like the optic nerve. SRS is typically reserved for tumors less than 3 cm in diameter or a volume up to 10 cc, with local control and toxicity depending on the size of tumor [24, 25].
- The complications after SRS include cranial nerve deficits, headaches, hemiparesis, seizure, cyst formation and stroke.

- Proton beam therapy has been used for benign meningioma, with excellent 5 year and 10 year local control rates of 88–100% with acceptable toxicity [26, 27].
- *In patients with NF1 syndrome, radiation therapy may lead to increased risk of malignant conversion or cerebral vasculopathies leading to complication like Moya Moya syndrome.*

30.8.2 Dose of RT

- The most common dose prescribed is 50–54 Gy in fraction of 1.8–2 Gy per day, both in the post-operative and the definitive setting [28].
- The dose for stereotactic radio surgery (SRS) is 13–16 Gy in single fraction.

30.8.3 Grade 2 (Atypical) and Grade 3(Anaplastic) Meningiomas

- Post-operative radiotherapy is an important part of adjuvant treatment in Grade 2 and Grade 3 meningiomas.
- Irrespective of the extent of resection, adjuvant radiotherapy is the standard of care for malignant meningioma, considering the recurrence rates of 60–90% at 5 years after surgery and a 5 year overall survival of around 20–50%, as reported in multiple retrospective data [28, 29].
- The dose of radiotherapy for malignant meningiomas should be at least 60 Gy to achieve a durable control [30, 31].
- In completely resected atypical meningioma, the role of radiotherapy is debatable. The benefit of radiotherapy has to be outweighed with the risk of long-term toxicity. In absence of any randomized trials, clinical decision has to be take in individual patient basis, and as per reports of multiple retrospective data analysis and population based studies [32, 33], most recent of which have reported increased survival associated with gross total resection and adjuvant radiotherapy.
- *The results of the ongoing EORTC/ROAM-1308 trial, comparing radiation* versus *observation after surgery in atypical meningioma, will shed more light regarding the optimal management of these meningiomas.*
- The dose of radiotherapy usually prescribed is between 54 and 60 Gy in 1.8 and 2 Gy per fraction.

30.8.4 Spinal Meningiomas

- Spinal meningiomas most commonly occur in the thoracic spine.
- Pathological features are similar to that of intracranial meningiomas.
- They can usually be resected completely, and even at recurrence, surgery is preferred.
- RT is considered when these meningiomas are not further resectable.

30.8.5 Recurrent Tumors

- Recurrence can be local (in and around previous surgical or irradiated site), spinal metastasis that develop by spread through CSF (more common in higher grade meningiomas) and rarely distant extra-cranial metastasis, usually associated with malignant meningiomas, with incidence as high as 10% in some retrospective series.
- The most curative form of treatment for recurrent tumors is surgical resection, if feasible. Radiotherapy (including re-irradiation) may also be effective in prolonging the progression free survival.
- There is no role of systemic therapy in recurrent meningiomas although somatostatin analogues like octreotide, hydroxyurea may be tried.

30.9 Summary of Adjuvant Treatment

1. In whom RT must be given	• Grade 1 recurrent tumors • Grade 2 tumors after Simpson 4/5 resection • All grade 3 tumors, irrespective of extent of resection • In-operable tumors
2. In whom RT can be avoided safely	• Grade 1 tumors after Simpson 1 resection • Incidental meningiomas which are asymptomatic
3. Gray area (at discretion of physician and patient's choice)	• Grade 2 meningioma post Simpson 3/4 resection • Recurrent grade 1 meningioma in elderly • Grade 1/ 2 meningiomas in patients with NF1/ NF2

30.10 Late Effects of Radiotherapy

- In view of the excellent local control and long-term survival, especially for benign meningiomas, critical consideration of late effects is important.
- There is a very small, but finite amount of risk of malignant transformation to higher grade after radiotherapy, with an increase in incidence after radiosurgery [34].
- Another rare risk is that of secondary tumors, that have been best described in the long-term follow up of those treated with radiotherapy for pituitary tumors, with a risk of 2.4% to 2.7% at 20 to 30 years [35, 36], with the most common being meningiomas, in most reports.
- A recent analysis showed 0% incidence of second malignancy at 15 years of follow up after SRS for benign tumors [34]. The risk of neurocognitive decline is not well described specifically for meningioma, but can be extrapolated from experience regarding the same after irradiation in low grade brain tumors, and this has been shown to be related to dose to bilateral hippocampi [37].
- The risk of hypopituitarism is also related to dose received [38].

30.11 Techniques of Radiotherapy

- Planning for radiotherapy is usually done by 3D-CRT, IMRT or VMAT based on orientation and size of tumor, with typical energy used being 6MV photons.
- Image guidance during treatment in the form of orthogonal KV X-rays and volume based cone beam CT's may also be used during treatment.
- Image co-registration and planning CT scans with diagnostic MRI scans helps in accurate delineation of the tumor and critical structures at risk.
- GTV delineated on T-1 weighted post contrast sequences of the MRI.
- GTV is equal to CTV in benign tumors, however a margin of 0.5–1 cm may be used if dural tail is not covered, if there is any uncertainty based on the MRI or in case of atypical or anaplastic meningioma. PTV margin of 3–5 mm is given.

30.12 Recent Advances

- Two major classes of meningiomas have emerged based on molecular classification on the basis of mutation profile: Those driven by NF2 inactivation and those with non-NF2 driver gene alteration such as mammalian target of rapamycin (mTOR) and Hedgehog, Wingless (WNT)/b-catenin, notch, transforming growth factor-b receptor(TGF-bR), mitogen-activated protein kinase (MAPK) and phospholipase c pathway alterations.
- With chemotherapy having minimal role in the treatment of meningiomas, targeted therapy based on their molecular characteristics may open new treatment avenues for meningiomas.
- An Alliance consortium group trial (Clinical Trials. Gov. identifier: NCT02523014) is underway investigating targeted agents in progressive meningiomas.

References

1. Louis DN, Perry A, Reifenberger G, von Deimling A, Figarella-Branger D, Cavenee WK, et al. The 2016 World Health Organization classification of tumors of the central nervous system: a summary. Acta Neuropathol. 2016;131(6):803–20.
2. Palma L, Celli P, Franco C, et al. Long-term prognosis for atypical and malignant meningiomas: a study of 71 surgical cases. J Neurosurg. 1997;86(5):793.
3. Ostrom QT, Gittleman H, Truitt G, et al. CBTRUS statistical report: primary brain and other central nervous system tumors diagnosed in the United States in 2011-2015. Neuro-Oncology. 2018;20:iv1–iv86.
4. Liu Y, Li F. Zhu S, et al; clinical features and treatment of meningiomas in children: report of 12 cases and literature review. Pediatr Neurosurg. 2008;44(2):112.
5. Claus EB, Bondy ML. Schildkraut JM, et al; epidemiology of intracranial meningioma. Neurosurgery. 2005;57(6):1088.
6. Braganza MZ, Kitahara CM, de González AB, et al. Ionizing radiation and the risk of brain and central nervous system tumors: a systematic review. Neuro-Oncology. 2012;14(11):1316–24.

7. Lin MC, Lee CF, Lin CL, et al. Dental diagnostic X-ray exposure and risk of benign and malignant brain tumors. Ann Oncol. 2013;24(6):1675–9.
8. Davis F, Il'yasova D, Rankin K, et al. Medical diagnostic radiation exposures and risk of gliomas. Radiat Res. 2011;175(6):790–6.
9. Goutagny S, Kalamarides MJ. Meningiomas and neurofibromatosis. J Neuro-Oncol. 2010;99(3):341.
10. Perry A, Giannini C, Raghavan R, et al. Aggressive phenotypic and genotypic features in pediatric and NF2-associated meningiomas: a clinicopathologic study of 53 cases. J Neuropathol Exp Neurol. 2001;60(10):994.
11. Caroll RS, Zhang J, Dashner K, et al. Androgen receptor expression in meningiomas. Neurosurgery. 1995;82(3):453.
12. Wiemels J, Wrensch M, Claus EB. Epidemiology and etiology of meningioma. J Neuro-Oncol. 2010;9(3):307.
13. Muskens IS, Wu AH, Porcel J, et al. Body mass index, comorbidities, an hormonal factors in relation to meningioma in an ethnically diverse population: the Multiethnic Cohort. Neuro-Oncology. 2019;21(4):498.
14. Louis DN, Perry A, Reifenberger G. The 2016 World Health Organization classification of tumors of the central nervous system: a summary. Acta Neuropathol. 2016;131(6):803–20.
15. Sahm F, Schimprf D, Stichel D. DNA-methylation based classification nand grading system for meningiomas: a multicentre, retrospective analysis. Lancet Oncol. 2017;18(5):682.
16. Juratli TA, Mc Cabe D, Nayyar N. DMD genomic deletions characterize a subset of progressive/higher-grade meningiomas with poor outcome. Acta Neuropathol. 2018;136(5):779.
17. Brastianos PK, Horowitz PM, Santagata S. Genomic sequencing of meningiomas identifies oncogenic SMO and AKT1 mutations. Nat Genet. 2013;45(3):285–9.
18. Nanda A, Bir SC, Maiti TK. Relevance of Simpson grading system and recurrence-free survival after surgery for World Health Organization Grade 1 meningioma. J Neurosurg. 2017;126(1):201.
19. Simpson D. The recurrence of intracranial maningiomas after surgical treatment. J Neurol Neurosurg Psychiatry. 1957;20(1):22.
20. Goldsmith BJ, Wara WM, Wilson CB. Post-operative irradiation for subtotally resected meningiomas. A retrospective analysis of 140 patients treated from 1967 to 1990. J Neurosurg. 1994;80(2):195–201.
21. Taylor BW Jr, Marcus RB Jr, Friedman WA, et al. The meningioma controversy: post-operative radiotherapy. Int J Radiat Oncol Biol Phys. 1988;15(2):299–304.
22. Soyuer S, Chang EL, Selek U, et al. Radiotherapy after surgery for benign cerebral meningioma. Radiother Oncol. 2004;71:85–90.
23. Sughrue ME, Rutkowski MA, Aranda D, et al. Treatment decision making based on published natural history and growth rate of small meningiomas. J Neurosurg. 2010;113:1036–42.
24. DiBiase SJ, Kwokzy Y, Yovino S, et al. Factors predicting local tumor control after gamma knife stereotactic radiosurgery for benign intracranial meningiomas. Int J Radiat Oncol Biol Phys. 2004;60:1515–9.
25. Pollock BE, Stafford SL, Link MJ, et al. Single fraction radio-surgery for presumed intracranial meningiomas: efficacy and complication from 22 years' experience. Int J Radiat Oncol Biol Phys. 2012;70:639–45.
26. Slater JD, Loredo LN, Chung A, et al. Fractionated proton therapy for beingn cavernous sinus meningioma. Int J Radiat Oncol Biol Phys. 2012;83:e633.
27. Arvold ND, Niemierko A, Broussard GP, et al. Projected second tumor risk and dose to neurocognitive structures after proton versus photon radiotherapy for benign meningioma. Int J Radiat Oncolo Biol Phys. 2012;83:495–500.
28. Rosenberg LA, Prayson RA, Lee J, et al. Long term experience with World Health Organization grade 3 (malignant) meningiomas at a single institution. Int J Radiat Oncol Biol Phys. 2009;74(2):427.
29. Sughrue ME, Sanai N, Shangari G, et al. Outcome and survival following primary and repeat surgery for WHO grade 3 meningiomas. J Neurosurg. 2010;113(2):202.

30. Hug EB, Devries A, Thornton AF, et al. Management of atypical and malignant meningiomas: role of high-dose, 3D conformal radiation therapy. J Neuro-Oncol. 2000;48(2):151.
31. Dziuk TW, Woo S, Butler EB, et al. Malignant meningioma: an indication for initial aggressive surgery and adjuvant radiotherapy. J Neuro-Oncol. 1998;37(2):177.
32. Aizer AA, Arvold ND, Catalano P, et al. Adjuvant radiation therapy, local recurrence, and the need for salvage therapy in atypical meningioma. Neuro-Oncology. 2014;16(11):1547–53.
33. Rydzewski NR, Lesniak MS, Chandler JP, et al. Gross total resection and adjuvant radiotherapy most important predictors of improved survival in patients with atypical meningioma. Cancer. 2018;124(4):734.
34. Pollock BE, Link MJ, Stafford SL, et al. Risk of radiation-induced tumor or malignant transformation after single fraction intracranial radiosurgery: results based on a 25-year experience. Int J Radiat Oncol Biol Phys. 2017;97:919–23.
35. Minniti G, Traish D, Ashley S, et al. Risk of second brain tumor after conservative surgery and radiotherapy for pituitary adenoma: update after an additional 10 years. J Clin Endocrinol Metab. 2005;90:800–4.
36. Breen P, Flickinger JC, Kondziolka D, et al. Radiotherapy for nonfunctional pituitary adenoma: analysis of long term tumor control. J Neurosurg. 1998;89:933–8.
37. Gondi V, Hermann BP, Mehta MP, et al. Hippocampal dosimetry predicts neurocognitive function impairment after fractionated stereotactic radiotherapy for benign or low-grade adult brain tumors. Int J Radait Oncol Biol Phys. 2013;85:34–54.
38. Darzy KH. Radiation-induced hypopituitarism. Curr Oin Endocrinol Diabetes Obes. 2013;20:342–53.

Pituitary Tumors: Diagnosis and Management

31

Renu Madan, Arun Yadav, and Shikha Goyal

31.1 Introduction

Sellar and suprasellar regions contain several diverse and closely approximated tissues, abnormal growth of which may lead to several neoplastic and non-neoplastic lesions. Pituitary tumors are the most common cause of sellar-sprasellar mass and represent around 10–15% of primary intracranial tumors. A slight female preponderance is seen, with presentation commonly in the age group of 45–55 years. Majority of the pituitary tumors are benign. Pituitary adenoma (PA) is the commonest pituitary tumor accounting for around 10% of all intracranial tumors [1]. Pituitary carcinoma is rare. Classification of various pituitary gland tumors has been shown in Table 31.1 [2].

31.2 Pituitary Anatomy and Physiology

Normal pituitary gland (master gland) is a pea-sized structure, situated in the sella turcica of sphenoid bone. It has two different embryological origins. Adenohypophysis (anterior pituitary, composed of anterior and intermediate lobes) is derived from rathke's pouch while neurohypophysis (posterior pituitary/ lobe) and pituitary stalk are derived from the diencephalon. Adenohypophysis is responsible for most of the endocrine function of pituitary gland. Anterior lobe produces prolactin (PRL), adrenocorticotropic hormone (ACTH), follicle-stimulating hormone (FSH), luteinising hormone (LH), growth hormone (GH), and thyroid-stimulating hormone (TSH) while intermediate lobe produces melanocyte-stimulating

R. Madan (✉) · A. Yadav · S. Goyal
Department of Radiotherapy and Oncology, PGIMER, Chandigarh, India

S. Mallick et al. (eds.), *Evidence based practice in Neuro-oncology*,
https://doi.org/10.1007/978-981-16-2659-3_31

Table 31.1 WHO 2017 classification of pituitary tumors

Pituitary adenoma	• Somatotroph adenomas • Lactotroph adenomas • Thyrotroph adenomas • Corticotroph adenomas • Gonadotroph adenomas • Null cell adenomas • Plurihormonal and double adenomas
Tumors of the posterior pituitary	• Pituicytoma • Granular cell tumor of the neurohypophysis • Spindle cell oncocytoma • Sellar ependymoma
Neuronal and paraneuronal tumors	• Gangliocytoma and mixed gangliocytoma-pituitary adenoma • Neurocytoma • Paraganglioma • Neuroblastoma
Craniopharyngioma	• Adamantinomatous craniopharyngioma • Papillary craniopharyngiom
Mesenchymal tumors	• Meningioma • Schwannoma • Chordoma • Solitary fibrous tumor/hemangiopericytoma
Others	• Pituitary carcinoma • Pituitary blastoma • Germ cell tumors • Hematological tumors • Secondary tumors

hormone (MSH). The posterior lobe stores and releases vasopressin and oxytocin produced by the hypothalamus. Pituitary stalk connects pituitary gland and hypothalamus forming hypothalamic-pituitary axis. The gland is closely related to the optic chiasma, anterior cerebral arteries, floor of the third ventricle and cavernous sinuses that contain internal carotid artery and cranial nerves III, IV, V, and VI. Tumor extension to these regions can lead to cranial nerves palsies and visual defects (Fig. 31.1).

31.3 Classification of Pituitary Adenoma

PA can be broadly classified based on their secretory activity and size. Based on the size, these can be devided into microadenoma (<1 cm) and macroadenoma (>1 cm). The term Pico adenoma is used for tumors less than 3 mm in size [3]. Tumors more than 4 cm are termed as giant adenomas (6–10% of all pituitary adenoma) [4]. Based on their secretory activity, PA can be classified as functioning PA (FPA) or non-functioning PA(NFPA). Prolactinoma is the commonest PA followed by NFPA. Other functioning adenoma are GH-secreting adenoma, ACTH-secreting adenoma, and TSH-secreting adenoma.

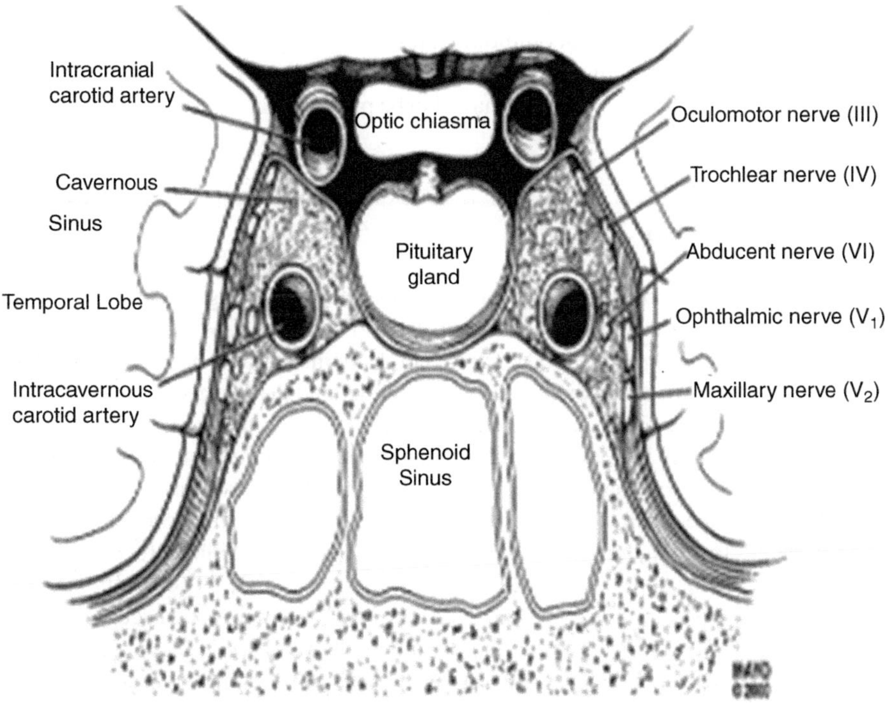

Fig. 31.1 Depicts the Pituitary gland and structures in relation to it

31.4 Etiology and Pathogenesis

Pathogenesis of pituitary tumors is not well understood. Most pituitary tumors are believed to be sporadic in origin; however, syndromic association is seen in approximately 5% of the patients. The genes responsible for syndromic association include (1) multiple endocrine neoplasia I (MEN I), (2) cyclin dependent kinase inhibitor 1 B (CDKN), (3) protein kinase, cAMP dependent regulatory 1A (PRKAR)—Carney complex (CNC)—characterised by endocrine overactivity, schwannomas, skin pigmentation, and myxoma, (4) aryl hydrocarbon receptor-interacting protein (AIP) [5]. No causal relationship has been found with smoking, alcoholism, or trauma.

31.5 Clinical Presentation

Clinical features depend on tumor extent and secretory status. FPA are usually detected on evaluation for features of hyperpituitarism (acromegaly, cushing's syndrome, galactorrhea) and are usually microadenomas. NFPA are either detected incidentally during imaging for unrelated reasons or present with pressure symptoms over normal pituitary or over visual pathway manifesting as vision loss that can include bitemporal or homonymous hemianopia, supero-inferior field effects and central scotoma, and uncommonly cranial nerve dysfunction and seizures. Signs and symptoms of hypopituitarism such as hypogonadism, adrenal insufficiency and rarely diabetes insipidus may also be present. Rarely patients can present with pituitary apoplexy.

31.6 Evaluation

Detailed physical examination, laboratory assessment and imaging should be done to assess visual, endocrinological and neurological abnormalities. Laboratory investigation includes complete blood count, blood chemistry and urine examination. Hormonal analysis depends on patient's age, gender and symptoms and may include TSH, free thyroxine (T4), ACTH, cortisol, PRL, somatomedin C or insulin-like growth factor-1 (IGF-1), LH, FSH, and testosterone. Dynamic contrast enhanced T1 weighted magnetic resonance imaging (MRI) of the brain or sella is the imaging modality of choice. Thin slices (1 mm) should be obtained in the coronal/axial and sagittal planes to define tumor extent and its relation to surrounding structures (optic pathway, cavernous sinuses, and anterior cerebral vessels). PA is generally hypointense on T1W MRI and hyperintense on T2W MRI. Contrast enhanced computed tomography (CT) scan of the brain can be done when there is a suspicion of bone destruction or MRI is not feasible. Role of functional imaging, i.e., somatostatin receptor scintigraphy or fluorodeoxyglucose (FDG) and 11 C-methionine (11 C-Met) positron emission tomography (PET) has been explored by some groups but its utility remains unknown [6]. Susceptibility weighted imaging (SWI) can be used to detect any hemorrhagic component.

31.7 Classification

Grade	Criteria
A. *Hardy's classification of pituitary adenoma* [7]	
Grade I	Microadenoma (<10 mm diameter)
Grade II	Macroadenoma (>10 mm diameter), within Sella
Grade III	>10 mm, focal sellar erosions, outside the Sella
Grade IV	Infiltrates sphenoid and cavernous sinuses, compresses optic nerves, cranial nerves and/or invades adjacent brain parenchyma
B. *Knosp's grading of parasellar invasion* [8]	
Grade 0	Adenoma does not extend beyond the medial carotid line
Grade 1	Adenoma extends beyond the medial line, but doesn't reach the median line, or intercarotid line
Grade 2	Adenoma extends to the lateral aspects of internal carotid artery (ICA)
Grade 3a	Adenoma extends beyond lateral aspects of ICA and into the superior cavermons sinus compartment
Grade 3b	Adenoma extends beyond lateral aspects of ICA and into the inferior cavenous sinus compartment
Grade 4	Adenoma totally wraps around the intracavernous carotid artery

31.8 Treatment

Goals of the treatment are to control over-secretion of hormones, to correct hormone deficiencies and to reduce tumor size while minimising potential morbidity such as hypopituitarism and vision changes. Three main treatment modalities are surgery, medical management, and radiotherapy (RT). Observation is an option for selected group of patients.

Observation NFPA patients without hypopituitarism, mass effect or visual abnormalities can be considered for observation. Follow-up clinical examination and MRI should be done to rule out progression as approximately 10% of microadenoma and 23% of macroadenoma may progress without treatment.

Surgery Surgery with or without RT remains the first line of treatment for most of the PA including apoplexy and prolactinoma resistant to medical treatment. Debulking of the tumor provides histological diagnosis, rapidly reduces mass effect and hormone hypersecretion, arrests progression of neurological deficits and vision loss. Surgical management yields a long-term control of 80–90% in microadenoma and 25–50% in macroadenomas. Transsphenoidal surgery (TSS) approach is appropiate for majority of the patients; however, if the tumor extends to frontal or temporal lobe, transcranial route is preferred. Complications of TSS approach such as perioperative mortality, central diabetes insipidus, cerebrospinal fluid (CSF) rhinorrhoea, and meningitis occur only in 1–3% of the patients [9].

Medical Therapy Medical management alone, or in combination with surgery or RT, is important as hormonal excess or deficiency both can have a detrimental effect on quality of life of a patient.

Type of hormonal therapy depends on tumor type, laboratory investigations and clinical condition of the patient. Often, hydrocortisone and cortisone acetate are used for glucocorticoid replacement. For correction of hypothyroidism, L-thyroxine is used. Gonadal steroids oestrogen and progestin are used for women and testosterone for men, if deficient. For prolactinoma, medical management is the first treatment of choice that includes therapy with a dopamine agonist, such as bromocriptine or cabergoline. These drugs rapidly normalise PRL levels and reduce tumor size in 80–90% of patients [10]. For GH-secreting adenoma or acromegaly, medical treatment can be used as an adjunct if patient is not a candidate for surgery or has not responded to surgery or RT. Most commonly used drugs are first generation somatostatin analogues (octreotide and lanreotide) administered once a month which inhibit GH production or antagonise its action [11]. Normalisation of IGF-1 and GH at 1 year can be achieved in 25% and 38% patients, respectively [12]. Cabergoline can be added if patients do not show adequate response to somatostatin analogues. Pasireotide, a multisomatostatin receptor (SSTR)-ligand has a longer half-life than first generation somatostatin analogues and is used in resistant cases [13]. GH receptor antagonist pegvisomant, in combination with somatostatin analogue has also been tested in placebo controlled trial in resistant cases which has shown to normalise IGF-1 levels in

90–97% of patients [14]. Pegvisomant monotherapy can increase the tumor size due to lack of negative feedback as it does not inhibit GH secretion. Other limitations of pegvisomant therapy are high cost and need for lifelong therapy.

For ACTH-secreting adenoma or cushing's disease, medical treatment is considered for those who do not respond to surgery or RT. Steroidogenesis inhibitors, i.e., ketoconazole, aminoglutethimide, metyrapone, mitotane, and etomidate are used for hormonal normalisation. Ketoconazole is a well tolerated which results in control of cortisol excess in up to 70% of the patients [15]. Pasireotide has also shown biochemical control in approximately 25% of the patients [16]. Glucocorticoid receptor inhibitor mifepristone may reduce the systemic side effects of cortisol [17]. Bilateral adrenalectomy may be considered in severe refractory disease [18].

TSH-secreting adenomas are exceedingly rare, and thus, treatment paradigm is not well defined. Surgery with or without RT is typically the first line of the treatment and medical treatment is considered only in resistant cases. Somatostatin analogues such as octreotide have been shown to normalise TSH levels in approximately 79% of patients [19]. A more detailed description of medical management is provided in the chapter on neuro-endocrinology.

Radiotherapy (RT) Typical indications of RT include incomplete resection of a large NFPA, cavernous sinus invasion, recurrent or progressive tumors, inoperable symptomatic or potentially symptomatic tumors, or refractory secreting tumors. RT techniques for PA have evolved from conventional two-dimensional (2D) to three-dimensional (3D) treatment, intensity modulated radiotherapy and stereotactic treatments, with either fractionated stereotactic radiotherapy (SRT) or stereotactic radiosurgery (SRS). Sophisticated RT techniques can provide a steeper dose gradient thus delivering higher dose to the tumor and minimising the dose to surrounding critical structures, thereby allowing for maximum tumor control and lesser long-term RT-induced side effects. Figure 31.2 depicts IMRT dose color wash for a prescription dose of 54 Gy in 30 fractions planned for large residual functioning pituitary adenoma. Figure 31.3 depicts dose volume histogram (DVH) of optic chiasm for the same patient.

Normalisation of excess hormone production is slow after RT as compared to surgery and medical therapy.

Fractionated RT vs SRS As the dose gradient is sharper in SRS, the likelihood of radiation injury to surrounding non-contiguous normal structures may be lesser. Response to RT depends on the type of the tumor. In general, either form of RT controls the tumor in approximately 90% of the patients at 5 years. Evidence so far does not distinguish between these two approaches based on efficacy or safety. A meta-analysis of eight studies has shown comparable results with fractionated RT and SRS [20]. Few retrospective studies have suggested that compared to fractionated RT, SRS is more effective in functioning adenoma, while both are equiefficacious in NFPA [21]. SRS, for now, only offers the time advantage of treatment in a single session against a longer 5 weeks course of fractionated RT, and may be considered as treatment of choice in tumors situated atleast 2 mm away from optic structures, or cavernous sinus involvement or persistent high serum hormone levels postoperatively. Fractionated RT is preferred over SRS when optic pathway is involved or closely apposed, or for large lesions (diameter >3 cm) where risk of radio necrosis with SRS is high.

SRS Optic apparatus is the most important dose-limiting structure in RT planning of PA. To minimise the risk of radiation related optic neuropathy, a minimum gap of 2 mm between the tumor and optic apparatus is crucial. SRS which is generally delivered as a single fraction of high dose, may result in more rapid biochemical control. However, 2–5 fractions of SRS can be used if there is a higher probability of radiation damage to nearby structures. In functioning PA, where the treatment goal is to achieve radiological control as well as to reduce endocrine oversecretion,

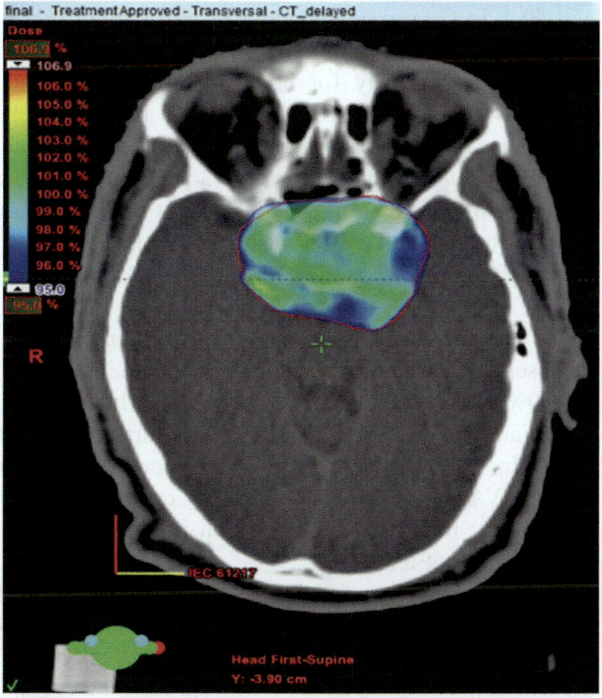

Fig. 31.2 Dose color wash of radiotherapy planning in a 30 years old patient with large residual FPA. CT/MR fusion was done for RT planning. Gross target volume (GTV) was delineated as seen on imaging and a planning target volume (PTV) margin of 3 mm was given. No margin was given for clinical target volume (CTV). She received 54 Gy/30 fractions by IMRT technique. Figure 31.2 is showing 95% dose color wash of PTV

Fig. 31.3 Dose volume histogram (DVH) of optic chiasm. Dmax-54 Gy, 0.02 cc of optic chiasm received 54 Gy

a higher dose of SRS (16–25 Gy) is generally required as compared to NFPA where a SRS dose of 13–14 Gy is adequate for long-term tumor control and acceptable neurological toxicity. (20) In a multicenter retrospective analysis of NFPA, SRS dose of <12 Gy was associated with poor outcome while a dose of >20 Gy did not provide additional tumor control as compared to doses between 12 and 20 Gy [22]. SRS is effective in achieving 5-year tumor control rate of 90–100% in NFPA.

Hypopituitarism is the most commonly reported toxicity after SRS with a 5-year incidence of increasing pituitary deficits ranging from 10% to 40% in different studies [23]. Risk of radiation related optic neuropathy is less than 3% if the maximum point dose to optic apparatus is 8–10 Gy, but significantly increases after 12 Gy [24]. Other cranial nerve deficits resulting from RT are rarely reported.

Fractionated RT Adaquete RT doses for NFPA and functioning PA are 45-50.4 Gy and 50.4-54 Gy respectively at 1.8 Gy/fraction [25]. Studies have shown that local control with conventional RT can vary from 80% to 94% and 75% to 90% at 10 and 20 years respectively for both functioning and nonfunctioning adenoma [26]. Normalisation of GH/IGF-1 occurs in approximately 50% of the patients after 5–10 years of RT for acromegaly [27]. Biochemical control in cushing disease can be seen in 50–80% patients within 3 years of fractionated RT [28]. RT in prolactinomas is indicated in those who do not respond to medical therapy. Around one third of the patients with prolactinomas respond to RT with normalisation of PRL after 5–10 years [7].

Most reported late toxicity of RT is hypopituitarism. Following conventional RT at doses of 45–50.4 Gy at 1.8 Gy per fraction, hypopituitarism can be seen in 30–50% of the patients after 5 years [28]. The reported incidence of vision loss due to radiation-induced optic neuropathy varies from 0 to 6% (<2% at RT dose of 45 Gy in 1.8 Gy daily fraction) [28]. Increased incidence of cerebrovascular accidents (CVA) has also been reported. Conventional RT for PA can increase the risk of CVA by four fold [8]. Radiation induced second cancer is another long-term complication of RT.

Target delieation in fractionated RT- Postoperative T1w MRI images are extremely helpful in target delineation. T2w MRI which is used to assess brain parenchyma infiltration and perilesion edema is generally not used. Any lesion which is visible on MRI is demarketed as GTV. Margin for CTV is generally not given except in the intracavernous portion of aggressive adenoma to encompass potential microscopic spread. A 3-5 mm margin is added for PTV for setup uncertainities. Reference- Minniti, G., Osti, M.F. & Niyazi, M. Target delineation and optimal radiosurgical dose for pituitary tumors. Radiat Oncol 11, 135 (2016) Particle radiation has been also been successfully used in the treatment of pituitary adenomas [29]. However, the potential advantage of proton therapy over photon remains unclear.

Pituitary Incidentaloma (PI) It is defined as an unexpected pituitary lesion (adenoma and cystic lesion), identified on brain imaging for an unrelated reason.

Although most of the pituitary incidentalomas are <1 cm in size and nonfunctioning, hormonal deficiencies or excess hormone secretion can also seen in these patients [30]. Studies have shown that the prevalence of microincidentaloma on MRI for an unrelated reason varies from 10% to 38% while that for macroincidentaloma is far less and varies from 0.16% to 0.3% [31]. Series of untreated PRL secreting microadenoma has shown that conversion rate to PRL macroadenoma was less than 5%. Endocrine society recommends that all patients of incidentaloma should be investigated for the presence of signs and symptoms of pituitary hormone dysfunction irrespective of their size [32]. All patients with macroincidentalomas or large solid microincidentalomas with a diameter of >6 mm should undergo laboratory investigation for the presence of hypopituitarism. For microincidentalomas of size >6 mm, assessment of PRL and IGF-I alone is sufficient [32]. Hormonal assessment is not required for cystic microincidentalomas and solid lesions <5 mm in diameter. Visual campimetry is necessary when the tumor extends till optic chiasm [33]. Treatment is always indicated for functioning macroadenomas depending on the type of tumor.

Pituitary Carcinoma It is a rare entity comprising of <0.5% of all pituitary tumors. Most pituitary carcinomas originate from adenohypophysis and secrete PRL and ACTH. Clinical presentation resembles that of PA. These tumors have a high propensity for CSF and systemic metastases. Mitotic count/Ki-67 index indicates aggressiveness of the tumor. Despite aggressive treatment with surgery and RT, overall prognosis remains poor with mean survival of 1.9 years. Studies have suggested that temozolomide (TMZ) is effective when used as a first-line chemotherapeutic agent after surgery [34]. Concurrent RT with TMZ may further improve the outcome.

Pituicytoma is a rare low-grade glial neoplasm which originates from neurohypophysis, and can be misdiagnosed preoperatively as a pituitary adenoma [35].

Pituitary Blastoma It is a rare aggressive malignant neoplasm of the pituitary gland mostly seen in children [36]. Pituitary blastomas are generally seen as a part of pleuropulmonary blastoma (PPB)-familial tumor, which is caused by heterozygous germline mutations in the DICER1gene.

Pituitary Apoplexy This is an uncommon clinical condition that occurs because of acute hemorrhage or infarction in the pituitary gland, often in PA patients. It affects 0.2–0.6% of the general population and 2–12% of patients with pituitary adenomas [37]. Headache is the most common symptom followed by vision loss, hypopituitarism, nausea, diplopia and altered consciousness [37].

Patients may present in acute (within 24–72 h of symptom onset) or subacute setting. The classic triad of acute pituitary apoplexy constitutes (1) severe, sudden-onset headache ("Worst headache of life!"), (2) acute hypopituitarism (addisonian crisis, hyponatremia), and (3) visual disturbances. If imaging shows acute hemorrhage or infarction a diagnosis of pituitary apoplexy should be made. Acute pituitary apoplexy is a medical emergency. A rapid treatment with hydrocortisone is often lifesaving. Once patient is medically stable, urgent resection of the hemorrhagic pituitary tumor should be done.

Several studies have compared conservative approach with surgical management in pituitary apoplexy. A systematic review has demonstrated a significantly higher recovery rate for ocular palsy and visual field in surgically treated patients, but no significant difference was seen in visual acuity or pituitary function recovery [38]. A large, multi-institution data registry (PASTOR—Pituitary Apoplexy Surgical Timing & Outcomes Registry) is prospectively analysing observation versus surgery in acute or sub-acute onset pituitary apoplexy. At present, urgency of surgical treatment depends on symptom severity and duration. Urgent surgery is the mainstay for the treatment of acute pituitary apoplexy, especially in those with vision loss. However, in milder cases, a conservative approach can be preferred.

31.9 Follow-up

Patients are followed up with clinical evaluation to see any change in pretreatment visual or hormonal symptoms. Hormonal evaluation is done 3 months after treatment and then at least once in 6 months. Ophthalmology evaluation (visual acuity, fields, fundus, oculomotor evaluation) should be conducted every 6 months for those with baseline abnormalities and every year in those who had normal vision at baseline. First follow-up MRI is generally recommended at 3–6 months after treatment and then every year for 5 years, and subsequently every 2–3 years in absence of progression.

31.10 Conclusions

Pituitary tumors are a common cause of morbidity and often require multidisciplinary management. Observation, surgery, medical management, and radiotherapy are treatment options that are tailored according to the clinical circumstances. With suitable therapy, long-term control with good quality of life is feasible for over 90% of the patients. Long-term follow-up with periodic imaging, hormonal and vision assessments is required to assess disease control and treatment sequelae.

References

1. Elston MS, McDonald KL, Clifton-Bligh RJ, Robinson BG. Familial pituitary tumor syndromes. Nat Rev Endocrinol. 2009;5(8):453–61.
2. Mete O, Lopes MB. Overview of the 2017 WHO classification of pituitary tumors. Endocr Pathol. 2017;28(3):228–43.
3. Bonneville J-F, Bonneville F, Cattin F. Magnetic resonance imaging of pituitary adenomas. Eur Radiol. 2005;15(3):543–8.

4. Iglesias P, Rodríguez Berrocal V, Díez JJ. Giant pituitary adenoma: histological types, clinical features and therapeutic approaches. Endocrine. 2018;61(3):407–21.
5. Crabtree JS, Scacheri PC, Ward JM, Garrett-Beal L, Emmert-Buck MR, Edgemon KA, et al. A mouse model of multiple endocrine neoplasia, type 1, develops multiple endocrine tumors. Proc Natl Acad Sci U S A. 2001;98(3):1118–23.
6. Bashari WA, Senanayake R, Fernández-Pombo A, Gillett D, Koulouri O, Powlson AS, et al. Modern imaging of pituitary adenomas. Best Pract Res Clin Endocrinol Metab. 2019;33(2):101278.
7. Tsagarakis S, Grossman A, Plowman PN, Jones AE, Touzel R, Rees LH, et al. Megavoltage pituitary irradiation in the management of prolactinomas: long-term follow-up. Clin Endocrinol. 1991;34(5):399–406.
8. van Varsseveld NC, van Bunderen CC, Ubachs DHH, Franken AAM, Koppeschaar HPF, van der Lely AJ, et al. Cerebrovascular events, secondary intracranial tumors, and mortality after radiotherapy for nonfunctioning pituitary adenomas: a subanalysis from the Dutch National Registry of Growth Hormone Treatment in Adults. J Clin Endocrinol Metab. 2015;100(3):1104–12.
9. Jane JA, Catalino MP, Laws ER. Surgical treatment of pituitary adenomas. In: Feingold KR, Anawalt B, Boyce A, Chrousos G, Dungan K, Grossman A, et al., editors. Endotext [internet]. South Dartmouth, MA: MDText.com, Inc.; 2000. Available from: http://www.ncbi.nlm.nih.gov/books/NBK278983/.
10. Gillam MP, Molitch ME, Lombardi G, Colao A. Advances in the treatment of prolactinomas. Endocr Rev. 2006;27(5):485–534.
11. Mehta GU, Lonser RR. Management of hormone-secreting pituitary adenomas. Neuro Oncol. 2017;19(6):762–73.
12. Melmed S, Cook D, Schopohl J, Goth MI, Lam KSL, Marek J. Rapid and sustained reduction of serum growth hormone and insulin-like growth factor-1 in patients with acromegaly receiving lanreotide Autogel therapy: a randomized, placebo-controlled, multicenter study with a 52 week open extension. Pituitary. 2010;13(1):18–28.
13. Grasso LFS, Auriemma RS, Pivonello R, Colao A. Somatostatin analogs, cabergoline and pegvisomant: comparing the efficacy of medical treatment for acromegaly. Expert Rev Endocrinol Metab. 2017;12(1):73–85.
14. Buchfelder M, van der Lely A-J, Biller BMK, Webb SM, Brue T, Strasburger CJ, et al. Long-term treatment with pegvisomant: observations from 2090 acromegaly patients in ACROSTUDY. Eur J Endocrinol. 2018;179(6):419–27.
15. Schteingart DE. Drugs in the medical treatment of Cushing's syndrome. Expert Opin Emerg Drugs. 2009;14(4):661–71.
16. Colao A, Petersenn S, Newell-Price J, Findling JW, Gu F, Maldonado M, et al. A 12-month phase 3 study of pasireotide in Cushing's disease. N Engl J Med. 2012;366(10):914–24.
17. Fleseriu M, Findling JW, Koch CA, Schlaffer S-M, Buchfelder M, Gross C. Changes in plasma ACTH levels and corticotroph tumor size in patients with Cushing's disease during long-term treatment with the glucocorticoid receptor antagonist mifepristone. J Clin Endocrinol Metab. 2014;99(10):3718–27.
18. Nagesser SK, van Seters AP, Kievit J, Hermans J, Krans HM, van de Velde CJ. Long-term results of total adrenalectomy for Cushing's disease. World J Surg. 2000;24(1):108–13.
19. Beck-Peccoz P, Brucker-Davis F, Persani L, Smallridge RC, Weintraub BD. Thyrotropin-secreting pituitary tumors. Endocr Rev. 1996;17(6):610–38.
20. Li X, Li Y, Cao Y, Li P, Liang B, Sun J, et al. Safety and efficacy of fractionated stereotactic radiotherapy and stereotactic radiosurgery for treatment of pituitary adenomas: a systematic review and meta-analysis. J Neurol Sci. 2017;372:110–6.
21. Scheick S, Amdur RJ, Kirwan JM, Morris CG, Mendenhall WM, Roper S, et al. Long-term outcome after fractionated radiotherapy for pituitary adenoma: the curse of the secretory tumor. Am J Clin Oncol. 2016;39(1):49–54.
22. Sheehan JP, Starke RM, Mathieu D, Young B, Sneed PK, Chiang VL, et al. Gamma Knife radiosurgery for the management of nonfunctioning pituitary adenomas: a multicenter study. J Neurosurg. 2013;119(2):446–56.

23. Minniti G, Flickinger J. The risk/benefit ratio of radiotherapy in pituitary tumors. Best Pract Res Clin Endocrinol Metab. 2020;33(2):101269. Available from: https://pubmed.ncbi.nlm.nih.gov/31053487/?from_term=The+risk%2Fbenefit+ratio+of+radiotherapy+in+pituitary+tumors&from_pos=1.
24. Pollock BE, Link MJ, Leavitt JA, Stafford SL. Dose-volume analysis of radiation-induced optic neuropathy after single-fraction stereotactic radiosurgery. Neurosurgery. 2014;75(4):456–60. discussion 460
25. Loeffler JS, Shih HA. Radiation therapy in the management of pituitary adenomas. J Clin Endocrinol Metab. 2011;96(7):1992–2003.
26. Sebastian P, Balakrishnan R, Yadav B, John S. Outcome of radiotherapy for pituitary adenomas. Rep Pract Oncol Radiother J Gt Cancer Cent Poznan Pol Soc Radiat Oncol. 2016;21(5):466–72.
27. Biermasz NR, Dulken HV, Roelfsema F. Postoperative radiotherapy in acromegaly is effective in reducing GH concentration to safe levels. Clin Endocrinol. 2000;53(3):321–7.
28. Minniti G, Osti M, Jaffrain-Rea ML, Esposito V, Cantore G, Maurizi Enrici R. Long-term follow-up results of postoperative radiation therapy for Cushing's disease. J Neuro-Oncol. 2007;84(1):79–84.
29. Wattson DA, Tanguturi SK, Spiegel DY, Niemierko A, Biller BMK, Nachtigall LB, et al. Outcomes of proton therapy for patients with functional pituitary adenomas. Int J Radiat Oncol Biol Phys. 2014;90(3):532–9.
30. Orija IB, Weil RJ, Hamrahian AH. Pituitary incidentaloma. Best Pract Res Clin Endocrinol Metab. 2012;26(1):47–68.
31. Vasilev V, Rostomyan L, Daly AF, Potorac I, Zacharieva S, Bonneville J-F, et al. Management Of Endocrine Disease: pituitary "incidentaloma": neuroradiological assessment and differential diagnosis. Eur J Endocrinol. 2016;175(4):R171–84.
32. Boguszewski CL, de Castro Musolino NR, Kasuki L. Management of pituitary incidentaloma. Best Pract Res Clin Endocrinol Metab. 2019;33(2):101268.
33. Chanson P, Raverot G, Castinetti F, Cortet-Rudelli C, Galland F, Salenave S, et al. Management of clinically non-functioning pituitary adenoma. Ann Endocrinol. 2015;76(3):239–47.
34. McCormack A, Dekkers OM, Petersenn S, Popovic V, Trouillas J, Raverot G, et al. Treatment of aggressive pituitary tumours and carcinomas: results of a European Society of Endocrinology (ESE) survey 2016. Eur J Endocrinol. 2018;178(3):265–76.
35. El Hussein S, Vincentelli C. Pituicytoma: review of commonalities and distinguishing features among TTF-1 positive tumors of the central nervous system. Ann Diagn Pathol. 2017;29:57–61.
36. Scheithauer BW, Horvath E, Abel TW, Robital Y, Park S-H, Osamura RY, et al. Pituitary blastoma: a unique embryonal tumor. Pituitary. 2012;15(3):365–73.
37. Briet C, Salenave S, Bonneville J-F, Laws ER, Chanson P. Pituitary apoplexy. Endocr Rev. 2015;36(6):622–45.
38. Tu M, Lu Q, Zhu P, Zheng W. Surgical versus non-surgical treatment for pituitary apoplexy: a systematic review and meta-analysis. J Neurol Sci. 2016;370:258–62.

Craniopharyngioma

32

Prashanth Giridhar

32.1 Introduction

Craniopharyngioma are rare embryonic malformations with low histological grade (I) that commonly occur in sellar and para-sellar region. They are associated with very good survival particularly in the pediatric age group (20 year OS: 87–95%) but treated patients are prone to long-term treatment related sequelae [1–5]. Adult craniopharyngioma are associated with a comparatively worse prognosis with increased risk of cerebrovascular mortality [6, 7]. Treatment therefore needs to be carefully tailored to balance survival with risk of toxicity.

32.2 Epidemiology

- Incidence of disease is 0.05–0.2 cases per 100,000 population every year.
- A bimodal age distribution is seen. First peak is 5–14 years age and second peak is 50–70 years age.
- No gender predilection seen.

32.3 Clinical Presentation [8]

1. Headache and vomiting due to raised intracranial tension.
2. Visual deficits due to compression of optic nerve/chiasma.
3. Endocrine disorders due to compression of pituitary and/or hypothalamus—decreased growth rate, hypothalamic obesity, polyuria and polydipsia are

P. Giridhar (✉)
Department of Radiation Oncology, National Cancer Institute, AIIMS, Haryana, India

S. Mallick et al. (eds.), *Evidence based practice in Neuro-oncology*,
https://doi.org/10.1007/978-981-16-2659-3_32

common manifestations. Growth hormone (75%) most commonly affected followed by gonadotropins (40%), ACTH (25%), and TSH (25%).

**Any clinical combination of headache, visual impairment, decreased growth rate, and/or polydipsia/polyuria should arouse suspicion of childhood craniopharyngioma in the differential diagnosis process.*

32.4 Radiology

32.4.1 Appearance on CT

Solid cystic mass seen in sellar and suprasellar region.

Most commonly they are a Suprasellar mass extending into sella.

Calcification seen in 90% cases.

32.4.2 Appearance on MRI

Signal intensity varies on MRI due to different protein concentrations in cystic fluid.

The solid component is iso-intense on T1 weighted imaging.

Radiological differentials include optic glioma, langerhans cell histiocytosis, intracranial germinoma, colloid cyst, and pituitary adenoma.

32.5 Management

- Surgical resection is the initial treatment of choice [8].
- *Tumor does not involve hypothalamus/ optic structures:* Attempt gross total resection with preservation of hypothalamus/ optic structures. Confirm GTR with CT after surgery.
- *Tumor involves hypothalamus/ optic structures:* Attempt subtotal resection with preservation of hypothalamus/optic structures.
- Surgical approach can be transcranial sub-frontal or trans-sphenoidal.
- Transcranial approach requires more neuro-vascular manipulation including optic nerves and brain retraction.
- Trans-sphenoidal approach is preferred. Risk of CSF fistula present.

32.6 Pathology

- The two types of craniopharyngioma are: Adamantinomatous and papillary.
- Adamantinomatous craniopharyngioma shows presence of squamous epithelium disposed in cords, nodules, and irregular trabeculae bordered by palisaded columnar epithelium. These islands of densely packed cells merge with loosely cohesive aggregates of squamous cells known as stellate reticulum.

- Essential features papillary craniopharyngioma include a monomorphous mass of well-differentiated squamous epithelium lacking surface maturation, the picket fence-like palisades and wet keratin.
- Most common genetic mutation: Beta catenin.

32.7 Adjuvant Therapy [8]

32.7.1 Radiation

- *If GTR done: No adjuvant radiotherapy.*
- *If STR done: Adjuvant radiation to dose of 54 Gy/30#/6 weeks.*
- Technique: Fractionated stereotactic radiotherapy.
- Target volume: Residual solid disease plus cyst forms gross tumor volume. 5 mm margin is given to form clinical target volume. PTV 2–3 mm if SRT done.
- 10 year PFS ranges from 78% to 95% with surgery plus radiotherapy.
- Data on proton beam therapy limited. Small studies show preservation of neuro-cognition.
- Single session gamma knife surgery may be considered if the following features are present:
  - <3 cm lesion
  - Only solid lesion
  - >3 mm away from brainstem and optic structures

    Dose: 12–15 Gy

32.7.1.1 Management of Recurrence

5–20% patients recur at 10 years.

Complete resection is possible in less than one fourth cases due to scarring from previous treatment.

Safe limited resection may be attempted.

If previously radiotherapy was not given—Radiotherapy to a dose of 54 Gy given.

If previously radiotherapy given and only cystic recurrence: Options include Ommaya reservoir placement and repeated cyst drainage; Intra-cystic instillation of interferon alpha; Instillation of beta emitters like Y90 and P32.

If previously radiotherapy given and solid recurrence: Close follow up after limited surgery.

32.7.2 Long-Term Sequelae

32.7.2.1 Pituitary Dysfunction

Pediatric craniopharyngioma: Growth hormone deficiency is most common leading to impaired growth. It is seen in 70–90% patients. In some patients with hypothalamic involvement, 'growth without GH' is seen leading to normal height.

Adult craniopharyngioma: Deficiencies in GH, ACTH, TSH, and gonadotropins seen in more than two-third patients after surgery. Diabetes insipidus presenting with polydipsia and polyuria develops in 70% patients.

32.7.2.2 Hypothalamic Dysfunction

Obesity, behavioural changes, disturbed circadian rhythm, disturbances in regulation of body temperature, thirst, heart rate, and/or blood pressure develop in more than half the number of patients.

Visual problems.

Neurovascular sequelae like hemiparesis and epilepsy in few patients.

References

1. Wisoff JH, Donahue BR. Craniopharyngiomas. In: Albright AL, Pollack IF, Adelson PD, editors. Principles and practice of pediatric neurosurgery. 2nd ed. New York, Stuttgart: Thieme; 2006. p. 560–77.
2. Müller HL. Childhood craniopharyngioma–current concepts in diagnosis, therapy and follow-up. Nat Rev Endocrinol. 2010;6:609–18.
3. Müller HL. Childhood craniopharyngioma: current controversies on management in diagnostics, treatment and follow-up. Expert Rev Neurother. 2010;10:515–24.
4. Müller HL. Childhood craniopharyngioma. Recent advances in diagnosis, treatment and follow-up. Horm Res. 2008;69:193–202.
5. Karavitaki N, Cudlip S, Adams CB, Wass JA. Craniopharyngiomas. Endocr Rev. 2006;27:371–97.
6. Tomlinson JW, Holden N, Hills RK, et al. Association between premature mortality and hypopituitarism. West Midlands Prospective Hypopituitary Study Group. Lancet. 2001;357:425–31.
7. Erfurth EM, Holmer H, Fjalldal SB. Mortality and morbidity in adult craniopharyngioma. Pituitary. 2013;16:46–55.
8. Müller HL, Merchant TE, Warmuth-Metz M, et al. Craniopharyngioma. Nat Rev Dis Primers. 2019;5(1):75.

Acoustic Neuroma

33

V. R. Anjali

33.1 Introduction

- Acoustic neuroma or vestibular schwannoma (VS) is a benign tumour arising from the Schwann cells in the nerve sheath covering the vestibulocochlear nerve. The nomenclature VS is preferred since the lesion is seen in superior or inferior branch of vestibular nerve and histology being schwannoma.
- It accounts for 5–10% of intracranial lesions and 80–90% of cerebellopontine angle (CPA) tumours [1].
- The mean age of presentation is 62 years.
- There is no gender predilection.
- Majority are sporadic (90%) and unilateral.
- Hereditary variants are associated with neurofibromin 2 (NF2) and is usually bilateral.
- Pathogenesis of both familial and sporadic (45%) VS has been associated with NF2 gene mutation [2].
- NF2 is located on chromosome 22, encodes for tumour suppressor protein moesin-ezrin-radixin–like protein merlin/schwannomin.
- Other risk factors
  - Ionising radiation
  - Acoustic trauma
- The non-vestibular schwannomas are usually seen in trigeminal nerve (CN V) and oculomotor nerve (CN III).

V. R. Anjali (✉)
Department of Radiation Oncology, All India Institute of Medical Sciences, New Delhi, India

S. Mallick et al. (eds.), *Evidence based practice in Neuro-oncology*,
https://doi.org/10.1007/978-981-16-2659-3_33

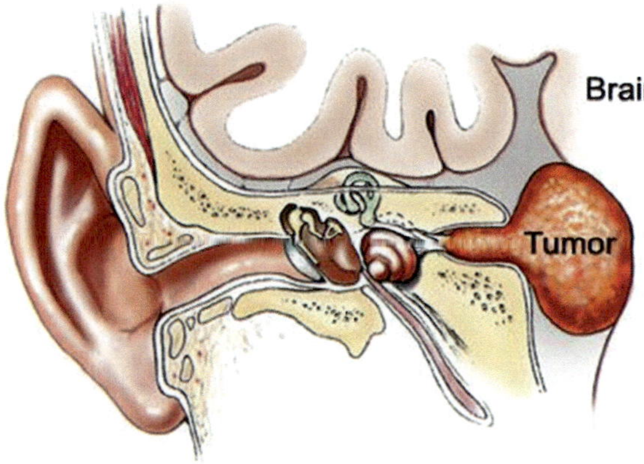

33.2 Symptoms

Symptoms are attributed to cranial nerve VIII involvement, compression on surrounding structures like cranial nerves, cerebellum, brainstem.

- Unilateral progressive sensorineural hearing loss:
  - due to impaired blood supply to cochlear nerve
  - Commonest and the earliest symptom
  - Slowly progressive
  - High frequency retro-cochlear sensorineural type
- Tinnitus
- Vertigo
- Disequilibrium
- Headache
- Trigeminal
- Facial nerve dysfunction

In 1917, the following stages of evolution of symptoms were proposed by Cushing et al.: (1) auditory and labyrinthine dysfunction; (2) occipitofrontal pain; (3) cerebellar ataxia; (4) cranial nerve involvement; (5) features of raised ICT; (6) dysphagia and dysarthria (7) brainstem compression causing respiratory depression and death [3].

In 1993, Selesnick et al. classified symptoms depending on location of tumour into four stages summarised in Table 33.1. The average duration of symptoms prior to the diagnosis disease for hearing loss is 3.9 years, vertigo 3.6 years, tinnitus 3.4 years, headache 2.2 years, disequilibrium 1.7 years, trigeminal nerve

Table 33.1 Symptoms depending on location of tumour

Stage	Symptoms
Intracanalicular	Hearing loss Tinnitus Vertigo
Cisternal	Worsening of hearing loss Vertigo diminishes Disequilibrium increases
Brainstem compression	Midfacial and corneal hypaesthesia Occipital headache Ataxia
Hydrocephalic	Trigeminal dysfunction worsens Gait deterioration Raised intracranial function Lower cranial nerve dysfunction Death due to tonsillar herniation

dysfunction 0.9 years and facial nerve dysfunction 0.6 years. At an average growth rate time for the tumour to grow from intracanalicular to cisternal stage is 1.5–2 years [4].

33.3 Investigation

1. Magnetic Resonance Imaging (MRI):
   (a) Thin slices of TI weighted MRI with gadolinium at CPA is the gold standard for diagnosis. It shows hypo to iso intensity with strong enhancement after gadolinium.
   (b) T2 weighted MRI shows heterogeneously hyperintense lesion.
   (c) Heavily T2 weighted sequences like CISS (constructive interference in steady state), FIESTA-C (fast imaging employing steady-state acquisition cycled phase) provide high resolution images of tumour and nerve.
2. Computed tomography (CT) scan:
   (a) Used in patients who have contraindication for MRI/claustrophobic. Widely available and cheaper than MRI. VS >1.5 cm has better visualisation with sensitivity of 100%. Sensitivity decreases for tumours<1 cm and is unable to detect intracanalicular VS.
3. Pure Tone Audiometry (PTA) and Speech discrimination score (SDS):
   (a) The hearing quality is assessed by the PTA and SDS.
   (b) The most important factors in preoperative decision making.
   (c) SDS was considered as a good predictor for preservation of functional hearing.

(d) Patients with relatively poor PTA thresholds, but with SDS preserved are attempted hearing-preservation surgery, because amplification can successfully rehabilitate the loss.
(e) If PTA is preserved but SDS is poor, patient won't be a candidate for hearing preserving surgery, since amplification is less useful.
(f) A patient with hearing equal to or better than 50 dB PTA and better than 50% SDS can be considered for hearing-preservation surgery.

4. Auditory brainstem response (ABR):
   (a) ABR retains 93.4% sensitivity for detecting VS of any size. The sensitivity is 95.6% for larger tumours (>1 cm) and 85.8% for smaller tumours (<1 cm) [5].

33.4 Grading System

Various grading systems for VS is available based on tumour size/anatomical location. Sterker and House used tumour size, Samii used anatomical location of tumour while Koos used both for tumour grading. Grading summarised in Table 33.2.

33.5 Histopathology

- Arise near the porus acusticus. This area represents transition point between glial and Schwann cells (Obersteiner-Redlich zone).
- On grossing areas of cystic degeneration, haemorrhage, calcification is found.
- On light microscopy, tumour is made up of spindle cells with elongated nuclei and fibrillary cytoplasm.
- Alternating zones of dense and sparse cellular areas, Antoni A and Antoni B are characteristic.

Table 33.2 Grading system

Sterker	House	Tumour size	Koos	Tumour location	Samii
Tube type	Intracanalicular	0	Grade I	Confined to internal acoustic canal	T1
Small	Grade 1	≤10 mm	Grade II	Super passing IAC	T2
	Grade 2	≤15 mm		Occupying CPA	T3a
Mild		≤20 mm			
	Grade 3	≤30 mm	Grade III	Occupy CPA and brain stem contact	T3b
Large	Grade 4	≤40 mm	Grade IV	Brain stem compression	T4a
Huge	Grade 5	>40 mm		Severe brain displacement Deformation of fourth ventricle	T4b

- Antoni A tissue composed of prominent, organised, interwoven course of elongated bipolar cells.
- Antoni B is composed of irregular conglomeration of cells around foci of cystic change, necrosis, old haemorrhage.
- Immuno histochemistry is diffusely positive for S100 and SOX10.

33.6 Management

VS is a slow growing benign tumour, but because of its critical location even small lesions can lead to symptoms. The main goal of treatment is local control of disease with preservation of cranial nerve function and maintaining good quality of life.

Treatment option includes:

- Conservative/Observation: For small tumours and elderly patients
- Radiotherapy: Stereotactic radiation
- Surgery

There is no consensus on management guidelines. Each treatment option carries its own toxicity/risk as well as safety profile/benefits. Tumour growth is the most important predictor of treatment strategy.

33.6.1 Conservative/Observation

Indication:

- For tumours <15 mm
- Intracanalicular VS without tinnitus
- Elderly patients with significant comorbidities

Observation with radiological surveillance and audiometry [6, 7].

Nearly 50% patients experience no growth or negative growth during observation [8]. Average tumour growth rate is 1.9 mm/year [8]. Maximum tumour growth is seen in first year.

33.6.1.1 Predicting Factors for Rapid Growth

- Large tumour size at diagnosis.
- Tinnitus at diagnosis (threefold) [9].
- IAC tumours grow significantly at a lesser rate than CPA tumours [10, 11].
- Extrameatal tumours.
- Compared to solid lesions, cystic VS has greater growth rate [12].

Radiological surveillance includes MRI at 6 months after first imaging then followed by yearly MRI for 5 years then every alternate year for next 4 years for minimum surveillance of 10 years.

33.6.1.2 Limitation

- Risk of rapid tumour progression
- Easily resectable /radio ablative lesion become complicated one
- Possibility of reduction in hearing by 40–60%
- Success is dependent upon patient's compliance

33.6.2 Surgery

Surgery is option for patients who are not a candidate for wait and watch strategy.

33.6.2.1 Indication

- Large tumours (Koos grade 4).
- Mass effect, brain stem compression or hydrocephalus.
- Cystic VS, surgery is considered in view of rapid and dramatic growth pattern of lesion.

33.6.2.2 The Surgical Approaches are Mainly Three Types (Table 33.3)

1. Middle cranial fossa approach
2. Retro sigmoid
3. Trans labyrinthine

The factors which influenced the type surgical approach are

1. Tumour size
2. Hearing status of patient/serviceable hearing
3. Age of the patient
4. Surgeons preference

Table 33.3 Surgical approaches

Middle cranial fossa approach	Small intracanalicular tumours <1.5 cm Normal/near normal hearing Younger patients (<65 years) Hearing preservation is goal
Retro sigmoid approach	Size >1.5 cm Intracanalicular lesion with>1 cm into CPA Good hearing Hearing preservation is aimed
Trans labyrinthine	Tumours>3 cm; poor hearing Hearing could not be preserved

(a) The MC and RS approach is preferred for patients with residual hearing and in patients which hearing preservation is planned.
(b) TL approach is done in large tumours with poor hearing, where preservation of hearing is not achieved.
(c) Facial nerve preservation outcome with NTR, STR and GTR are 97%, 96% and 77% respectively.
(d) For tumours size more than 2.5 cm, GTR shows a trend for worse facial nerve function due to nerve injury.
(e) So NTR or STR with functional preservation is preferred over gross total resection with nerve injury. Tumour regrowth was seen higher in NTR (2%) and STR (10%) groups as compared to GTR (0%), but the need for further treatment was extremely low.
(f) The local control after surgery ranges between 80% and 98%.
(g) The preservation of serviceable hearing is 50–75% at 5 years and 25–50% at 10 years [12]. Permanent damage to the facial nerve is about 10–20%.

33.6.3 Radiotherapy

Radiotherapy is noninvasive alternative treatment modality with good local control and better toxicity profile.

33.6.3.1 Indications

- Tumours up to 3 cm is treated with highly conformal RT.
- Residual or recurrent lesions post-surgery.
- Progressive lesion on observation.

Table 33.4 Radiation techniques

RT technique	SRS	CFSRT	HFSRT
Delivered using	Gammaknife Linear accelerator Cyberknife	Linear accelerator	Linear accelerator Cyberknife
Dose	10–13 Gy in single fraction prescribed to 50% isodose line	54–57.6 Gy @ of 1.8 Gy per fraction	20–25 Gy in 5 fractions/18 Gy in 3 fractions prescribed to 80% isodose line
Target volume	GTV—enhancing mass lesion on T1 weighted MRI with gadolinium. No CTV PTV—depending on setup error and institutional protocol		
Toxicity	Facial nerve neuropathy and trigeminal neuropathy 1–3% Loss of useful hearing 14–16% [13]		Facial nerve dysfunction 0–9% and trigeminal neuropathy 0–9.4%
Outcome in terms of local control	Excellent local control 92–98% at 5 years 92–98% at 10 years [14] No significant difference in LRC after SRS/CFSRT		90.5%–99.1% at 5 years [15, 16]

33.6.3.2 Radiation Techniques (Table 33.4)

- Radiotherapy options include stereotactic radiotherapy (SRS)
- Hypo fractionated stereotactic radiotherapy (HFSRT)
- Conventionally fractionated stereotactic radiotherapy (CFSRT)

33.6.3.3 Specific Highlights of Radiation

- Salvage SRS provides local control 85–95% at 5 years.
- SRS dose should not exceed 13 Gy as risk of toxicity increases without additional benefit on disease control.
- After SRS some tumours tend to increase in size for a transient period of first 1–3 years called the pseudo-progression.

33.6.3.4 Proton Therapy

- Results are not encouraging.
- Hearing-preservation rates are low compared to other RT techniques.
- Local tumour control 95.3% and 93.6% at 2 and 5 years, respectively.
- Cranial nerve preservation rates 90%.
- Patients who retained serviceable hearing was only 33% [17] with proton beam SRS and 42% with FSRT [18].

33.6.4 Systemic Therapy

Both familial and sporadic tumours have mutation of NF2 gene product Merlin. Merlin interact with tyrosine kinase receptor including epidermal growth factor family (EGF), platelet derived growth factor receptor (PDGFR) and c kit. Immunohistochemical analysis showed Her 2 was upregulated in 76% of sporadic and 94% of familial VS [19]. Erlotinib (oral EGFR inhibitor), Bevacizumab (humanised anti VEGF monoclonal antibody) are the systemic agents under trial for VS.

33.6.5 Quality of Life

Two prospective studies evaluated QOL of patients with vestibular schwannoma treated with microsurgical resection and SRS with follow up of 24 and 42 months. Both studies showed significantly better facial nerve and hearing outcomes from SRS group than from surgery [20, 21]. Another prospective study showed similar QOL for patients, after observation, radiotherapy, or surgery [22].

33.7 Treatment Summary in Flowchart

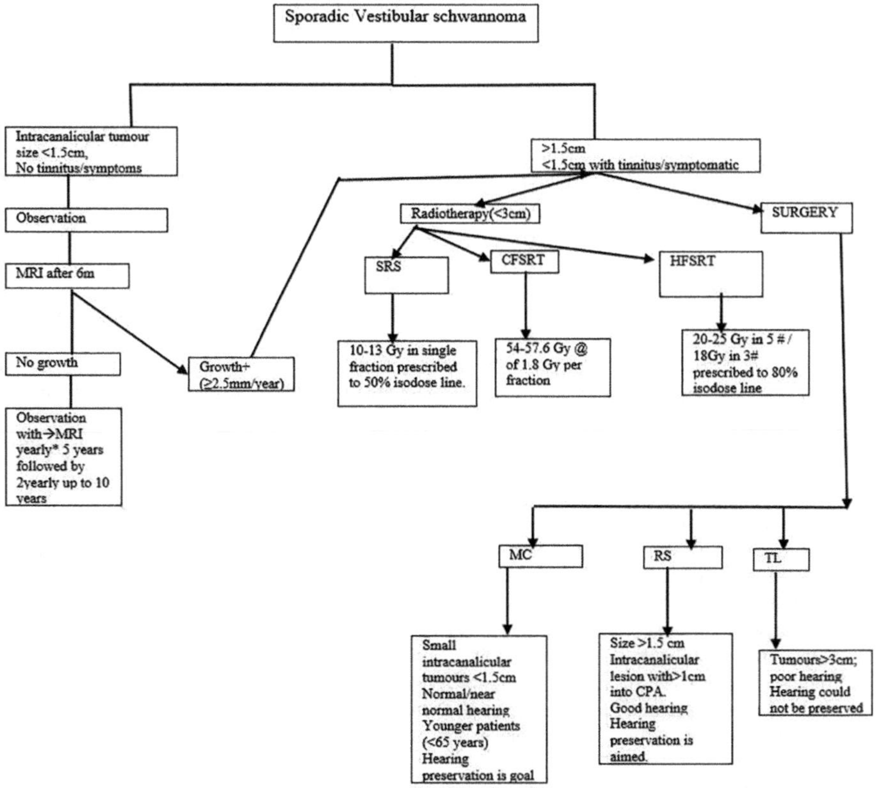

References

1. Hart RG, Davenport J. Diagnosis of acoustic neuroma. Neurosurgery. 1981;9(4):450–63.
2. Bikhazi PH, Lalwani AK, Kim EJ, et al. Germline screening of the NF-2 gene in families with unilateral vestibular schwannoma. Otolaryngol Head Neck Surg. 1998;119:1–6.
3. Cushing H. Tumors of the nervous acousticus. Philadelphia: WB Saunders; 1917.
4. Selesnick SH, Jackler RK. The changing clinical presentation of acoustic tumors in the MRI era. Laryngoscope. 1993;103:431–6.
5. Koors P, Thacker L, Coelho D. ABR in the diagnosis of vestibular schwannomas: a meta-analysis. Am J Otolaryngol. 2013;34:195–204.
6. Germano IM, Sheehan J, Parish J, et al. Congress of neurological surgeons systematic review and evidence-based guidelines on the role of radiosurgery and radiation therapy in the management of patients with vestibular schwannomas. Neurosurgery. 2018;82(2):E49–51.
7. Stangerup SE, Caye-Thomasen P, Tos M, et al. The natural history of vestibular schwannoma. Otol Neurotol. 2006;27(4):547–52.

8. Smouha EE, Yoo M, Mohr K, et al. Conservative management of acoustic neuroma: a meta-analysis and proposed treatment algorithm. Laryngoscope. 2005;115:450–4.
9. Agrawal Y, Clark JH, Limb CJ, et al. Predictors of vestibular schwannoma growth and clinical implications. Otol Neurotol. 2010;31(5):807–12.
10. Battaglia A, Mastrodimos B, Cueva R. Comparison of growth patterns of acoustic neuromas with and without radiosurgery. Otol Neurotol. 2006;27:705–12.
11. Raut VV, Walsh RM, Bath AP, et al. Conservative management of vestibular schwannomas-Vsecond review of a prospective longitudinal study. Clin Otolaryngol. 2004;29:505–14.
12. Carlson ML, Vivas EX, McCracken DJ, et al. Congress of neurological surgeons systematic review and evidence-based guidelines on hearing preservation outcomes in patients with sporadic vestibular schwannomas. Neurosurgery. 2018;82(2):E35–e39.
13. Combs SE, Engelhard C, Kopp C, et al. Long-term outcome after highly advanced single-dose or fractionated radiotherapy in patients with vestibular schwannomas - pooled results from 3 large German centers. Radiother Oncol. 2015;114(3):378–83.
14. Murphy ES, Suh JH. Radiotherapy for vestibular schwannomas: a critical review. Int J Radiat Oncol Biol Phys. 2011;79:985–97.
15. Anderson BM, Khuntia D, Bentzen SM, et al. Single institution experience treating 104 vestibular schwannomas with fractionated stereotactic radiation therapy or ste-reotactic radiosurgery. J Neuro-Oncol. 2014;116:187–93.
16. Tsai JT, Lin JW, Lin CM, et al. Clinical evaluation of CyberKnife in the treatment of vestibular schwannomas. Biomed Res Int. 2013;2013:297093.
17. Weber DC, Chan AW, Bussiere MR, et al. Proton beam radiosurgery for vestibular schwannoma: tumor control and cranial nerve toxicity. Neurosurgery. 2003;53(3):577–88.
18. Vernimmen FJ, Mohamed Z, Slabbert JP, et al. Long-term results of stereotactic proton beam radiotherapy for acoustic neuromas. Radiother Oncol. 2009;90(2):208–12.
19. Doherty J, Ongkeko W, Crawley B, et al. ErbB and Nrg: potential molecular targets for vestibular schwannoma pharmacotherapy. Otol Neurotol. 2008;29(1):50–7.
20. Myrseth E, Moller P, Pedersen PH, et al. Vestibular schwannoma: surgery or gamma knife radiosurgery? A prospective, nonrandomized study. Neurosurgery. 2009;64:654–63.
21. Pollock BE, Driscoll CL, Foote RL, et al. Patient outcomes after vestibular schwannoma management: a prospective comparison of microsurgical resection and stereotactic radiosurgery. Neurosurgery. 2006;59:77–85.
22. Di Maio S, Akagami R. Prospective comparison of quality of life before and after observation, radiation, or surgery for vestibular schwannomas. J Neurosurg. 2009;111:855–62.

Spinal Cord Tumors

34

BhanuPrasad Venkatesulu

34.1 Introduction

Spinal cord tumors constitute 3–4% of all the primary central nervous system tumors [1]. The spinal cord ends at L1–L2 in adults and L3–L4 in the infants. The termination of spinal cord is followed by filum terminale which attaches the dural sac to the coccygeal vertebra. They can be classified according to the WHO classification based on histological subtypes as well as clinical stratification based on the location of the tumors within the spinal cord layers. The spinal cord is embedded with a protective lining of outer duramater, middle arachnoid and inner piamater [2]. The tumors can be classified based on imaging into extradural, intradural–extramedullary, or intramedullary. The spinal cord tumors that originate from the spinal cord parenchyma are referred as intramedullary tumors; those that arise from duramater are referred as intradural–extramedullary tumors; and those that arise from structures outside the spinal cord coverings but within the spinal canal are referred as extradural. Around 33% of tumors are intramedullary in nature and rest are extramedullary in nature [3]. The spinal cord tumors are mostly benign with only 20% being malignant on presentation. The most common sites are the spinal cord (60%), followed by the spinal meninges (36%), and the cauda equina (4%). The most common malignant intramedullary tumors are gliomas. In adults, the most common tumor is meningioma and in children, the most common histology types are astrocytoma [4]. In this chapter, we will discuss the approach to the management of spinal cord tumors.

B. Venkatesulu (✉)
Department of Radiation Oncology, Stritch School of Medicine Loyola University Chicago, Loyola University Medical Center, Maywood, IL, USA

Edward Hines VA hospital, Maywood, IL, USA

S. Mallick et al. (eds.), *Evidence based practice in Neuro-oncology*,
https://doi.org/10.1007/978-981-16-2659-3_34

34.2 Clinical Symptoms

The most common symptoms include radicular pain that radiates along the nerve root distribution followed by paresthesia. The allied symptoms that usually happens with more advanced involvement includes motor deficits, sensory changes, and alteration in bowel and bladder habits. Some classic presentations that are pathognomic of specific sites of involvement include cauda quina syndrome that causes radicular pain, saddle anesthesia, absent ankle reflexes and motor wasting in the hamstring muscles; Conus medullaris syndrome involve the loss of bladder and bowel function.

34.3 Management (Table 34.1)

- Comprehensive history and a physical examination that involves a neurological examination assessing sensory, motor, reflexes, autonomic function domains will provide insight into the possible location of the tumor.
- Investigations involve CBC, BMP, PT-INR, MRI of brain and spine with contrast. Ependymomas, GBM, and anaplastic astrocytomas have propensity for CSF spread and hence imaging should involve brain in addition to the entire neuraxis. CT is important for extradural tumors that arise from the bony structures. The CSF analysis should be done only after MRI or CT if clinically indicated. MRI features of Astrocytoma: T1: iso/hypointense, T2: hyperintense, enhancement, asymmetric expansion of Spinal cord; Ependymoma: T1: iso/hypointense, mixed if cyst/hemorrhage, T2: hyperintense, peritumor edema, central/symmetric expansion of Spinal cord.

Table 34.1 shows the classification of spinal cord tumors based on histology and anatomical location

Intramedullary	Intradural–extramedullary	Extradural
Glial neoplasms Ependymoma Astrocytoma Ganglioglioma Glioblastoma *Non-glial neoplasms* Metastasis Primary lymphoma of spinal cord Primitive neuroectodermal tumor Solitary fibrous tumor Hemangioblastoma Paraganglioma	Meningioma Schwannoma Neurofibroma Solitary fibrous tumor Hemangiopericytoma Malignant peripheral nerve sheath tumor. Metastases from solid malignancies Drop metastases from primary brain tumors like medulloblastoma in children and Glioblastoma with primitive neuronal components in adults Primary neoplasms of the cauda equina region Myxopapillary ependymoma Paraganglioma	Metastasis Chordoma Lymphoma Plasmacytoma Langerhans cell histiocytosis

- Biopsy is indicated in metastatic disease with a solid primary only; the standard approach is to go for surgical resection directly for decompression of the spinal cord as well as histological confirmation.

34.4 Intramedullary Tumors

- Intramedullary tumors constitute one third of the primary spinal cord tumors with glial tumors being the most common among them.
- Ependymomas
  - The most common Glial intramedullary tumors with histological grades of Grade I myxopapillary ependymoma and sub-ependymoma; Grade II tumors are classic, clear cell, papillary and tanycytic and Grade III tumors are defined as anaplastic ependymomas.
  - The standard approach for management is complete en bloc excision with neuronavigational techniques including intraoperative ultrasonography and neurophysiologic monitoring. The need for adjuvant radiation depends upon grade, extent of disease and progression of disease after surgical resection [5].
  - Ependymomas are more frequently amenable to gross total excision than astrocytomas.
  - Myxopapillary ependymomas involve the cauda equina and have a propensity to involve the CSF. Encapsulated tumors are completely resectable with a low recurrence rate whereas non-encapsulated tumors are difficult to resect completely with high propensity for recurrence [6].
- Spinal cord astrocytomas.
  - In comparison to astrocytomas of the brain, the spinal cord astrocytoma is relatively low-grade Juvenile pilocytic astrocytomas, fibrillary astrocytomas (grade II) which are commonly seen compared to high grade (III, IV) tumors like anaplastic astrocytomas and glioblastoma multiforme. (grade IV) are infiltrative, and complete resection carries a significant risk of neurologic disability. In these cases, a subtotal resection or biopsy may be the only safe surgical option.
  - The standard approach for management is complete en bloc excision with neuronavigational techniques including intraoperative ultrasonography and neurophysiologic monitoring. The need for adjuvant radiation depends upon grade, extent of disease and progression of disease after surgical resection [7, 8].

34.5 Intradural–Extramedullary Tumors

Meningiomas have a slow growth pattern which can be usually excised completely with a minimal propensity for recurrence. They are more commonly seen in females with a high incidence in the cervico-thoracic vertebra. Neurofibromas are or Schwannomas or neuromas are nerve sheath tumors arise from the Schwann cell, are well encapsulated are usually completely excised. Some of these nerve sheath tumors can have an extradural component referred as dumbbell tumors [9].

34.6 Extradural Tumors

The majority of extradural tumors are metastatic and are treated on the lines of malignant spinal cord compression.

34.7 Chemotherapy

There is no clear evidence to support the role of chemotherapy in spinal cord tumors. Temozolomide can be used in spinal cord glioblastoma or grade III spinal astrocytomas. In recurrent spinal cord ependymomas Carboplatin and Etoposide has been tried with some success [10].

34.8 Radiation

- Most of the spinal cord histology treatments are based on extrapolation of data from Primary brain tumors as well as from retrospective series. Low-grade astrocytomas and ependymomas with complete resection do not need any adjuvant therapy.
- Incomplete resection as well as piecemeal excision of low-grade ependymomas and astrocytomas may warrant close follow up with need of surgery or salvage radiation at time of progression. There is no clear data to support routine post-op radiation therapy for incompletely excised low-grade astrocytomas as well as ependymomas.
- High-grade ependymomas and astrocytomas will need postoperative adjuvant radiation therapy in both completely resected as well as incompletely resected scenarios.
- 2D borders: The superior and inferior borders should involve the two vertebral bodies above and below the site of the tumor with the lateral borders encompassing the tips of lateral process of the vertebra.
- 3D volumes: The planning CT and MRI spine should be co-registered. Low-grade astrocytomas or ependymomas, contour the preop–gross tumor volume and a 1 cm margin to form the clinical target volume. The PTV margin is based on the institutional standard.
- The planning CT and MRI spine should be co-registered. High-grade astrocytomas or ependymomas, contour the preop–gross tumor volume and a 1.5-cm margin to form the clinical target volume. The PTV margin is based on the institutional standard. Myxopapillary ependymomas with conus medullaris involvement require a larger 1.5-cm CTV margin to the GTV is used and should encompasses the thecal sac.

- The dose schedule used for low-grade astrocytomas and ependymomas is f 50.4 Gy in 1.8 Gy fractions. High-grade astrocytomas can be treated to a dose of 54 Gy with 1.8 Gy daily fractions. In spinal cord tumors with CSF spread cranio-spinal irradiation to 36–45 Gy, with a boost to sites of gross tumor to 50.4–54 Gy should be considered [11–13].

References

1. Duong LM, McCarthy BJ, McLendon RE, Dolecek TA, Kruchko C, Douglas LL, et al. Descriptive epidemiology of malignant and nonmalignant primary spinal cord, spinal meninges, and cauda equina tumors, United States, 2004-2007. Cancer. 2012;118(17):4220–7. https://doi.org/10.1002/cncr.27390.
2. Grimm S, Chamberlain MC. Adult primary spinal cord tumors. Expert Rev Neurother. 2009;9(10):1487–95. https://doi.org/10.1586/ern.09.101.
3. Louis DN, Ohgaki H, Wiestler OD, Cavenee WK, Burger PC, Jouvet A, et al. The 2007 WHO classification of tumours of the central nervous system. Acta Neuropathol. 2007;114(2):97–109. https://doi.org/10.1007/s00401-007-0243-4.
4. Abdel-Wahab M, Etuk B, Palermo J, Shirato H, Kresl J, Yapicier O, et al. Spinal cord gliomas: a multi-institutional retrospective analysis. Int J Radiat Oncol Biol Phys. 2006;64(4):1060–71. https://doi.org/10.1016/j.ijrobp.2005.09.038.
5. Ferrante L, Mastronardi L, Celli P, Lunardi P, Acqui M, Fortuna A. Intramedullary spinal cord ependymomas--a study of 45 cases with long-term follow-up. Acta Neurochir. 1992;119(1–4):74–9. https://doi.org/10.1007/bf01541785.
6. Weber DC, Wang Y, Miller R, Villà S, Zaucha R, Pica A, et al. Long-term outcome of patients with spinal myxopapillary ependymoma: treatment results from the MD Anderson Cancer Center and institutions from the Rare Cancer Network. Neuro-Oncology. 2015;17(4):588–95. https://doi.org/10.1093/neuonc/nou293.
7. Garcés-Ambrossi GL, McGirt MJ, Mehta VA, Sciubba DM, Witham TF, Bydon A, et al. Factors associated with progression-free survival and long-term neurological outcome after resection of intramedullary spinal cord tumors: analysis of 101 consecutive cases. J Neurosurg Spine. 2009;11(5):591–9. https://doi.org/10.3171/2009.4.Spine08159.
8. Jallo GI, Freed D, Epstein F. Intramedullary spinal cord tumors in children. Childs Nerv Syst. 2003;19(9):641–9. https://doi.org/10.1007/s00381-003-0820-3.
9. Gezen F, Kahraman S, Canakci Z, Bedük A. Review of 36 cases of spinal cord meningioma. Spine. 2000;25(6):727–31. https://doi.org/10.1097/00007632-200003150-00013.
10. Brandes AA, Cavallo G, Reni M, Tosoni A, Nicolardi L, Scopece L, et al. A multicenter retrospective study of chemotherapy for recurrent intracranial ependymal tumors in adults by the Gruppo Italiano Cooperativo di Neuro-Oncologia. Cancer. 2005;104(1):143–8. https://doi.org/10.1002/cncr.21110.
11. Isaacson SR. Radiation therapy and the management of intramedullary spinal cord tumors. J Neuro-Oncol. 2000;47(3):231–8. https://doi.org/10.1023/a:1006470523052.
12. Wen BC, Hussey DH, Hitchon PW, Schelper RL, Vigliotti AP, Doornbos JF, et al. The role of radiation therapy in the management of ependymomas of the spinal cord. Int J Radiat Oncol Biol Phys. 1991;20(4):781–6. https://doi.org/10.1016/0360-3016(91)90023-w.
13. Chao ST, Kobayashi T, Benzel E, Reddy CA, Stevens GH, Prayson RA, et al. The role of adjuvant radiation therapy in the treatment of spinal myxopapillary ependymomas. J Neurosurg Spine. 2011;14(1):59–64. https://doi.org/10.3171/2010.9.Spine09920.

Brain Metastases

35

Prashanth Giridhar

35.1 Introduction

Brain metastases are the most common intracranial tumour in adults. Intracranial metastases include brain (parenchymal) metastases, dural metastases, cranio-spinal fluid (CSF) metastases, ocular and pituitary metastases. The present chapter will confine its discussion to brain (parenchymal) metastases. Brain metastases arise commonly from lung (20–56% patients), breast (5–20%), melanoma (7–16%), renal or colorectal primary. They are associated with poor outcomes due to various reasons including but not limited to poor drug penetration due to presence of blood–brain and blood–tumour barriers, associated neurological and cognitive deficit with its associated risk of aspiration and infections, raised intracranial tension and risk of herniation and associated widespread extracranial disease. Brain metastases have also been shown in recent studies to harbour different mutations in comparison to the primary disease. All these factors make treatment of brain metastases difficult. The chapter will summarise the standard medical, surgical and radiotherapy management along with future directions in the management of brain metastases.

35.2 Clinical Features

The symptoms occur due to brain compression and mass effect. *Mass effect is a phenomenon in which a focal lesion causes surrounding areas of brain structures to be compressed and injured due to the degree of space that leaking blood, cerebrospinal fluid, or edema takes up within the restricted skull space.*

P. Giridhar (✉)
Department of Radiation Oncology, National Cancer Institute, AIIMS, Haryana, India

S. Mallick et al. (eds.), *Evidence based practice in Neuro-oncology*,
https://doi.org/10.1007/978-981-16-2659-3_35

The radiographic features of mass effect include *brain tissues pushed away from mass, midline shift and CSF spaces being occluded hampering CSF flow*. E.g.: IV ventricle compression leading to hydrocephalus and III and lateral ventricle dilation.

Symptoms	Percentage patients with symptom (%)
Headache	50
Focal neurological deficit	20
Acute stroke like symptoms (*Intra-tumoral haemorrhage*)	10
Altered mental status/cognition	

35.3 Diagnosis

Contrast enhanced MRI Brain is investigation of choice for diagnosis.

- Radiological features on MRI typically include *T1 iso to hypo-intense lesions, T2 hyper-intense lesions, ring enhancement on T1 contrast imaging with peritumoral edema, location at grey-white matter junction and spherical well encapsulated shape.*
- MR spectroscopy shows *intra-tumoral choline peak and NAA depletion.*
- Metastases that haemorrhage include melanoma, renal cell carcinoma, choriocarcinoma, thyroid cancer, lung and breast cancers.

The differentials include.

Condition	Radiological findings	When to suspect clinically
Glioma	Epicentre on white matter extends to ependymal surface NAA present	Long term remission present in primary cancer and patient has no disease elsewhere
Meningioma	Usually extra-axial homogenous enhancement Dural tail	Slow development of symptoms Prior history of radiotherapy
Abscess	Central restricted diffusion Dual rim sign	Usually associated with fever
Subacute stroke	Gyriform enhancement typical Vascular territory	Other risk factors present

35.4 Biopsy Is Indicated if Primary Cancer Is Unknown and Patient Presents with Brain Metastases on Imaging

35.4.1 Standard Management

Primary treatment	Adjuvant treatment	Evidence	Comments
Single brain metastasis			
Surgery (or) stereotactic radiosurgery (SRS) *Surgery preferred if:* Patient with a neurological deficit 4 cm or larger tumour Tumour with midline shift Central necrosis/bleeding *SRS preferred* In tumours less than 4 cm	*Post-surgery:* Surgical cavity SRS *(if resection cavity < 5 cm)* *If larger cavity* Whole brain radiotherapy may be considered *Post SRS* No adjuvant treatment	Brown et al. [1]; level 1 evidence; phase 3 RCT Patchell et al. [2]; level 3 evidence; phase 2 RCT Aoyoma et al. [3], Chang et al. [4]; level 1 evidence	*Post-operative SRS preserves cognition in comparison to WBRT* *No difference in survival* *WBRT improves survival in comparison to surgery alone* *Adding WBRT to SRS worsens cognition with no effect on survival*
2–3 brain metastases			
Surgery (or) stereotactic radiosurgery (SRS) *Surgery preferred if:* Patient with a neurological deficit *3 cm or larger tumour/>15 cc* Tumour with midline shift Central necrosis/bleeding *SRS preferred* *In tumours less than 3 cm and 10 cc volume*	*Post-surgery:* Whole brain radiotherapy may be considered *Post SRS* No adjuvant treatment	Patchell et al.; level 3 evidence; phase 2 RCT Brown et al. [5], Chang et al.; level 1 evidence; phase 3 RCT	*BRT improves survival in comparison to surgery alone* *Adding WBRT to SRS worsens cognition with no effect on survival*
4 or more metastases			
Whole brain radiotherapy	*No adjuvant treatment*		30 Gy/10#/2 weeks or 20 Gy/5#/1 week is the standard treatment. Multiple studies trying altered fractionation failed

Primary treatment	Adjuvant treatment	Evidence	Comments
Non-small cell lung carcinoma: Observation may be considered		UARTZ [6] trial; level 1 evidence; phase 3 RCT	No difference in survival when patients were observed Evidence primarily from Caucasian population. May not be applicable to Asian population with higher rates of EGFR mutations in NSCLC
RPA class III/GPA score <2 Observation may be considered			Median survival is 2–3 months

35.5 Radiotherapy Management Protocols

35.5.1 Stereotactic Radiosurgery

Max tumour diameter	Assigned dose
<2.0 cm	24 Gy
2.1–3.0 cm	18 Gy
3.1–4.0 cm	15 Gy
Volume	*Dose*
1–4 cc:	24 Gy
4–10 cc	20 Gy
10 cc	18 Gy
For large lesions consider 8–10 Gy × 3 fractions or 6–7 Gy × 5 fractions. Try to keep V12 <10 cc	

35.6 Dose Prescription in Multiple Metastases and Post-op SRS

Multiple lesions	
1–4 cc	22 Gy
4–10 cc	20 Gy
For radio-resistant histology:	
Renal cell carcinoma, melanoma, sarcoma	30 Gy in 3 fractions
Postoperative cases	
<10 cc	16 Gy
10–15 cc	14 Gy
>15 cc	12 Gy

35.7 Volumes for Contouring

SRS	Volumes	
GTV	Enhancing portion of the metastatic lesion on the contrast enhanced CT or MRI scan Surrounding edema are not considered part of the target volume	MRI:T1 CONTRAST WITH 3D sequence, straight cut not oblique, delayed (90 s), 1 mm slice thickness, no gap
PTV	GTV + 3 mm	
Postoperative	*Postoperative volumes*	
CTV	CTV should include the entire contrast-enhancing surgical cavity using the T1-weighted Gadolinium enhanced axial MRI scan, excluding edema determined by MRI CTV should include entire surgical tract seen on post-operative CT or MRI If the tumour was in contact with the dura preoperatively, CTV should include a 5–10 mm margin along the bone flap beyond the initial region of preoperative tumour contact If the tumour was not in contact with the dura, CTV should include a margin of 1–5 mm along the bone flap If the tumour was in contact with a venous sinus preoperatively, CTV should include a margin of 1–5 mm along the sinus	
PTV	CTV + 3 mm	

35.8 Premedication

Tab Dexamethasone 8 mg 1 h before.

Tab Ondansetron 8 mg 1 before.

Antiepileptics not given if no seizure history present.

Stop targeted agent 1 day prior of BRAF positive in melanoma patient it increases skin reactions.

Taper steroid earliest as tolerated.

35.9 Dose Coverage and Constraints

Dose coverage	GTV should get 90–95% of prescribed dose PTV should get 85–90% of prescribed dose Prescription dose 80% isodose line Hotspots are preferred at the Centre of GTV (110–120%) Gradient index (G I) should be less than 3 Conformity index (C I)-1

Brain	V12 < 10 cc
Optic pathway Dmax (0.1 cc)	8 Gy
Cochlea	9 Gy
Brainstem (not medulla) Dmax (0.1 cc)	15 Gy
Spinal canal (inc. medulla) Dmax (0.1 cc)	14 Gy
Lens Dmax (0.1 cc)	1.5 Gy

35.10 Medical Management in Brain Metastases

Steroids E.g.: Dexamethasone	Provides short term symptomatic improvement; can be considered as therapy alone in NSCLC
Anti-epileptics	Indicated only if patient presents with seizures
Chemotherapy	Not indicated in initial treatment. Multiple agents tried in subsequent treatment on progression
Targeted therapy	Not indicated in initial treatment. Multiple agents tried in subsequent treatment on progression

All the above described treatments provide reasonable local control and moderate improvements in survival. Yet patients with brain metastases are prone to progression and subsequent demise with a median survival of about 6 months. In the quest to improve treatment outcomes, researchers have tried to improve drug penetrability of blood–brain barrier as well as understand the different unique molecular aberrations in brain metastases.

35.11 Blood–Brain Barrier

The BBB consists of endothelial cells, two basement membranes (Endothelial and astrocytic), embedded pericytes and astrocytic terminal processes (end feet).

The endothelial cells have low transcytosis rates, *high expression of efflux pumps* and they are bound by *tight junctions*.

The astrocytes also play a major role in maintaining the blood–brain barrier by helping in coupling between endothelial cells, pericytes and neurons to form a neurovascular unit, ionic regulation via end feet ion and *aquaporin channel expression*, and harbouring of specific transporter to transport glucose and proteins.

Due to this BBB, only small uncharged molecules can diffuse without a specific transporter.

There is another, less tight barrier the blood–CSF barrier which is formed by the choroid plexus epithelial cells that harbour tight junctions and choroid plexus capillaries that contain gap junctions and fenestrations.

How do cancers disrupt the BBB and form brain metastases [7]? Figures 35.1 and 35.2.

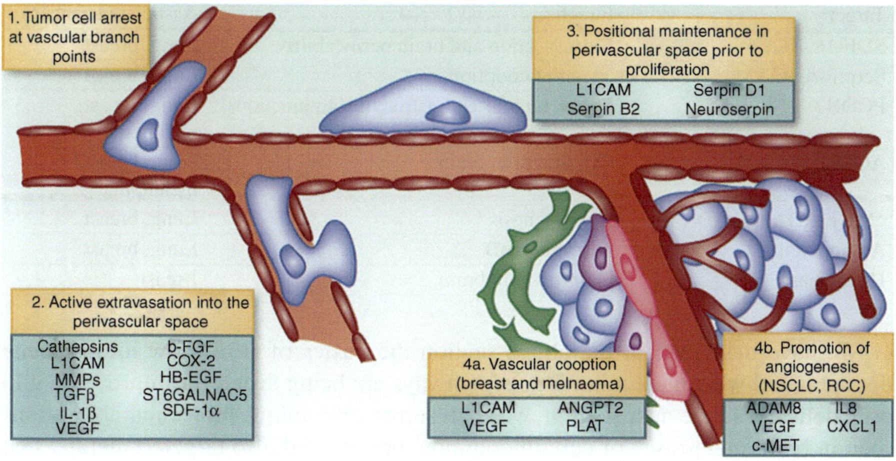

Fig. 35.1 Tumour cell penetration through blood–brain barrier (adapted from Peter E Fecci et al.)

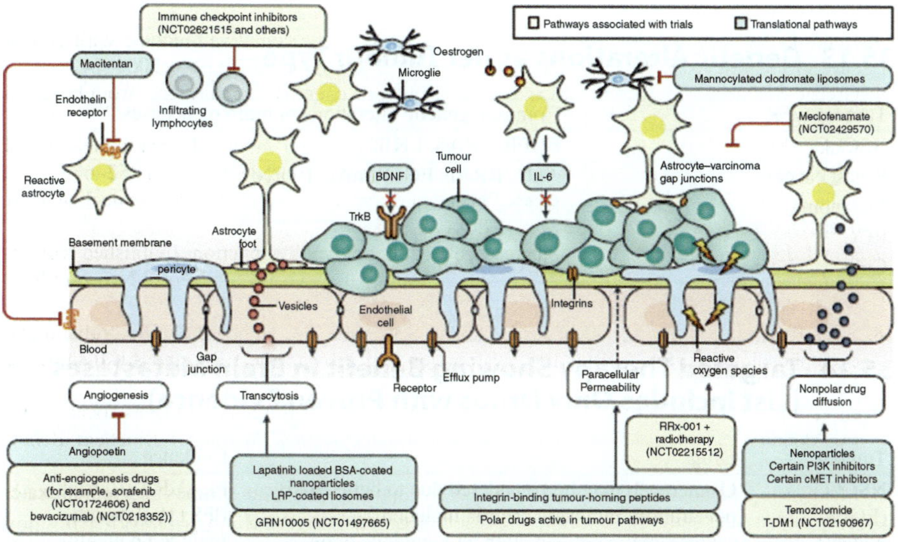

Fig. 35.2 Schematic representation of BBB and experimental treatments being tried to improve drug delivery. Figure adapted from Achal Singh Achrol et al. [8]

35.12 Molecular Targets of Metastases Involved in BBB Disruption and Their Function [7]

Target	Function	Associated cancer
ST6GALNAC5, COX2, HBEGF	Adhesion to brain endothelial cells and extravasation	Breast

Target	Function	Associated cancer
SDF-1/CXCR4	Extravasation and brain permeability	Lung, breast
Serpins/L1CAM	Vascular cooption	Lung, breast
PCDH7	Forms tumour cell-astrocyte gap junctions	Lung, breast, melanoma
ACSS2	Outgrowth in brain	Lung, breast, melanoma
cMET, ADAM8, ADAM9	Angiogenesis	Lung, breast
Angiopoietin	Disrupts BBB	Lung, breast
Hexokinase	Outgrowth in brain	Breast

We have described in the above section the basics of *BBB*, how tumour cells disrupt it and what molecular targets/pathways are being tested for improvement in drug delivery. In the next section, we shall summarise unique molecular aberrations seen in brain metastases of certain common cancers and also targeted therapy that have shown benefit. These therapies should not at present replace standard treatment with surgery and/or radiotherapy.

35.13 Genetic Alterations as per Tumour Type

Tumour type	Common genetic alterations in brain metastases
NSCLC	EGFR, kRAS, LKB1
Breast cancer	Akt1, BRAF, EGFR, myc, PTEN
Melanoma	PTEN, Akt1
RCC	PI3KCA, PTEN
Colo-rectal cancer	kRAS, nRAS

35.14 Targeted Therapy Showing Benefit in Brain Metastases (List Includes Only Drugs with Proven Evidence)

Tumour	Drug	Evidence
NSCLC (EGFR mutant)	Osimertinib may be considered due to better BBB penetration. Inclusion criteria included asymptomatic patients with NO previous brain directed therapy	FLAURA [9] trial; PFS: 18.9 months Included patients without brain metastases also
NSCLC (ALK positive)	Alectinib may be considered. Patients could have received prior RT or crizotinib	Global ALEX [10] trial; CNS RR: 81%
Breast cancer	Lapatinib and capecitabine. If no prior CNS directed therapy and ERBB2 positive	LANDSCAPE [11] trial; PFS: 5.5 months
Breast cancer	Neratinib and capecitabine. If prior CNS directed therapy given and ERBB2 positive	TBCRC [12] 022 trial; PFS: 1.9 months
Melanoma	BRAF V600E mutation; prior CNS directed therapy given	COMBI MB [13] trial; PFS: 7.2 months

References

1. Brown PD, Jaeckle K, Ballman KV, et al. Effect of radiosurgery alone vs radiosurgery with whole brain radiation therapy on cognitive function in patients with 1 to 3 brain metastases: a randomized clinical trial. JAMA. 2016;316:401–9.
2. Patchell RA, Tibbs PA, Regine WF, et al. Postoperative radiotherapy in the treatment of single metastases to the brain—a randomized trial. JAMA. 1998;280:1485–9.
3. Aoyama H, Shirato H, Tago M, et al. Stereotactic radiosurgery plus whole-brain radiation therapy vs stereotactic radiosurgery alone for treatment of brain metastases. JAMA. 2006;295:2483–91.
4. Chang EL, Wefel JS, Hess KR, et al. Neurocognition in patients with brain metastases treated with radiosurgery or radiosurgery plus whole-brain irradiation: a randomised controlled trial. Lancet Oncol. 2009;10:1037–44.
5. Brown PD, Asher AL, Ballman KV, et al. NCCTG N0574 (Alliance): a phase III randomized trial of whole brain radiation therapy (WBRT) in addition to radiosurgery (SRS) in patients with 1 to 3 brain metastases. ASCO Meet Abstr. 2015;33:LBA4.
6. Mulvenna P, Nankivell M, Barton R, et al. Dexamethasone and supportive care with or without whole brain radiotherapy in treating patients with non-small cell lung cancer with brain metastases unsuitable for resection or stereotactic radiotherapy (QUARTZ): results from a phase 3, non-inferiority, randomised trial. Lancet. 2016;388(10055):2004–14.
7. Fecci PE, Champion CD, Hoj J, et al. The evolving modern management of brain metastasis. Clin Cancer Res. 2019;25(22):6570–80.
8. Achrol AS, Rennert RC, Anders C, et al. Brain metastases. Nat Rev Dis Primers. 2019;5(1):5.
9. Soria JC, Ohe Y, Vansteenkiste J, et al. Osimertinib in untreated EGFR-mutated advanced non-small-cell lung Cancer. N Engl J Med. 2018;378(2):113–25.
10. Peters S, Camidge DR, Shaw AT, et al. Alectinib versus crizotinib in untreated ALK-positive non-small-cell lung cancer. N Engl J Med. 2017;377(9):829–38.
11. Bachelot T, Romieu G, Campone M, et al. Lapatinib plus capecitabine in patients with previously untreated brain metastases from HER2-positive metastatic breast cancer (LANDSCAPE): a single-group phase 2 study. Lancet Oncol. 2013;14(1):64–71.
12. Freedman RA, Gelman RS, Anders CK, et al. TBCRC 022: a phase II trial of neratinib and capecitabine for patients with human epidermal growth factor receptor 2-positive breast cancer and brain metastases. J Clin Oncol. 2019;37(13):1081–9.
13. Davies MA, Saiag P, Robert C, et al. Dabrafenib plus trametinib in patients with BRAF(V600)-mutant melanoma brain metastases (COMBIMB): a multicentre, multicohort, open-label, phase 2 trial. Lancet Oncol. 2017;18(7):863–73.

Paediatric CNS Tumours 36

Prashanth Giridhar

36.1 What Will This Chapter Cover?

1. Basic paediatric CNS tumour epidemiology.
2. Signs/symptoms of presentation of paediatric CNS tumours.
3. Genetic syndromes associated with paediatric CNS tumours.
4. Molecular aberrations in common paediatric CNS tumours.
5. Radiological findings in common paediatric CNS tumours.
6. Standard management of paediatric CNS tumours.
7. Newer drugs for targeted therapy in paediatric CNS tumours.
8. Follow up and toxicity management in paediatric CNS tumours.

36.2 Introduction

CNS tumours are the most common type of solid tumours that develop in children. They account for 20–25% of all paediatric malignancies. The incidence is highest in children aged 1–4 years and lowest in those aged 10–14 years. The known risk factors are certain genetic syndromes (described later) and history of ionising radiation. The most common types of paediatric CNS tumours in children are pilocytic astrocytoma, other low-grade glioma, malignant glioma and medulloblastoma. The distribution of major paediatric CNS tumours is shown in Fig. 36.1. Survivorship of paediatric CNS tumours has improved significantly over the past two decades with the average 5-year survival being 75.3% today. Due to this improved survivorship extreme caution should be taken while choosing various treatment modalities so as

P. Giridhar (✉)
Department of Radiation Oncology, National Cancer Institute, AIIMS, Haryana, India

S. Mallick et al. (eds.), *Evidence based practice in Neuro-oncology*,
https://doi.org/10.1007/978-981-16-2659-3_36

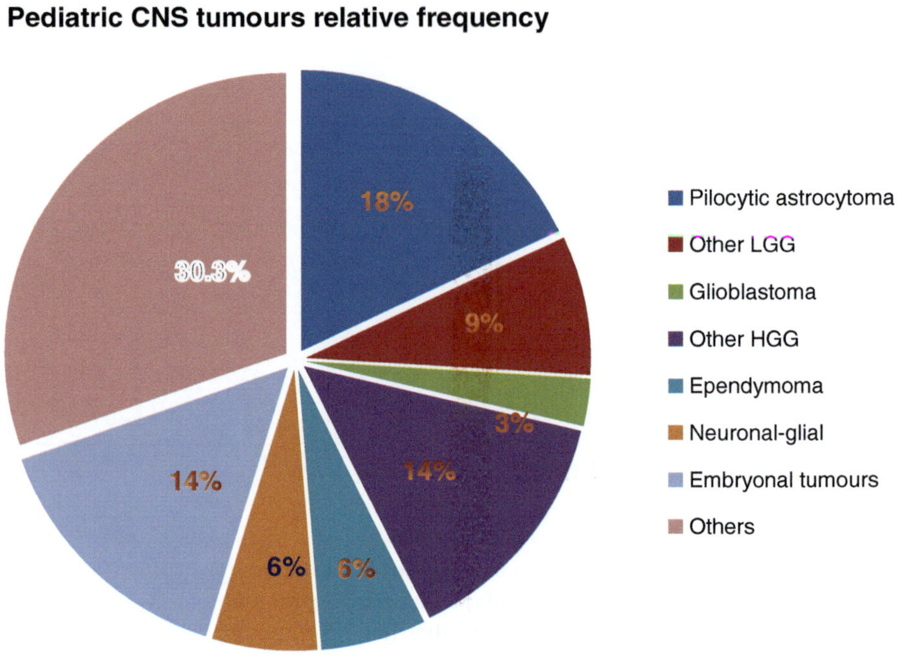

Fig. 36.1 Distribution of paediatric CNS tumours

to minimise long-term treatment related sequelae. But, this improvement in survival is not seen across all histological subtypes with glioblastoma and diffuse midline glioma still having 5 year survival rates of <20%. This has in recent years led to research to identify genetic mutations and drivers of oncogenesis, proliferation and progression in paediatric CNS tumours. Interestingly, the molecular aberrations found in paediatric CNS tumours seem to be totally unique and not the same as of adult CNS tumours of matching histological variants. This has further led to exploration and use of newer targeted therapy and the preliminary results of many such studies seem promising. The present chapter will summarise clinical and radiological features, standard management, genetic mutations and newer targeted therapy in these tumours.

The following table describes signs and symptoms seen in paediatric CNS tumours (Table 36.1).

Many paediatric tumours are associated with certain genetic syndromes. Table 36.2 summarises the most common syndromes.

The genetic mutations in paediatric CNS tumours are different from adults. They are summarised in Table 36.3.

Table 36.1 Signs and symptoms of presentation

Cause/Location	Symptoms
Elevated intracranial pressure	[a]Vomiting (early morning despite empty stomach), headache, personality changes, cranial nerve VI palsy, papilledema and visual impairment, sunset phenomenon, [b]macrocephaly in neonates
Cerebellar tumour	Unsteady gait, scanning speech, ataxia, nystagmus, intention tremor, dysdiadochokinesis
Brain stem tumour/ infiltration	Horizontal ophthalmoplegia, cranial nerve palsies, spastic palsies, long tract signs
Cerebellopontine tumour/ extension	Facial paralysis, hearing loss, torticollis
Cerebral hemispheric tumour	Cerebral seizures (e.g., complex partial seizures), pareses, paralyses, sensory impairments
Suprasellar tumour/ chiasma/hypothalamus tumours	Vision loss, narrowing of visual field, nystagmus
Tumour of hypophyseal and hypothalamic region	Short stature, diabetes insipidus, disordered pubertal development, eating disorders
Diencephalic tumours/ infiltration	Diencephalic syndrome: Cachexia in infants who sometimes appear euphoric; Normal intake
Pineal/midbrain tumour	Parinaud syndrome: Upward gaze palsy, convergence retraction nystagmus, light-near dissociation, and bilateral lid retraction

The above signs and symptoms also depend on rate of growth with generally high-grade tumours being more symptomatic than lower grade

Some features like the Diencephalic syndrome, development of macrocephaly, short stature, and disordered pubertal development are unique to paediatric CNS tumours and not seen in adults

[a]More than 50% of all CNS tumours in children occur in the posterior cranial fossa or in the brain stem. A tumour in this region may obstruct CSF drainage and thus increase the intracranial pressure

[b]In neonates and young infants classical features of raised intracranial tension may be absent. Instead they may have macrocephaly and failure to thrive

Table 36.2 Genetic syndromes associated with paediatric CNS tumours

Syndrome	Gene (Chromosome locus)	CNS Tumours	Clinical findings
Li–Fraumeni syndrome	TP53 (17p13)	Astrocytoma; cPNET; Medulloblastoma, choroid plexus tumour	
Familial adenomatous polyposis (Turcot syndrome)	APC (5q21)	Medulloblastoma	Multiple adenomatous colonic polyps
Cowden syndrome	PTEN (10q23)	Dysplastic cerebellar gangliocytoma	Trichilemmoma, thyroid disorders, gastrointestinal polyps, macrocephaly
Tuberous sclerosis	TSC1 (9q34) TSC2 (16p13)	Sub-ependymal giant cell astrocytoma (SEGA)	Neurodevelopmental defects, renal angiomyolipoma, cardiac rhabdomyoma, Ashleaf spots, shagreen patches

(continued)

Table 36.2 (continued)

Syndrome	Gene (Chromosome locus)	CNS Tumours	Clinical findings
von Hippel–Lindau syndrome	VHL (3p25)	Hemangioblastoma	Clear cell RCC, Pheochromocytoma, pancreatic neuro-endocrine tumour
Neurofibromatosis 1	NF1 (17q11)	Optic pathway glioma and other astrocytoma	Café au lait spots, Neurofibroma, scoliosis, behavioural issues
Neurofibromatosis 2	NF2 (22q12)	Bilateral vestibular schwannomas, peripheral schwannomas, meningioma, astrocytoma, spinal ependymoma,	Café au lait spots, Neurofibroma
Goltz-Gorlin syndrome	PTCH (9q31)	Medulloblastoma	Odontogenic keratocysts, palmar and/or plantar pits, and ectopic calcifications of the falx cerebri

Patients with neurofibromatosis and CNS tumour have an increased risk of vasculopathy (Moya Moya disease) and cardiovascular accident on treatment with radiotherapy. *Radiotherapy should be deferred in such patients*

Table 36.3 Genetic mutations in paediatric CNS tumours

Type	Variant/grade	Genetic mutations	Comments
Low-grade glioma (Grade I/II)	*Pilocytic astrocytoma* (I) Sub-ependymal giant cell astrocytoma (I) *Diffuse astrocytoma* (II) Pleomorphic xanthoastrocytoma (II) Oligodendroglioma (II) Ganglioglioma (I)) Dysembryoplastic Neuroepithelial tumour (I) Angiocentric glioma (I) Polymorphous low-grade Neuroepithelial tumour of the Young (PLNTY)	*KIAA1549-BRAF (70–80%)* BRAF V 600 E (5%) TSC 1/2 SNV (85%–95%) *BRAF p.V600E (20–40%)* MYBL1 alteration (5–10%) KIAA1549-BRAF (5–10%) BRAF p.V600E (80–90%) FGFR1-TKD duplication (10–20%) FGFR1 SNV (10–20%) BRAF p.V600E (5–10%) BRAF p.V600E (40–50%) KIAA1549-BRAF (10–15%) FGFR1-TKD duplication (20–30%) FGFR1 SNV (20–30%) MYB (80–90%) BRAF p.V600E (30–40%) FGFR2/3 fusions (30–40%)	*1p/19q co-deletion seen only in 5% cases*

(continued)

Table 36.3 (continued)

Type	Variant/grade	Genetic mutations	Comments
High-grade glioma (hemispheric or non-brain stem)	*High-grade glioma* (hemispheric or non-brain stem) Includes glioblastoma and anaplastic PXA as well	*NTRK fusion (15%)* *BRAF V600E (5–10%)* G34V/R histone mutant (7–15%) Hypermutant glioblastoma (MMR deficiency, POLE mutation; 5–10%) IDH mutant (5%)	5 year OS: 43% 2 year OS: 67% 2 year OS: 27% 5 year OS: 30% 2 year OS: 59%
High-grade glioma (brain stem)	*High-grade glioma (brain stem)*	*H3.3K27M (25–30%)* H3.1K27M (5–10%) G34V/R histone mutant (5–10%) BRAFV600E (5–10%) Wild type (50%)	Median OS: 12 months Median OS: 15 months Median OS: 42 months Median OS: 85 months Median OS: 15 months
Ependymoma	Supratentorial	RELA- C11ORF95 fusion YAP1 fusion	5 year OS: 60–80% 5 year OS: 100%
Ependymoma	Posterior fossa	PF-A – May have 1q gain PF-B	1q gain present: 5 year EFS: 35% 1q gain absent: 5 year EFS: 80% 5 year OS: 90%
Medulloblastoma	WNT variant SHH variant Group 3 Group 4	CTNNB1 mutation SMO mutation MYC, GLI mutation CDK6 mutation	5 year OS: 90% 5 year OS: If MYC amp; 50%
Atypical rhabdoid teratoid tumour	Group 1 Group 2 Group 3	SHH TYR MYC	

Imaging in Paediatric CNS Tumours CEMRI is used in the diagnosis of brain tumours. The diagnosis is made by using location of lesion, differentiating intra-axial from extra-axial lesions, enhancement patterns, and presence of edema supported by the clinical picture. An extra-axial lesion has the following features:

- CSF cleft sign.
- Displaced sub-arachnoid vessels.
- Cortical grey matter between mass and white matter.
- Displaced and expanded sub-arachnoid space.
- Bony reaction.
- Dural tail.

Location of tumour:

Supratentorial (25–40%)	*Infratentorial (45–60%)*
Low-grade glioma (10–20%)	*Medulloblastoma (20–30%)*
High-grade glioma (5–10%)	*Low-grade glioma (10–20%)*
Ependymoma (2–5%)	*Diffuse intrinsic pontine glioma (diffuse midline glioma) (5%)*
Ganglioglioma (1–5%)	*Ependymoma (5%)*
Oligodendroglioma (1–5%)	Others (5%)
Germ cell tumour (1%)	
Choroid plexus tumours (1%)	
Meningioma (1%)	
Others (5%)	

Table 36.4 describes the common radiological findings in paediatric CNS tumours.

Table 36.4 Radiological findings of paediatric CNS tumours

Subtype	Imaging findings (CEMRI)
Pilocytic astrocytoma	T1: Enhances with contrast medium; T2: Hyper intense Well circumscribed Minimal peritumoral edema Often cystic
Optic pathway glioma	T1: Hyper intensity in posterior lobe of hypophysis T2: Hyper intense Well circumscribed D/D: Craniopharyngioma/germ cell tumour. These do not show T2 hyper intensity in posterior lobe of hypophysis Look for NF1 features
Pleomorphic xanthoastrocytoma	Location: Supratentorial, temporal lobe T1: Solid and cystic components; solid component vivid enhancement Leptomeningeal involvement in 70% cases T2: Hyper intense solid component
Oligodendroglioma	Location: Supratentorial; frontal lobe T1: Variable enhancement T2: Typically hyper intense Calcification seen as area of blooming
Ganglioglioma	Location: Supratentorial; temporal lobe Circumscribed cyst with mural nodule T1: Variable contrast enhancement T2: Hyper intense solid component
DNET	Location: Cortical well circumscribed; scallops the cortical bone; no erosion T1: Heterogeneous enhancement T2: Hyper intense bubbly appearance
Glioblastoma	T1: *Garland-like* contrast uptake T2: Inhomogeneous Peritumoral edema

(continued)

Table 36.4 (continued)

Subtype	Imaging findings (CEMRI)
Diffuse intrinsic pontine glioma (included in diffuse midline glioma in new WHO classification)	T1: Variable contrast enhancement Basilar artery typically enclosed >50% pons involved usually
Medulloblastoma (SHH variant)	Location: Cerebellar hemispheres Intense contrast enhancement
Medulloblastoma (Wnt variant)	Location: Cerebellopontine angle
Medulloblastoma (Group 3)	Location: Cerebellar vermis Intense contrast enhancement
Medulloblastoma (Group 4)	Location: Cerebellar vermis Poor contrast enhancement
Atypical teratoid rhabdoid tumour	T1: Iso or hypo intense Heterogeneous enhancement Variable contrast uptake Cysts/necrosis common Marked peritumoral edema
Ependymoma including grade III	T1: Strong contrast uptake usually T2: Variable, inhomogeneous 50% have calcifications Commonly cystic May protrude into foramen magnum
Craniopharyngioma	T1: Cysts with/without solid components; hypo-intense or hyper intense if colloid present T2: Hypo or hyper intense Calcifications may be seen Do not show T2 hyper intensity in posterior lobe of hypophysis
Germ cell tumour	T1: Homogenous contrast uptake in pineal or hypophysis region Do not show T2 hyper intensity in posterior lobe of hypophysis

36.3 Management of Various Paediatric CNS Tumours

36.3.1 Low-Grade Glioma

- Low-grade glioma (LGG) is a heterogeneous group of tumour as shown in Table 36.3.
- They encompass tumours of astrocytic, oligodendroglial and mixed glial-neuronal histology of grade I and II.
- The majority of low-grade glioma in children is indolent and *do not undergo malignant transformation unlike adult LGG.*
- Pilocytic astrocytoma is very common but *diffuse astrocytoma is rare*. In adults diffuse astrocytoma is common but pilocytic variant rare.
- Usual age of occurrence of pilocytic astrocytoma is 5–9 years.

Table 36.5 Rare variants of low-grade glioma

Histology	Grade	Age at presentation	Location	Comments
Pilomyxoid astrocytoma	II	18 months	Hypothalamus	Cerebrospinal dissemination common (14%)
Pleomorphic xanthoastrocytoma	II	18 years	Temporal lobe	BRAF V600E common
Ganglioglioma	I	9.5 years	Temporal lobe	
Oligodendroglioma	II	14 years	Frontal lobe	1p 19q co-deletion rare

- Most common locations of LGG include:
  Cerebellum (15–25% cases).
  Hemispheric (Cerebral) glioma (10–15%).
  Gliomas of deep midline structure (10–15%).
  Optic pathway glioma (5%).
  ssBrainstem glioma (5%)—Dorsal exophytic, Cervicomedullary, Focal types
- Children with Neurofibromatosis type 1 account for the majority (over 70%) of the optic pathway/hypothalamic glioma. 15–20% of children with neurofibromatosis type 1 will develop an optic pathway/hypothalamic glioma.
- Rare variants of low-grade glioma are described in Table 36.5.

36.3.2 Investigations

Contrast enhanced MRI Brain; Contrast enhanced MRI spine also in case of Pilomyxoid variant.

Pituitary function tests/Visual acuity testing if radiotherapy planned.

36.3.3 Treatment

36.3.3.1 At Initial Diagnosis

Assess for operability. If operable, attempt to achieve gross total resection (GTR). Functional brain mapping with functional MRI and magnetoencephalography, microneurosurgical techniques, intraoperative neuro-monitoring and imaging have enabled resection of previously inoperable tumours while minimising surgical morbidity.

Inoperable sites: Hypothalamus, Optic pathway, certain parts of Brainstem, Thalamus.

If GTR achieved, no adjuvant treatment needed.

If subtotal resection (STR) achieved, no adjuvant treatment needed in most cases except brainstem lesions where adjuvant chemotherapy or radiotherapy may be considered (As even minimal progression may lead to severe deficits).

36.3.3.2 Inoperable Sites

Age <5 years and all cases of Neurofibromatosis: Chemotherapy; Deliver Radiation when child crosses 5 years age or *on progression of disease (RANO criteria) with no targetable mutations.*

Age >5 years: Radiation therapy (54 Gy/30#/6 weeks).

Chemotherapy Multiple studies have tried to ascertain efficacy of chemotherapy. In general, chemotherapy achieves a progression free survival of 40–50% in in non-NF1 patients and 60–70% in NF1 patients. We summarise some trials of chemotherapy in paediatric LGG and also present the commonly used chemotherapy regimen (Tables 36.6 and 36.7).

36.3.4 Radiation Therapy

36.3.4.1 Indications

Inoperable disease and child >5 years.

Adjuvant treatment in child >5 years with brainstem LGG and STR.

Progressive disease after Surgery and chemotherapy in child <5 years with no targetable mutations.

Type of Treatment Proton therapy if available as multiple small studies has shown reduction in acute and late toxicity including neurocognitive decline, visual symptoms and endocrinopathy. Photon therapy if proton therapy not available.

Table 36.6 Chemotherapy in non-NF1 patients

Study name/type	HIT-LGG 1996 single arm	COG randomised	SIOP single arm	SIOP randomised
Chemotherapy used	Carboplatin/VCR	Carboplatin/VCR vs TPCV	Carboplatin/VCR	Carboplatin/VCR vs carboplatin/VCR/etoposide
No. of patients	161	274	166	497
PFS (5 year)	47%	45% ±3%	40% ±11%	46%
OS	90%	86% ±2%	88% ±1%	90%

Table 36.7 Chemotherapy in NF1 patients

Study name/Type	HIT-LGG 1996 single arm	COG single arm	SIOP I single arm	SIOP single arm
Chemotherapy used	Carboplatin/VCR	Carboplatin/VCR	Carboplatin/VCR	Carboplatin/VCR
No. of patients	55	127	44	284
PFS (5 year)	68%	69%	60%	N.A
OS	N.A	98%	100%	N.A

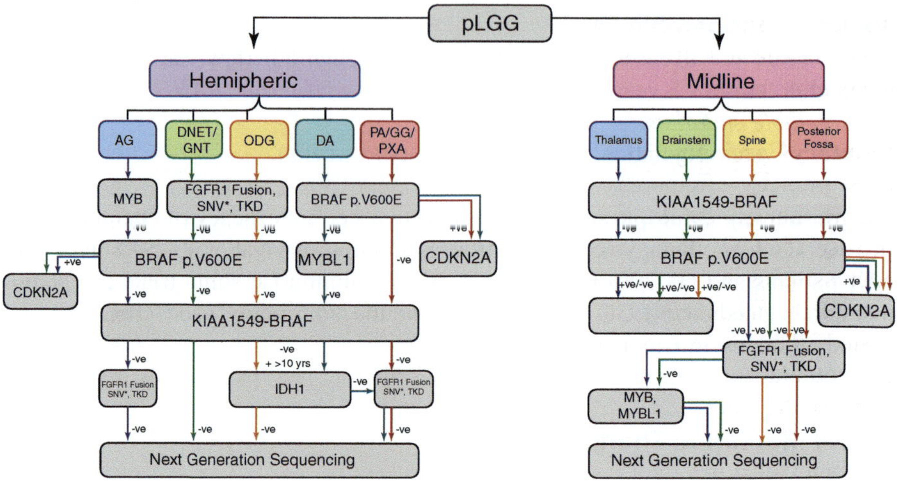

Fig. 36.2 Possible Flowchart for molecular diagnosis of paediatric low-grade glioma

Dose Fractionation 54 Gy/30#/6 weeks.

Treatment Volumes Gross disease/residuum on MRI plus 1 cm isotropic margins.

Progressive Disease On progression after surgery/chemotherapy in children <5 years or after surgery/chemotherapy/radiation in children >5 years, consideration for targeted therapy should be given. The molecular aberrations in various LGG is summarised in Table 36.3. We present a possible flow chart (Fig. 36.2) for identification of mutations in LGG in the absence of Next generation sequencing.

The following table provides a list of some promising newer drugs being tested in trials (Table 36.8).

36.3.5 High-Grade Glioma

- Paediatric high-grade gliomas (pHGG) are a heterogeneous group of tumours classified by the World Health Organization (WHO) as Grade III and Grade IV.
- The survival remains poor with 5-year survival rates being fairly stagnant over the last decades at about 20%.
- The OS for pHGG by subtype is 25.2%, and 20.4% at 5 and 10 years for anaplastic astrocytoma (AA) and 21.8%, and 18.1% for glioblastoma (GBM); *however, these assessments do not take into account any molecular characterisation.*
- *The various molecular aberrations and their survival is summarised in Table 36.3.*
- *Median age for hemispheric glioma is 12 years and brainstem glioma 10 years.*
- In infants, hemispheric high-grade glioma is more common than midline glioma.
- In older children, brainstem glioma is more common.

Table 36.8 Experimental drugs for low-grade glioma

Drug	Class	Dose/regimen	Possible indications	Side effects	Highest level evidence	Response rates/PFS	CNS penetration	Comments
Dabrafenib	BRAF inhibitor	Age < 12 years: 5.25 mg/kg/day Age > 12 years: 4.5 mg/kg/day in 2 divided doses, Oral	BRAF V600E mutant relapsed/ refractory low-grade glioma	Fatigue Rash Dry skin Arthralgia Pyrexia Vomiting	Phase II (Kieren et al)	ORR: 44%	Restricted Effluxed by p-glycoprotein and BCRP	Risk of cutaneous SCC may be present BBB breakage with RT may improve efficacy NOT TO BE USED IN KIAA1549-BRAF fusion positive low-grade glioma
Vemurafenib	BRAF inhibitor	240–480 mg twice daily. Start at lower dose, Oral	BRAF V600E mutant relapsed/ refractory low and high-grade glioma	Maculopapular rash Arthralgia Fatigue Alopecia Skin papilloma QT prolongation	Case series		Restricted Effluxed by p-glycoprotein and BCRP	Risk of cutaneous SCC may be present BBB breakage with RT may improve efficacy NOT TO BE USED IN KIAA1549-BRAF fusion positive low-grade glioma

(continued)

Table 36.4 (continued)

Drug	Class	Dose/regimen	Possible indications	Side effects	Highest level evidence	Response rates/PFS	CNS penetration	Comments
Trametinib	MEK 1/2 inhibitor	0.015 mg/kg to 0.03 mg/kg daily dose	KIAA1549-BRAF fusion positive relapsed/ refractory low and high-grade glioma	Diarrhoea Rash Paronychia	Case series		Restricted Effluxed by p-glycoprotein and BCR	
Selumetinib	MEK 1/2 inhibitor	25 mg/m^2 twice daily dose, Oral	KIAA1549-BRAF fusion positive or BRAF V600E mutation low-grade glioma relapsed/ refractory (both NF and non-NF)	Rash Mucositis Raised amylase Diarrhoea Hypoalbuminemia	Phase II (Fangusaro et al)	ORR: 36–40%	CSF penetration limited	Phase III trials on going Binimetinib and Cabometinib are similar drugs where phase 1 and 2 studies on

36.3.6 Investigations

CE MRI Brain is investigation of choice.

Pituitary function tests/Visual acuity testing if radiotherapy planned.

36.3.7 Treatment

36.3.7.1 At Initial Diagnosis

Maximal safe resection.

36.3.7.2 Adjuvant Treatment

Children >3 years: Radiation therapy 54 Gy/30#/6 weeks.

Children <3 years: No consensus. Chemotherapy is generally ineffective. No data to support efficacy of Temozolomide in paediatric population.

36.3.7.3 Inoperable Disease (Diffuse Intrinsic Pontine Glioma; Thalamic Glioma)

Radiation therapy (60 Gy/33#/6.5 weeks).

36.3.7.4 Progressive Disease

Re-surgery if possible.

If surgically inoperable consider reirradiation.

If further progression or not suitable for both, consider trial of targeted therapy and inclusion in clinical trial.

The following table shows the list of promising new drugs (Table 36.9).

36.3.8 Medulloblastoma

- MB accounts for approximately 20% of all paediatric CNS tumours.
- Despite the conventional aggressive multimodal therapeutic approach involving surgery, radiotherapy, and chemotherapy, approximately one-third of patients with MB die from their disease.
- The internationally accepted subgroups of MB have been termed wingless (WNT), sonic hedgehog (SHH), Group 3, and Group 4, which have been adopted in the 2016 WHO classification of central nervous system tumours (Table 36.10).

Investigation (Pre-surgery) CE MRI Brain is the initial investigation of choice.

36.3.9 Treatment

- *Surgical resection is the initial management.*
- Using micro-operative techniques, intraoperative imaging and the ultrasonic aspirator facilitate gross total resection with minimal neurological deficits.

Table 36.9 Experimental drugs for high-grade glioma

Vemurafenib	BRAF inhibitor	240–480 mg twice daily. Start at lower dose, Oral	BRAF V600E mutant relapsed/ refractory low and high-grade glioma	Maculopapular rash Arthralgia Fatigue Alopecia Skin papilloma QT prolongation	Case series		Restricted Effluxed by p-glycoprotein and BCRP	Risk of cutaneous SCC may be present BBB breakage with RT may improve efficacy Not To Be Used In KIAA1549-BRAF fusion positive low-grade glioma
Trametinib	MEK 1/2 inhibitor	0.015 mg/kg to 0.03 mg/kg daily dose	KIAA1549-BRAF fusion positive relapsed/refractory low and high-grade glioma	Diarrhoea Rash Paronychia	Case series		Restricted Effluxed by p-glycoprotein and BCRP	
Entrectinib	NTRK/ ROS1/ ALK inhibitor	<12 years: Not defined yet >12 years: 400–600 mg/ day, Oral	NTRK/ROS1/ ALK Fusion positive relapsed/refractory High-grade glioma	Fatigue Loss of taste Elevated creatinine Weight gain Hyperglycemia	Phase I/IIB study (STARTRK-NG trial-early results)	100% in initial results	Good CNS penetration	Acts even with NTRK resistance mutations Larotrectinib is a similar drug in phase 1 trials

Table 36.10 Molecular classification of medulloblastoma

Molecular subtype	Molecular aberration	Percentage cases (%)	Age group	Metastases at presentation (%)	Potential targeted agents for relapsed MB	Prognosis
WNT	CTNNB1 gene mutation	10	All age groups	5–10		Very good
SHH	PTCH1, SMO gene mutation	30	All age groups	15–20	SMO inhibitor, Aurora kinase inhibitor	Intermediate
Group 3	MYC amplification, GLI mutation	25	Infants	45–50	BET inhibitor, HDAC inhibitor WEE inhibitor	Very poor
Group 4	No specific mutation	35	All age groups	35–40	HDAC inhibitor	Intermediate

- Children with medulloblastoma typically present with features of raised ICP due to obstructive hydrocephalus.
- Routine pre-operative ventriculo-peritoneal (VP) shunt should generally be avoided (It can lead to "*reverse herniation*" of the superior vermis into the quadrigeminal cistern and occasionally seeding of the tumour into the peritoneal cavity).
- For surgery, patient is positioned in *prone position.*
- A standard *midline suboccipital craniectomy or craniotomy* can be used for access to the tumour. A good approximation of the deep fascia is the most important aspect in preventing a postoperative CSF leak.
- When the tumour infiltrates the brainstem, the neurosurgeon should not be aggressive in this location.
- Postoperative cerebellar mutism is a relatively common complication. It develops over 48–72 h after resection of the tumour and can persist from weeks to months, with associated findings of dysmetria, hypotonia, dysphagia, hemiparesis, and increased mood lability.
- Other common temporary postoperative complications are ataxia, hemiparesis, and VI cranial nerve palsy.

36.3.10 Work Up: (After Surgery)

CBC/LFT/KFT

CEMRI Brain and spine (2 weeks after surgery)

CSF cytology (3 weeks after surgery. Perform after MRI. 2 mL volume; to reach lab within 2 h; DO NOT REFRIGERATE)

Pituitary function tests (S. GH, S. TSH, S. ACTH, S. Prolactin, S. LH, S. FSH)
Ophthalmology review
Pure tone audiometry
Thyroid function tests (S.T3, S.T4)
2-D Echocardiography
DTPA renal function scan (If suspicion of kidney disorder)
Neurocognitive function assessment at baseline

36.3.11 Risk Stratification for Adjuvant Therapy

Modified Chang risk classes:

Standard risk	Age >3 years; M0 disease; <1.5 × 1.5 cm^2 residual ds
High risk	Age <3 years; M+ disease; larger residual ds

36.3.12 Adjuvant Therapy

Standard risk	Maximal safe resection f/b craniospinal irradiation (CSI)
High risk	Maximal safe resection f/b CSI with intrathecal methotrexate (if M+ disease) f/b adjuvant chemotherapy (MRI repeated after CSI before adjuvant chemotherapy)

36.3.13 Dose Prescription

Standard risk	CSI: 30 Gy/17#/3.5 weeks f/b posterior fossa/tumour bed boost till 55.8 Gy	In case of myelosuppression (Grade 3 or higher); do not stop RT; Continue PF boost; Give G-CSF injection (5 μg/kg/day) and platelet transfusions (<20,000) as necessary
High risk	CSI: 36 Gy/20#/4 weeks f/b posterior fossa/tumour bed boost till 55.8 Gy	In case of myelosuppression (Grade 3 or higher); do not stop RT; continue PF boost; give G-CSF injection (5 μg/kg/day) and platelet transfusions (<20,000) as necessary
Few spinal drop metastases	Local boost till 50.4 Gy	

[1]In many centres, dose reduced CSI of 23.4 Gy with concurrent Vincristine followed by adjuvant chemotherapy (Packer et al) has become standard. Though the trial showed good disease control (5-year event-free survival 81%), Grade 3/4 haematological toxicity was seen in >90% patients. We therefore do not yet advocate this regimen on a routine basis. The regimen is shown below

[2]For standard risk MB, tumour bed boost and posterior fossa boost provide equivalent disease control rates with possibility of reducing cochlear and brainstem dose

[3]Proton therapy reduces risk of acute and late effects. It should be considered standard if available

Table 36.11 Chemotherapy regimens

Regimen	Drug	Dosage
A		
Day 0	CCNU	75 mg/m^2 by mouth
Day 1	CDDP	75 mg/m^2 intravenously
Day 1, 7, 14	VCR	1.5 mg/m^2; max 2 mg intravenous bolus, maximum of eight doses
B		
Day 0	CDDP	75 mg/m^2 intravenously
Day 1, 7, 14	VCR	1.5 mg/m^2; max 2 mg, intravenous bolus
Day 21, 22	Cyclo	1000 mg/m^2 intravenously over 60 min daily

CCNU lomustine, *CDDP* cisplatin, *VCR* vincristine, *max* maximum, *Cyclo* cyclophosphamide, *min* minute

36.3.14 Reduced Dose CSI Regimen

Craniospinal irradiation: 23.4 Gy/12#/2.5 weeks f/b posterior fossa boost till 55.8 Gy with concurrent vincristine (1.5 mg/m^2, Max dose: 2 mg). CSI is followed by 8 cycles of adjuvant chemotherapy (one of two regimens) 6 weeks after CSI (Table 36.11).

36.3.15 Radiation Therapy Volumes

CSI Volumes	Volumes	Comments
Brain CTV	Entire brain with meninges up to foramen magnum Carefully include cribriform plate, optic nerves, temporal lobes	
Spine CTV	Entire spinal canal with exiting nerve roots Lower end of thecal sac to be identified on MRI and contoured	Identify thecal sac ending on sagittal cut of T2 weighted MRI Seen as grey structure surrounded by white
Boost CTV (high risk)	Entire posterior fossa	
Boost CTV (standard risk)	Tumour bed + residual ds + surgical cavity +1.5 cm	Brainstem is a natural barrier; if involved then to be included
PTV	CTV + 5 mm	

36.3.16 Adjuvant Chemotherapy

36.3.16.1 Indications

Reduced dose CSI in standard risk MB (Explained above).

In children <3 years to defer CSI till 3 years of age.

All high risk MB after CSI.

There is no standard chemotherapy after CSI is completed. Most evidence is from small prospective studies. The different regimens that can be considered are described below:

Children <3 years	Carboplatin: 600 mg/m^2 day 1 i.v
	Cyclophosphamide:1000 mg/m^2 day 1 i.v
	Etoposide: 100 mg/m^2 day 1–3 i.v
The above regimen is given 4 weekly for a maximum of 12 cycles. This regimen is commonly used in low middle income countries	

Alternative regimen in children <3 years: The below regimen is more aggressive and provides a 5 year PFS of 60–80%.

Week 1	Week 3	Week 5	Week 7
Methotrexate (2 mg/day, intraventricular), day 1–4 Cyclophosphamide (800 mg/m^2/day, intravenous), day 1–3 Vincristine (1.5 mg/m^2, intravenous), day 1	Methotrexate (2 mg/day, intraventricular), day 1–2 Methotrexate (5 g/m^2 intravenous), 24 h Vincristine (1.5 mg/m^2, intravenous), day 1	Methotrexate (2 mg/day, intraventricular), day 1–2 Methotrexate (5 g/m^2 intravenous), 24 h Vincristine (1.5 mg/m^2, intravenous), day 1	Methotrexate (2 mg/day, intraventricular), day 1–4 Carboplatin (200 mg/m^2, intravenous), day 1–3 Etoposide (150 mg/m^2, intravenous), day 1–3
One two-month cycle of intraventricular and systemic chemotherapy consisted of four treatments. The second and third cycles of chemotherapy started at weeks 10 and 19, respectively. Treatment finished if patients in complete remission after three cycles of chemotherapy			

36.3.17 Relapsed Medulloblastoma

Localised Recurrence Surgery plus consider reirradiation.

Disseminated Disease Trial of targeted therapy should be given or inclusion in clinical trial considered. List of drugs mentioned in table after the atypical teratoid rhabdoid tumour (Table 36.12).

Table 36.12 Experimental drugs in medulloblastoma

Drug	Class	Dose/Regimen	Possible Indications	Side effects	Highest level evidence	Response rates/PFS	CNS penetration	Comments
Sonidegib	Selective hedgehog (Hh) inhibitor of smoothened	680 mg/m^2 once daily, oral	Recurrent/ progressive SHH type Medulloblastoma	Increased creatinine Muscle spasm Alopecia Hyperglycaemia Anaemia	Phase I/II trial (Kieran et al)	ORR: 50%	Good CNS penetration	Phase II trials on Vismodegib phase II trials ongoing
Alisertib	Aurora kinase inhibitor	80 mg/m^2 oral; 7 days on/14 days off	Atypical rhabdoid teratoid tumour/ Group 3 MB	Neutropenia Leukopenia Anaemia Thrombocytopenia	Case series (Wetmore et al)			Phase II trials ongoing Haematological toxicity in 50% patients

36.3.18 Atypical Teratoid Rhabdoid Tumour (ATRT)

- Rare embryonal tumour accounting for 1–2% of CNS tumours in children.
- Peak incidence of children aged less than 3 years.
- ATRT has a median survival of only 8.5 months.
- Two-thirds of these neoplasms occur in the cerebellum, commonly at the cerebellar-pontine angle (CPA), with a potential of multifocal or disseminated disease in 20% of patients at the time of diagnosis.
- Currently accepted diagnostic criteria include biallelic loss and/or negative immunohistochemistry staining of SMARCB1 or SMARCA4 and their respective gene products hSNF/INI1/BAF47 and BRG1.

Investigations CE MRI Brain and spine (*Leptomeningeal spread seen in 25% cases*).

36.3.18.1 Management

Maximal safe resection (Attempt gross total resection).

36.3.18.2 Adjuvant Therapy

No standard therapy exists. *The following treatment may be considered based on Dana Farber cancer institute study which showed a 2 year OS of 70%. The regimen is aggressive with toxic deaths reported.*

36.3.18.3 Radiation Therapy

- Patients with M0 stage at presentation: Focal RT. 54 Gy/30#/5.5 weeks.
- For infratentorial tumours, a 1.5-cm margin and for supratentorial tumours, a 1.0-cm margin.
- For patients older than 3 years with M+ disease at diagnosis, craniospinal irradiation was prescribed. The total dose to the craniospinal axis was 36 Gy in 1.8 Gy fractions, and primary sites of disease received a total dose of 54 Gy. Spinal or brain metastases could be boosted, each with an additional 5.4 Gy in three fractions of 1.8 Gy.
- Patients with evidence of residual disease ≤2.5 cm on post-RT imaging studies could receive stereotactic radiosurgery.

36.4 Pre-irradiation Induction Therapy

Week 1: vincristine 2 mg/m^2 intravenously (IV) on days 1, 8, and 15; cisplatin 90 mg/m^2/day IV on day 1; doxorubicin 30 mg/m^2/day IV as continuous infusion on days 2 and 3; cyclophosphamide 300 mg/m^2/day IV as continuous infusion on days 2, 3, and 4.

Week 4: vincristine 2 mg/m^2 IV on days 1, 8, and 15; cisplatin 90 mg/m^2/day IV on day 1; doxorubicin 30 mg m^2/day IV as continuous infusion on days 1 and 2; etoposide 100 mg/m^2/day IV on days 1, 2, and 3.

36.5 Chemoradiation Induction Therapy

Weeks 7 and 10: vincristine 2 mg/m^2 IV on days 1, 8, and 15; cisplatin 90 mg/m^2/day IV on day 1; cyclophosphamide 600 mg/m^2/day IV as continuous infusion on day 2; etoposide 100 mg/m^2/day IV on days 1, 2, and 3; *radiation therapy*.

36.6 Post-radiation Induction Therapy

Week 13: vincristine 2 mg/m^2 IV on day 1; doxorubicin 30 mg/m^2/day IV as continuous infusion on days 1 and 2; cyclophosphamide 300 mg/m^2/day IV as continuous infusion on days 1, 2, and 3.

Week 16: vincristine 2 mg/m^2 IV on day 1; cyclophosphamide 300 mg/m^2/day IV on days 1, 2, and 3; dactinomycin 0.015 mg/kg/day intravenous push (IVP) on days 1 through 5.

36.7 Maintenance Therapy

Weeks 19 and 23: temozolomide 200 mg/m^2/day orally for conformal radiation therapy (150 mg/m^2/day for craniospinal irradiation) on days 1 through 5; dactinomycin 1.2 mg/kg/day IVP on day 1.

Weeks 27 and 33: vincristine 2 mg/m^2 IV on day 1; doxorubicin 30 mg/m^2/day IV as continuous infusion on days 1 and 2; cyclophosphamide 300 mg/m^2/day IV as continuous infusion on days 1, 2, and 3.

Week 30, 36, and 42: vincristine 2 mg/m^2 IV on days 1 and 5; cyclophosphamide 300 mg/m^2/day IV on days 1, 2, and 3; dactinomycin 0.015 mg/kg/day IVP on days 1 through 5.

Week 39: vincristine 2 mg/m^2 IV on day 1; doxorubicin 30 mg/m^2/day IVP on days 1 and 2; cyclophosphamide 300 mg/m^2/day IV on days 1, 2, and 3; dexrazoxane 300 mg/m^2/day IV on days 1 and 2.

36.8 Continuation Therapy with Doxorubicin (Patients Who Have Received <18 Gy RT to Thoracic Spine/Mediastinum)

Weeks 45 and 51: vincristine 2 mg/m^2 IV on day 1; doxorubicin 30 mg/m^2/day IVP on days 1 and 2; cyclophosphamide 300 mg/m^2/day IV on days 1, 2, and 3; dexrazoxane 300 mg/m^2/day IV on days 1 and 2.

Week 48: vincristine 2 mg/m^2 IV on days 1 and 5; cyclophosphamide 300 mg/m^2/day IV on days 1, 2, and 3; dactinomycin 0.015 mg/kg/day IVP on days 1 through 5.

36.9 Continuation Therapy without Doxorubicin (Patients Who Have Received >18 Gy RT to Thoracic Spine/Mediastinum)

Weeks 44, 48, and 51: vincristine 2 mg/m^2 IV on days 1 and 5; cyclophosphamide 300 mg/m^2/day IV on days 1, 2, and 3; dactinomycin 0.015 mg/kg/day IVP on days 1 through 5.

36.10 Intrathecal Therapy

Methotrexate 15 mg/m^2 (maximum 15 mg); hydrocortisone 30 mg/m^2 (maximum 30 mg); cytarabine 60 mg/m^2 (maximum 60 mg)—in preservative-free normal saline via filter on day 1 of weeks 1, 2, 4, 7, 13, 19, 27, 33, 39, 45, and 51; leucovorin 15 mg/m^2 IV/orally × 1 dose 24 h after methotrexate dose.

For patients with initially positive CSF cytology, weekly intrathecal therapy is administered until two consecutive CSF cytology analyses are negative for malignant cells. Once negative CSF cytology is achieved, patients continue on with CNS intrathecal therapy as prescribed above.

*Other rare tumours including ependymoma, germ cell tumours, pineal tumours, craniopharyngioma and pituitary adenoma are explained separately in other chapters

Primary Central Nervous System Lymphoma

37

Supriya Mallick and Prashanth Giridhar

Primary central nervous system lymphoma (PCNSL) accounts for nearly 3–4% of intracranial neoplasm.

PCNSL has two distinct types (1) Immune-competent and (2) immune-compromised.

It is a disease of elderly population and rare in pediatric age group.

37.1 Types

Immune-competent group—Mostly found in >60 years age and etiology unknown, nearly 20–25% multifocal.

In immune-compromised group (associated with HIV/AIDS, organ transplant) it is mainly because of EBV infection, mostly multifocal (50–60%).

Prognosis of HIV-related PCNSL is poor, with median survival varying from 2 to 4 months, but patients treated with chemotherapy do better (median survival: 1.5 years).

37.2 Presentation

Usually presents with features of raised intracranial tension, seizure and headache.

Systemic symptoms may include fever, night sweats, or weight loss.

However, patients may have diplopia, coma at presentation as well.

S. Mallick (✉) · P. Giridhar
Department of Radiation Oncology, National Cancer Institute, AIIMS, Haryana, India

S. Mallick et al. (eds.), *Evidence based practice in Neuro-oncology*,
https://doi.org/10.1007/978-981-16-2659-3_37

37.3 Workup

Patients with suspicion of PCNSL should not be advised steroids prior to imaging as it may reduce chances of diagnosis by 50%.

- Contrast enhanced MRI of the brain, spine screening MRI
- CSF cytology
- Ophthalmic evaluation
- Whole body CT/PET CT
- Bone marrow biopsy
- Viral markers
- A stereotactic biopsy should be obtained

37.4 Radiologic Characteristics

37.4.1 CT

Multiple patchy peri-ventricular lesions, most lesions are hyper attenuating (70%)

Immune-compromised cases show multiple lesions

37.4.2 MRI

T1: Hypo-intense

T1 post contrast: high-grade lesion show intense homogeneous enhancement, however, enhancement is less on lower grade lesions, Immune-compromised cases may have peripheral ring enhancement

T2: Iso to hypointense

DWI/ADC: Diffusion restriction, ADC values lower than a normal brain

MR Spectroscopy Choline peak and markedly decreased NAA.

37.5 Differential Diagnosis

GBM

Abscess

ADEM

37.6 Histology

- Most common—diffuse large B-cell lymphoma (DLBCL).
- Nearly 10% are Burkitt's lymphoma, indolent B-cell lymphomas.

- T cell lymphoma comprises less than 5%.
- EBV may be seen associated with DLBCL in non-immune-compromised patients.

37.7 Prognostic Factors

Prognosis is dependent on many factors.

Factors for poor prognosis: Age >60, ECOG PS >1, elevated LDH, elevated CSF protein, deep regions of the brain.

MSKCC has made RPA stratifications based on two factors KPS and age.

	Variables	Median OS
RPA class I	Age ≤50	8.5 years
RPA class II	Age ≥50 and KPS ≥70	3.2 years
RPA class III	Age ≥50 and KPS <70	1.1 years

IELSG Stratification Five variables included. Risk groups: 0–1 factors, 2–3 factors, 4–5 factors.

Variable	Criteria	Risk groups
Old age	>60 years	0–1: Low risk
Poor performance	PS 2 or more	
Elevated serum LDH	Present	2–3: Moderate risk
High CSF protein	Present	
Deep lesions	Present	4–5: High risk

Treatment:

- Surgery, if debulking in nature is not helpful and should be offered for situations where immediate cyto-reduction is aimed.
- *Historically whole-brain radiation* used to be the standard with 1–1.5 years median OS.
- *RTOG 9310 trial reported an improved survival with combination of chemotherapy plus radiation compared with radiation alone*. A high-dose methotrexate-based regimen produced a high response rate before RT was administered. High-dose methotrexate combined with whole-brain irradiation (WBRT) is an effective therapeutic approach to PCNSL, but neurotoxicity is a delayed risk of this approach. Table 37.1 summarizes trials that have explored this combination approach.

The above trials confirmed that high-dose methotrexate-based chemotherapy followed by consolidation with whole-brain radiotherapy provided a 5 year

Table 37.1 Trials of primary CNS lymphoma investigating combination chemotherapy and radiotherapy

Author/trial	Design	Results	Toxicity
DeAngelis et al./RTOG 9310 [1]	Five cycles of methotrexate 2.5 g/m^2, vincristine, procarbazine, and intraventricular methotrexate (12 mg) followed by whole-brain radiotherapy (RT) of 45 Gy/ 25#/ 5 weeks High-dose cytarabine after RT	Overall OS 36.9 months Median OS 50.4 vs 21.8 months for patients <60 vs >60 5-year OS 32%, no difference if CR after chemo or not	15% patients experienced severe delayed neurologic toxicity
O'Brien et al. TROG trial Phase II [2]	Methotrexate 1 g/m^2 on days 1 and 8, followed by whole-brain irradiation (45–50.4 Gy)	Median survival 36 months 5-year survival estimate was 37%	Neurotoxicity-30% Risk of neurotoxicity at 7 years was 58% for patients aged >60 years
Bessell EM et al. CHOD-BVAM I/II [6]	One cycle of CHOD and two of BVAM In the first trial, all 31 patients received 45-Gy whole-brain RT (CHOD/BVAM I) In the second, RT dose was reduced to 30.6 Gy if there was a complete response (CR) after chemotherapy (CHOD/BVAM II)	The 5-year overall survival (OS) was 36% 3-year relapse risks of 29% and 70% for CHOD/BVAM I and II, respectively 3-year OS, 92% v 60% for patients receiving 45 or 30.6 Gy	
Poortmans PM et al. EORTC 20962 [7]	Two cycles of MBVP (MTX 3 g/m^2 days 1 and 15, teniposide 100 mg/m^2 days 2 and 3, carmustine 100 mg/m^2 day 4, methylprednisolone 60 mg/m^2 days 1 to 5, and two intrathecal injections of MTX 15 mg, cytarabine 40 mg, and hydrocortisone 25 mg) followed by 40 Gy of RT	Median OS 3.8 years 3-year OS 58%	Grade 3–4 neutropenia 78%; treatment-related deaths 10%
Ferreri et al. Phase II [8]	Four cycles methotrexate 3.5 g/m^2 or methotrexate 3.5 g/m^2 on day 1 plus cytarabine 2 g/m^2 twice a day on days 2–3. Followed by whole-brain irradiation	CR rate 18% vs 46% Overall response rate of 40% and 69%	Grade 3–4 hematological toxicity was more common in the combination group (92%)
Houillier C et al. ANOCEF-GOELAMS trial [9]	Immunocompetent patients (18 to 60 years of age) Arm (1) Thiotepa/busulfan/ cyclophosphamide Arm (2) WBRT 40/20	2-year PFS chemo 87% vs WBRT 63%	Cognitive toxicity worse after WBRT

survival of approx. 40% but with a risk of acute hematological toxicity and late neurological toxicity. To reduce risk of cognitive decline, two approaches have been tried. The first approach has been to avoid whole-brain radiotherapy from treatment protocol. This approach has largely failed with higher risk of progression and lower survival. Table 37.2 summarizes the important trial evaluating withdrawal of radiotherapy.

The second approach has been to use autologous stem cell transplant (ASCT) for consolidation in place of whole-brain radiotherapy. Recent trials seem to show that ASCT provides equal or better survival outcomes with preserved neuro-cognitive function. This approach has not yet become widely accepted and practiced due to limited availability of resources. Table 37.3 summarizes the trials evaluating ASCT versus WBRT in PCNSL.

Table 37.2 Trial of primary CNS lymphoma investigating withdrawal of radiotherapy

Thiel E et al. Non-inferiority trial Primary endpoint: OS, defined non-inferior as lower 95% CI not below 0.9 Designed to have 60% power	Immuno-competent patients Arm A: HD-MTX + WBRT WBRT: 45 Gy/30F, 1.5 Gy daily fraction Arm B: HD-MTX	*Per protocol* Median overall survival was 32.4 months (WBRT) vs 37.1 months Median PFS was 18.3 months (WBRT) vs 11.9 months *ITT analysis* WBRT improved PFS (HR = 0.79, *P* = 0.041) On subgroup analysis, WBRT improved PFS in patients without CR (HR = 0.69, *P* = 0.011), but not with CR	More radiologic neurotoxicity 76% vs 46%, clinical neurotoxicity 49% vs 26%, in WBRT group

Table 37.3 Trials evaluating role of autologous stem cell transplant versus radiotherapy in PCNSL

Attributes	ANOCEF [3] study	IELSG–32 [4] study	OSHO 53 [5] study
No. of patients			
WBRT	70	59	6
ASCT	70	59	13
2 year PFS			
WBRT	58% (95% CI: 47–71%)	80% (95% CI: 70–90%)	80% (95% CI: Std error not reached)
ASCT	70% (95% CI: 59–82%)	69% (95% CI: 59–79%)	90% (95% CI: Std error not reached)
2 year OS			
WBRT	75% (95% CI: 65–87%)	85% (95% CI: 75–95%)	35% (95% CI: Std error not reached)
ASCT	66% (95% CI: 55–79%)	71% (95% CI: 60–82%)	60% (95% CI: 31–88%)

37.8 Recurrent PCNSL

No standard treatment exists.

Patients who responded initially to methotrexate should be re-challenged with methotrexate.

Patients with isolated extra-CNS relapses should be managed with anthracycline-based chemotherapy with or without ASCT.

Patients should be enrolled in phase 1 and phase 2 trials.

References

1. DeAngelis LM, Seiferheld W, Schold SC, Fisher B, Schultz CJ. Radiation Therapy Oncology Group Study 93-10. Combination chemotherapy and radiotherapy for primary central nervous system lymphoma: Radiation Therapy Oncology Group Study 93–10. J Clin Oncol. 2002;20(24):4643–8.
2. O'Brien PC, Roos DE, Pratt G, et al. Trans-Tasman Radiation Oncology Group. Combined-modality therapy for primary central nervous system lymphoma: long-term data from a phase II multicenter study (Trans-Tasman Radiation Oncology Group). Int J Radiat Oncol Biol Phys. 2006;64(2):408–13.
3. Houillier C, Taillandier L, Dureau S, et al. Intergroupe GOELAMS–ANOCEF and the LOC network for CNS lymphoma. Radiotherapy or autologous stem-cell transplantation for primary CNS lymphoma in patients 60 years of age and younger: results of the intergroup ANOCEF-GOELAMS randomized phase II PRECIS study. J Clin Oncol. 2019;37(10):823–33.
4. Ferreri AJM, Cwynarski K, Pulczynski E, et al. International Extranodal Lymphoma Study Group (IELSG). Whole-brain radiotherapy or autologous stem-cell transplantation as consolidation strategies after high-dose methotrexate-based chemoimmunotherapy in patients with primary CNS lymphoma: results of the second randomisation of the International Extranodal Lymphoma Study Group-32 phase 2 trial. Lancet Haematol. 2017;4(11):e510–23.
5. Kiefer T, Hirt C, Späth C, et al. Ostdeutsche Studiengruppe Hämatologie und Onkologie. Long-term follow-up of high-dose chemotherapy with autologous stem-cell transplantation and response-adapted whole-brain radiotherapy for newly diagnosed primary CNS lymphoma: results of the multicenter Ostdeutsche Studiengruppe Hamatologie und Onkologie OSHO-53 phase II study. Ann Oncol. 2012;23(7):1809–12.
6. Bessell EM, López-Guillermo A, Villá S, et al. Importance of radiotherapy in the outcome of patients with primary CNS lymphoma: an analysis of the CHOD/BVAM regimen followed by two different radiotherapy treatments. J Clin Oncol. 2002;20(1):231–6.
7. Poortmans PM, Kluin-Nelemans HC, Haaxma-Reiche H, et al. European Organization for Research and Treatment of Cancer Lymphoma Group. High-dose methotrexate-based chemotherapy followed by consolidating radiotherapy in non-AIDS-related primary central nervous system lymphoma: European Organization for Research and Treatment of Cancer Lymphoma Group Phase II Trial 20962. J Clin Oncol. 2003;21(24):4483–8.
8. Ferreri AJ, Reni M, Foppoli M, et al. International Extranodal Lymphoma Study Group (IELSG). High-dose cytarabine plus high-dose methotrexate versus high-dose methotrexate alone in patients with primary CNS lymphoma: a randomised phase 2 trial. Lancet. 2009;374(9700):1512–20.
9. Thiel E, Korfel A, Martus P, et al. High-dose methotrexate with or without whole brain radiotherapy for primary CNS lymphoma (G-PCNSL-SG-1): a phase 3, randomised, non-inferiority trial. Lancet Oncol. 2010;11(11):1036–47.

Part IV

Miscellaneous

38 Neurocognition in Neurooncology

Prashanth Giridhar

38.1 What Is Neurocognition?

Neurocognition includes processes having to do with the ability to think and reason. This includes the ability to concentrate, remember things, process information, learn, speak and understand.

38.2 Major Neurocognitive Domains

- Verbal learning and memory
- Visual learning and memory
- Speech and language
- Visuospatial processing
- Attention and concentration
- Executive functions
- Sensory perceptual functions
- Motor speed and strength
- Emotion and personality functioning
- Academic skills

Table 38.1 summarizes various cognitive domains and tests used to assess it.

Various researchers have tried to identify the discrete anatomical basis of these functions. Functional MRI studies have tried to analyse data to compartmentalize each domain to a specific Brodmann area. The results of such studies have shown that each component of cognition involves various Brodmann areas, and there seems

P. Giridhar (✉)
Department of Radiation Oncology, National Cancer Institute, AIIMS, Haryana, India

S. Mallick et al. (eds.), *Evidence based practice in Neuro-oncology*,
https://doi.org/10.1007/978-981-16-2659-3_38

Table 38.1 Neurocognitive domains assessed in neuropsychological evaluation [1]

Cognitive domain	Cognitive functions	Impairments	Test used in assessment (preferred marked in bold)
Verbal learning and memory	Learning, encoding, storing and retrieval of verbal information, including working memory, immediate free and cued recall, short-term and long-term free and cued recall and recognition of verbal information	Difficulty in encoding, retaining and retrieval of verbal information	**Hopkins Verbal Learning Test–Revised (HVLT-R)**; Wechsler Memory Scale; California Verbal Learning Test; Wide Range Assessment of Memory and Learning
Visual learning and memory	Learning, encoding, storing and retrieval of visual information, including working memory, immediate free and cued recall, short-term and long-term free and cued recall and recognition of visual stimuli	Difficulty in encoding, retaining and retrieval of visual information	**Brief Visuospatial Memory Test-Revised**; Wide Range Assessment of Memory and Learning
Speech and language	Oral and written abilities to comprehend information, repeat information and express information; word finding and naming	Wording finding difficulties, paraphasias, dysarthria, dysfluency, communication difficulties	**Controlled Oral Word Association (COWA) (also known as FAS Test)**, Boston Naming Test, Clinical Evaluation of Language Fundamentals-4
Visuospatial processing	Abilities and skills involved in making sense of the visual world, including shapes, angles, details, the meaning of forms, reproduction of what one sees, locating objects in space and spatial analysis of parts related to more complex objects	Visual field alterations, diplopia, neglect or inattention, impairment in spatial orientation, judgement and relationships	WAIS-IV processing speed index, **the Trail Making Test Part A**

Table 38.1 (continued)

Cognitive domain	Cognitive functions	Impairments	Test used in assessment (preferred marked in bold)
Attention and concentration	Ability to focus awareness on a given stimulus or task, to concentrate on a stimulus or task long enough to accomplish a goal and to shift awareness, if appropriate. Attentional processes include alertness/arousal, focused attention, selective attention, divided attention and sustained attention or vigilance	Diminished ability to focus awareness and concentration, difficulty with focused searching, problems with filtering irrelevant from relevant information, disrupted interactions and completion of tasks	**Conners Continuous Performance Test, Test of Variables of Attention**, the WAIS-IV working memory index, Digit Vigilance and the Paced Auditory Serial Addition Test
Executive functions	Higher order cognitive functions necessary for purposeful, goal-directed behaviour, including planning, conceptualizing, organizing, reasoning, insight, problem solving, evaluative judgement and effective execution of action, particularly in novel situations	Simplified or confused mental tracking of actions, problems with organizing and following plans, difficulties with reasoning, decision-making and concept formation, perseveration of speech and action	**Trail making test**, the Wisconsin Card Sorting Test; the Wisconsin Card Sorting Test
Sensory perceptual functions	Somatosensory function, sensory perception and discrimination, including tactile, pressure and localization, right-left discrimination	Somatosensory alterations (especially lateralized or confined to one limb), right-left disorientation	
Motor speed and strength	Gross and manual fine-motor abilities and grip strength	Lateralized weakness or clumsiness, problems with fine motor coordination, diminished ability in performing fine-motor manual skills	Finger tapping test
Emotion and personality functioning	Current affective and emotional states and stable traits, personality characteristics	Affective changes, diminished emotional control with temper outbursts, increased irritability, personality changes	**EMOTICOM neuropsychological test battery**

(continued)

Table 38.1 (continued)

Cognitive domain	Cognitive functions	Impairments	Test used in assessment (preferred marked in bold)
Academic skills	Reading, reading comprehension, writing skills, spelling, computation and functional academic skills that might be relevant to independent activities of daily living and work situations (e.g., handling finances, reading rate and retention)	Alterations in reading and reading comprehension, writing of words, letters and numbers, manipulating numbers, calculations	Schools Cognitive Assessment Battery (S-CAB)

to be significant interplay between various areas. The chapter will summarize the various causes of neurocognitive decline in neurooncology and management.

38.2.1 Causes for Neurocognitive Decline in Brain Tumours [2]

- Tumour
- Cancer treatments including but not limited to surgery, chemotherapy and radiotherapy
- Depression, anxiety and fatigue due to cancer diagnosis
- Hormonal imbalance, e.g. hypothyroidism
- Low serum folate/vitamin B12
- Use of anti-epileptics
- Use of steroids for raised ICT

38.3 Tumour-Induced Neurocognitive Decline

38.3.1 Depending on Location of Tumour [3]

As mentioned above, exact localization of all cognitive functions is difficult, but the table below based on cognitive deficits in post-surgery patients may help in identifying the cause of a particular deficit.

Cognitive domain	Involved areas of brain (bold: most strongly associated)
Overall composite score	All cognitive functions except information speed and memory seem to involve only right hemisphere. Information speed and memory involve both hemispheres

Cognitive domain	Involved areas of brain (bold: most strongly associated)
Executive function	Right hemisphere: Frontal pole, insular cortex, **inferior frontal gyrus**, superior and middle temporal gyrus, inferior fronto-occipital fasciculus
Visual construction	Right hemisphere: **Insular cortex**, inferior frontal gyrus, middle temporal gyrus, inferior fronto-occipital fasciculus, planum polare, amygdala
Information speed	Right hemisphere: **Frontal pole**, insular cortex, inferior frontal gyrus, superior and middle temporal gyrus, inferior fronto-occipital fasciculus Left hemisphere: Superior frontal gyrus, cingulate gyrus, **cingulum**
Memory (*from multiple sources*)	Pre-frontal cortex, **hippocampus (both hemispheres)**, cingulum, amygdala
Attention (multiple areas including cerebellum involved; **data not clear**)	Prefrontal cortex bilaterally in the middle frontal/premotor gyri, extending down the medial prefrontal wall to the anterior cingulate. Orbitofrontal cortex and the right inferior frontal gyrus

38.3.2 Depending Also on Rate of Growth of Tumour

High-grade glioma has more deficit than low-grade glioma.

38.4 Treatment-Associated Neurocognitive Decline

38.4.1 Impact of Surgery

- Glioma surgery aims to improve survival by maximizing tumour removal while preserving integrity of brain function.
- The impact of surgery on cognitive decline is less studied. The outcomes of such studies have also been variable due to multiple reasons: varied time points of assessment, varied tests used for assessment, impact of adjuvant treatment on cognition, surgical ischaemia.
- Cognitive alterations after resective surgery are prevalent in *15–50% of patients.*
- *Attention and information processing speed* are the most frequently declined cognitive domains.
- Brain regions most vulnerable for cognitive decline after surgery are located in the *right hemisphere.*

38.4.2 Impact of Radiotherapy [4–6]

- Neurocognitive decline occurs in *50–90% of patients* receiving radiotherapy. In a few cases, it is progressive and debilitating.

- The neurocognitive domains most affected by radiation are *verbal and non-verbal memory, executive function, attention and information processing speed.*
- Radiation induced neurocognitive decline has a *biphasic pattern*. Initially, there is a temporary decline followed by improvement at 4 months after radiotherapy. This, in a few cases, is followed by a progressive cognitive decline after 12 months or later.
- *Higher dose per fraction* leads to higher risk of neurocognitive decline.
- Concurrent high-dose *methotrexate* (both intrathecal and intravenous) leads to higher cognitive decline.
- Putative mechanisms involved in radiation-induced cognitive dysfunction:
  - Loss of neural stem cells in hippocampus.
  - Alterations in NMDA receptor subunit composition.
  - Disrupted arc expression in hippocampal neurons.
  - Oxidative stress/neuro-inflammation.

38.4.3 Impact of Chemotherapy/Targeted Therapy [7–13]

- The course of chemotherapy-induced cognitive decline is largely unknown.
- The neurocognitive domains most affected by chemotherapy are *memory, executive function and information processing speed.*
- It is difficult to ascertain the effect of individual drugs on cognition, the reason being that these drugs are used in combination regimens as well as with radiotherapy in most cases.
- *Human studies* evaluating effect of individual chemotherapy on cognition is *sparse*.
- Most of the data is from animal studies. The following table summarizes data available on common chemotherapy used in brain tumours.

Drug	Animal studies	Human studies	Cognitive domain affected	Putative mechanism
Methotrexate	Rodent studies show decline with both systemic and intrathecal methotrexate	**Data available from ALL survivors for both IT and IV methotrexate**	Recognition and spatial memory	Folate depletion in CNS
Vincristine	Rodent studies show decline	Not available for individual drug	Spatial memory	Disruption of microtubules in neurons

Drug	Animal studies	Human studies	Cognitive domain affected	Putative mechanism
Temozolomide	Rodent studies show decline	**Small studies available 25%–50% incidence of decline** No difference on cognition between standard dose and dose dense regimens	Psychomotor speed, attention and executive function. *Minimal effect on memory*	Unknown
Cyclophosphamide	Rodent studies show decline	Not available for individual drug	Impaired working and spatial memory	ERK and Akt signal activation affecting synaptic plasticity
Carboplatin	Rodent studies show decline	Not available for individual drug		Decreased dopamine and serotonin release in neurons
Etoposide	Not available	Not available for individual drug	Not available	Not available
Bevacizumab	Not available	**RTOG 0825 showed decline in global cognitive function**	Speech and language, visuo-spatial processing and executive functions	Reduced synaptic plasticity in hippocampus by affecting VEGF
Lomustine		*CeTeG/NOA-09 did not show detrimental effect of Lomustine addition*	No effect	
Selumetinib	Not available	Not available. **Phase III trial ongoing**	Not available	Not available

38.4.4 Impact of Anti-Epileptics [14]

- Common anti-epileptics used in brain tumours include levetiracetam, phenytoin, carbamazepine and sodium valproate.
- They are used commonly in patients *presenting with seizures* or for a short duration after surgery.

- *Not to be used prophylactically* in patients who do not have seizures.
- *Interaction with steroids and chemotherapy* need to be kept in mind.
- The effect of various anti-epileptics on cognition is mentioned in the following table.

Drug	Risk of effect	Cognitive domain affected
Phenytoin	**High**	Attention, memory, visuo-motor functions and processing speed
Levetiracetam—Preferred drug in brain tumours	**No risk.** *Most studies show improvement in cognition on use*	
Carbamazepine	**Moderate**	Processing speed and attention
Sodium valproate	**No risk**	1% risk of parkinsonism

38.4.5 Impact of Steroids [15]

- Steroid use is common in brain tumours.
- *RANO criteria includes steroid* use as one of the criteria for response assessment.
- *Dexamethasone is preferred* due to lack of mineralocorticoid activity.
- *Memory loss* is seen with prolonged use.
- Prolonged exposure to glucocorticoids can inhibit glucose utilization, which increases concentration of glutamate and leads to excitotoxic neuronal death.
- Reduced synaptic plasticity in hippocampal neurons and reduced CNS development following prolonged use of glucocorticoids has been reported.

38.4.6 Impact of Tumour-Treating Fields on Cognition [16]

- As per results from EF 14 study, TTF has minimal effect on cognition. These results should be viewed with caution as cognition was not evaluated with robust tests and follow-up is short.

38.4.7 Management of Neurocognitive Decline

There are no standard treatment guidelines for the management of cognitive decline in cancer. Most treatment modalities including but not limited to drugs and behavioural therapy have shown only mild to moderate efficacy. We describe a management algorithm for cognitive decline in neurooncology.

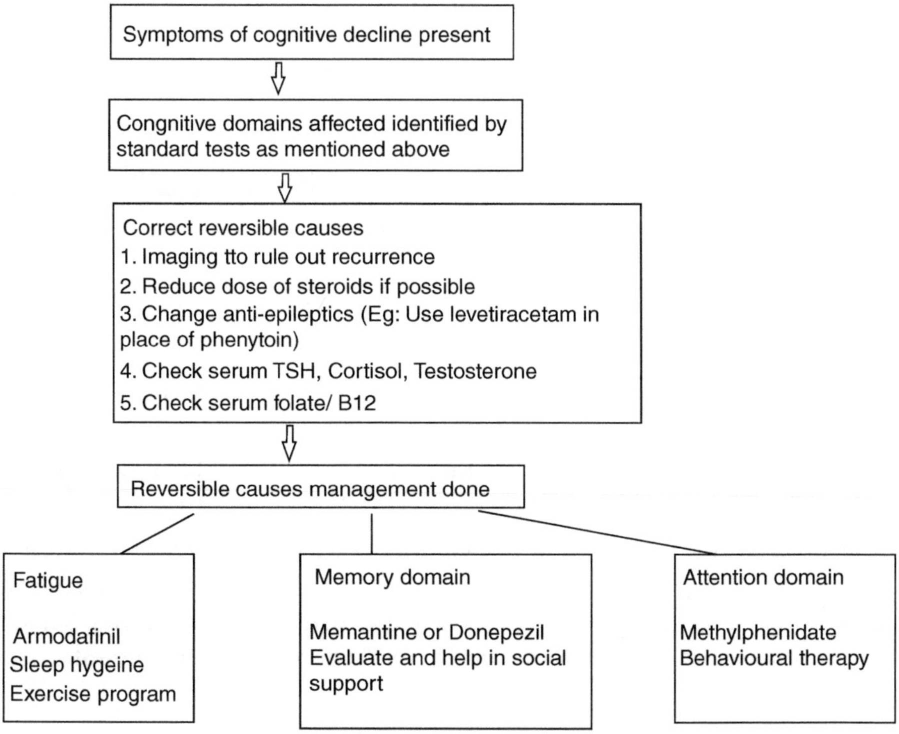

Drug	Class	Dosage/ regimen	Side effects	Important drug interactions
Memantine	NMDA antagonist	5 mg daily orally (7 mg if extended release tablets)	Dizziness, headache, constipation	Monitor closely if acetazolamide given
Methylphenidate	CNS stimulant (mechanism unknown)	20 mg daily orally before breakfast	Insomnia, mood swings, nasopharyngitis, back pain	Contraindicated with linezolid, monitor closely if carbamazepine, amlodipine, enalapril, ramipril, telmisartan being taken
Donepezil	AchE inhibitors	5 mg daily orally at bedtime	Nausea, diarrhoea, insomnia, infection	Monitor closely if carbamazepine given
Armodafinil	CNS stimulant	150 mg daily orally 1 h before work	Headache, nausea	Contraindicated with linezolid

References

1. Pendergrass JC, Targum SD, Harrison JE. Cognitive impairment associated with cancer: a brief review. Innov Clin Neurosci. 2018;15(1–2):36–44.
2. Taphoorn MJ, Klein M. Cognitive deficits in adult patients with brain tumours. Lancet Neurol. 2004;3(3):159–68.
3. Hendriks EJ, Habets EJJ, Taphoorn MJB, et al. Linking late cognitive outcome with glioma surgery location using resection cavity maps. Hum Brain Mapp. 2018;39(5):2064–74.
4. Greene-Schloesser D, Robbins ME. Radiation-induced cognitive impairment--from bench to bedside. Neuro-Oncology. 2012;14(Suppl 4):iv37–44.
5. Makale MT, McDonald CR, Hattangadi-Gluth JA, et al. Mechanisms of radiotherapy-associated cognitive disability in patients with brain tumours. Nat Rev Neurol. 2017;13(1):52–64.
6. Attia A, Page BR, Lesser GJ, Chan M. Treatment of radiation-induced cognitive decline. Curr Treat Options in Oncol. 2014;15(4):539–50.
7. Wefel JS, Schagen SB. Chemotherapy-related cognitive dysfunction [published correction appears in Curr Neurol Neurosci Rep. 2012 Jun;12(3):340]. Curr Neurol Neurosci Rep. 2012;12(3):267–75.
8. Salas-Ramirez KY, Bagnall C, Frias L, Abdali SA, Ahles TA, Hubbard K. Doxorubicin and cyclophosphamide induce cognitive dysfunction and activate the ERK and AKT signaling pathways. Behav Brain Res. 2015;292:133–41.
9. Li Y, Vijayanathan V, Gulinello M, Cole PD. Intrathecal methotrexate induces focal cognitive deficits and increases cerebrospinal fluid homocysteine. Pharmacol Biochem Behav. 2010;95(4):428–33.
10. Madhyastha S, Somayaji SN, Rao MS, et al. Hippocampal brain amines in methotrexate-induced learning and memory deficit. Can J Physiol Pharmacol. 2002;80(11):1076–84.
11. Shabani M, Larizadeh MH, Parsania S, et al. Profound destructive effects of adolescent exposure to vincristine accompanied with some sex differences in motor and memory performance. Can J Physiol Pharmacol. 2012;90(4):379–86.
12. Fathpour P, Obad N, Espedal H, et al. Bevacizumab treatment for human glioblastoma. Can it induce cognitive impairment? Neuro-Oncology. 2014;16(5):754–6.
13. Fangusaro J, Onar-Thomas A, Young Poussaint T, et al. Selumetinib in paediatric patients with BRAF-aberrant or neurofibromatosis type 1-associated recurrent, refractory, or progressive low-grade glioma: a multicentre, phase 2 trial. Lancet Oncol. 2019;20(7):1011–22.
14. Park SP, Kwon SH. Cognitive effects of antiepileptic drugs. J Clin Neurol. 2008;4(3):99–106.
15. Dietrich J, Rao K, Pastorino S, Kesari S. Corticosteroids in brain cancer patients: benefits and pitfalls. Expert Rev Clin Pharmacol. 2011;4(2):233–42.
16. Zhu JJ, Demireva P, Kanner AA, et al. Health-related quality of life, cognitive screening, and functional status in a randomized phase III trial (EF-14) of tumor treating fields with temozolomide compared to temozolomide alone in newly diagnosed glioblastoma. J Neuro-Oncol. 2017;135(3):545–52.

Suggested Reading

Schagen SB, Klein M, Reijneveld JC, et al. Monitoring and optimising cognitive function in cancer patients: present knowledge and future directions. EJC Suppl. 2014;12(1):29–40.

Syndromes Associated with Brain Tumors

39

Subhas Pandit and Abish Adhikari

39.1 Introduction

- Many familial cancer predisposing syndromes (CPS) are associated with increased incidence of nervous system tumors. Approximately 1–5% of brain tumors are due to genetic syndromes that confer an increased risk of developing tumors of the nervous system [1].
- Most CPS involves loss of function of the "tumor suppressor gene" and is transmitted to the offspring in an autosomal dominant pattern.
- Individuals with CPS have inherited mutation in one allele of tumor-suppressor gene, which is usually silent due to normal functioning of the second allele. In those with "germline mutation," only a single mutation in a functional second allele causes complete loss of function of tumor suppressor gene leading to carcinogenesis. This is also called the Knudson's two-hit theory [2].
- Neurocutaneous syndromes (also referred to as Phakomatoses) are a group of genetic and acquired disorders that are characterized by variable multisystem involvement, typically affecting the central nervous system (CNS), eyes, and skin, all of which have an ectodermal origin.
- Red flag signs of CPS include cancer in early age, multiple synchronous or metachronous tumors, rarer histologies, congenital anomalies, certain skin disorders, and clustering of certain types of cancer among family members.
- CPS are often underdiagnosed. Early diagnosis is helpful in clinical care and also in risk reduction for patients and their family.
- Genetic counseling is useful in high-risk patients and family members. It helps mitigate the risk of late diagnosis of these cancers and poor outcome. It also recommends screening, genetic tests, and preventive measures if available.

S. Pandit (✉) · A. Adhikari
Kathmandu Cancer Center, Bhaktapur, Nepal
e-mail: dr.subhas@kccrc.org

S. Mallick et al. (eds.), *Evidence based practice in Neuro-oncology*,
https://doi.org/10.1007/978-981-16-2659-3_39

- Before undergoing genetic tests, medical, psychological, and social consequences of positive, negative, or ambiguous results on individual and family members are explained for informed decision-making. This is also followed by post-test counseling.

Common syndromes associated with brain tumors are discussed below.

39.1.1 Neurofibromatosis Type 1

- Most common in the list of syndromes associated with brain tumors.
- Neurofibromatosis type 1 (NF1) is an autosomal dominant condition affecting around one in 3000 live births[1].
- The *NF1* gene is located in locus 17q11.2 and codes for a protein, neurofibromin. This protein has tumor suppressor function by inactivating RAS signaling. Germline mutation in *NF1* gene causes unopposed RAS activity causing various benign and malignant tumors.
- They are prone to develop multiple CNS and peripheral nervous system benign and malignant tumors as well as malignant tumors in other organs (GIST, pheochromocytoma, rhabdomyosarcoma, myeloid leukemia, breast cancer, etc.).
- Most common feature is cafe-au-lait macules (light brown spots common in groin and underarms) which typically present before the age of 1 year. Skinfold freckling and Lisch nodules (iris hamartoma) develop in childhood and cutaneous neurofibroma in adulthood.
- *Neurofibroma* are benign tumor composed of schwann cells, fibroblasts, mast cells, macrophages, endothelial cells, pericytes, and perineural cells.
- NF1 patients have fivefold increased risk of developing gliomas. About 15–20% develop low grade (WHO Grade I) gliomas, 80% of which are located in the optic pathway, and 15% in the brainstem.
- Optic pathway gliomas (OPG) tend to arise in infancy and are rare after childhood.
- Although radiological appearance is similar to sporadic OPG, presence of foci of abnormal signal intensity (FASI) and bilateral optic nerve involvement is suggestive of NF1.
- Only a minority of children with OPG will require treatment. Conservative approach with repeated ophthalmologic examination, imaging, and endocrine evaluation is the standard of care. Treatment is initiated only when clinical as well as radiological progression is noted.
- Firstline treatment is usually chemotherapy. Surgical excision is not feasible. Radiotherapy is usually avoided as it is at higher risk of radiation-induced vasculitis. NF1-associated OPG have better outcomes compared to sporadic OPG.
- Screening MRI brain is not routinely recommended for all NF1 patients. Ophthalmological surveillance including visual fields is recommended throughout childhood.

39.1.2 Neurofibromatosis Type 2

- Familial tumor predisposition syndrome typically characterized by development of bilateral vestibular schwannomas, multiple spinal or peripheral schwannomas, meningiomas, and ependymoma.
- Previously known as central "neuromatosis" as it demonstrates predilection for "central" intracranial and spinal lesions.
- Results from loss of function of *NF2* gene located in chromosome 22 (locus 22q12.2) which codes for protein product merlin.
- Commonly associated with bilateral vestibular schwannoma, "plaque like" dermal plexiform schwannoma, eye lesions (posterior subcapsular lenticular opacities, juvenile cortical cataracts), multiple meningiomas, and ependymomas.
- Estimated incidence is 1 in 33,000 people worldwide.
- The symptoms of this disease typically become apparent during puberty or early adulthood. Average age of onset is 18–24 years.

39.1.3 Tuberous Sclerosis Complex

- Multisystem neurocutaneous disorder characterized by skin lesions, seizure, and cellular overgrowth (hamartoma) typically in the brain, heart, and kidneys.
- Caused by mutation in either *TSC1* or *TSC2* gene leading to deregulation of mTOR signaling pathway. This leads to abnormal cell growth and proliferation in multiple organs.
- Cutaneous features are seen in 90% of patients, typically hypo-melanotic macules (**ash leaf spots**).
- Other manifestations of TSC are renal (angiomyolipoma), cardiac (intracardiac rhabdomyoma), ocular (retinal abnormality), pulmonary (lymphangioleiomyomatosis).
- Neurological involvement is the most common cause of mortality and morbidity. These include intractable epilepsy, intellectual disability, autism spectrum disorder, and behavioral problems.
- Structural brain abnormalities that are typically found include cortical tubers (pathognomonic), subependymal nodules, and subependymal giant cell astrocytoma (SEGA).
- mTOR inhibitors like everolimus have been approved for the management of TSC-related conditions.

39.1.4 Von Hippel–Lindau Syndrome

- Mutation in tumor suppressor gene VHL (3p25) which increases hypoxia-inducible factor and high level of VEGF-a, PDGF-b, and TGF-a.
- Incidence of one in 31,000–36,000 worldwide.

- Predisposition to hemangioblastoma, retinal angiomas, renal cell carcinoma, and lymphangiomatosis.
- Brain tumors associated are usually hemangioblastomas.

39.1.5 Multiple Hamartoma (Cowden) Syndrome

- Characterized by mucocutaneous papillomatous lesions, multiple hamartomas, and higher risk for follicular thyroid, breast, and endometrial cancers [3].
- Incidence is approximately one in 250,000 worldwide.
- Germline mutation in *PTEN* gene (locus 10q23) leads to deregulation of PI3K-AKT and mTOR signaling pathway.
- Patients are at an increased risk of glial tumors, including gliosarcoma.
- Dysplastic gangliocytoma of the cerebellum (also called Lhermitte–Duclos disease) is a very rare, benign tumor commonly manifested in the third and fourth decades of life. Symptoms may include headache, nausea, cerebellar dysfunction, hydrocephalus, ataxia, and visual disturbances.

39.1.6 Brain Tumor Polyposis (Turcot) Syndrome

- Characterized by the association of gastrointestinal polyposis and neuroepithelial tumors of the central nervous system.
- Type 1 is associated with hereditary non-polyposis-related colorectal cancer (HNPCC) also known as Lynch syndrome. Mutation is seen in mismatch repair genes including *MLH1*, *MSH2*, *MSH6*, and *PMS2*. It is more commonly associated with gliomas particularly glioblastoma.
- Type 2 has been associated with germline mutations in the *APC* gene (locus 5q22.2). Unlike Type 1 TS, it is associated with familial adenomatous polyposis (FAP). It also has a high association with medulloblastomas and less commonly with astrocytomas and ependymomas.
- Medulloblastoma is typically wingless (WNT) subtype and is seen in younger age (<10 years).
- Other symptoms include cafe-au-lait spots in skin, multiple lipomatosis, and basal cell carcinoma of skin.

39.1.7 Nevoid Basal Cell Carcinoma Syndrome (Gorlin Syndrome)

- Prevalence is one in 57,000–164,000 worldwide [4].
- Germline mutation in *PTCH1* gene (locus 9q22.3-q31).
- Medulloblastoma is the most common associated CNS tumor. It is mostly sonic hedgehog subtype and is reported in up to 5% of individuals within the first 2 years of life.

- Additional intracranial findings include dural and/or falcine calcifications, bridging of the sella, eye anomalies, macrocephaly, and congenital facial malformations such as cleft lip/palate, frontal bossing, coarse facies, and moderate or severe hypertelorism.
- Skin symptoms include multiple basal cell carcinoma which is very common. Other presentations include odontogenic keratocyst, cardiac fibroma, and ovarian fibroma.

39.1.8 Li–Fraumeni Syndrome

- Cancer predisposition syndrome associated with high risks for a diverse spectrum of malignancies.
- Mutation of *p53* gene (locus 17p13.1), essential for regulating DNA repair.
- Commonly associated malignancies are of breast, osteosarcoma, soft tissue sarcoma, lungs, and adrenocortical tumors.
- CNS tumors are seen in 9–14% patients. Glioblastomas and astrocytomas are the most common. Other associated tumors are ependymomas, choroid plexus carcinomas, supratentorial primitive neuroectodermal tumors, and medulloblastoma.
- The lifetime risk of cancer in individuals with LFS is ≥70% for men and ≥90% for women.

Other less common cancer predisposing syndromes associated with brain tumors include:

- Ataxia telangiectasia
- Carney complex
- DICER1 syndrome
- Familial glioma
- Familial melanoma-astrocytoma syndrome
- Goldenhar syndrome
- Muir–Torre syndrome
- Rhabdoid tumor predisposition syndrome
- Rubinstein–Taybi syndrome
- Werner's syndrome

References

1. Malmer B, Iselius L, Holmberg E, et al. Genetic epidemiology of glioma. Br J Cancer. 2001;84(3):429–34.
2. Knudson A. Mutation and cancer: statistical study of retinoblastoma. Proc Natl Acad Sci U S A. 1971;68:820–3.

3. Huang S, Zhang G, Zhang J. Similar MR imaging characteristics but different pathological changes: a misdiagnosis for Lhermitte-Duclos disease and review of the literature. Int J Clin Exp Pathol. 2015;8(6):7583–7.
4. Lo Muzio L, Nocini PF, Savoia A, et al. Nevoid basal cell carcinoma syndrome: clinical findings in 37 Italian affected individuals. Clin Genet. 1999;55(1):34–40.

Endocrine Management in Neurooncology

40

Geethu Antony

40.1 Introduction

The pituitary which is situated in the sella turcica is often referred to as the "master gland" because, together with the hypothalamus, it orchestrates the complex regulatory functions of many other endocrine glands including thyroid, adrenals and gonads. The pituitary gland consists of predominant adenohypophysis (anterior and intermediate lobe) and neurohypophysis (posterior lobe). Functional anterior pituitary contains five main cell types: somatotrope cells produce growth hormone (GH), lactotrope cells produce prolactin (PRL), thyrotrope cells produce thyroid-stimulating hormone (TSH), gonadotrope cells produce gonadotropins (follicle-stimulating hormone, FSH and luteinizing hormone [LH]) and corticotrope cells produce adrenocorticotropic hormone (ACTH). The pituitary stalk serves as an anatomic and functional connection to the hypothalamus. Hypothalamic neural cells synthesize specific releasing and inhibiting hormones which regulate the anterior pituitary hormones. The hypothalamus is also responsible for essential metabolic homeostatic mechanisms including water and electrolyte metabolism and regulation of body weight by controlling food intake, energy expenditure and energy storage. Arginine vasopressin (AVP) and oxytocin are produced in the hypothalamus and stored in posterior pituitary.

Any tumours in the sellar region can disrupt the hypothalamo–pituitary (HP) axis leading to hormone deficiencies or excess. This can develop due to the compressive effects of tumour itself or as an adverse effect of surgery or radiation. Functioning pituitary adenomas, most common cause of sellar mass, result in specific hormone excess syndromes which is discussed in detail later.

G. Antony (✉)
Department of Endocrinology, Mar Sleeva Medicity, Palai, Kerala, India

S. Mallick et al. (eds.), *Evidence based practice in Neuro-oncology*,
https://doi.org/10.1007/978-981-16-2659-3_40

40.2 Hypothalamo-Pituitary Axis Dysfunction

HP axis dysfunction is commonly observed in survivors of central nervous system (CNS) tumours and in those HP region was exposed to radiation [1]. The causes of HP axis dysfunction are detailed in Table 40.1. Novel targeted chemotherapy agents like immune system modulators can cause hormone deficiencies. The presentation of HP axis dysfunction varies according to tumour location and treatment modalities. Patients experiencing direct HP injury related to local tumour growth or surgical resection generally present with multiple hormone disorders at the time of tumour diagnosis or shortly after surgery. In contrast, patients with radiation-induced dysfunction are diagnosed with one or multiple hormone disorders often sequentially and over a period of time extending from a few months to several decades. The various endocrine manifestations of HP axis damage are detailed in Table 40.2.

40.2.1 Effect of Sellar Tumours on Hypothalamo–Pituitary Function

Pituitary hormone deficiencies occur due to direct pressure effects of the expanding mass on hormone secreting cells or parasellar pressure effects that attenuate synthesis or secretion of hypothalamic hormones, with resultant pituitary failure. Majority of sellar masses are pituitary adenomas (90%). Macroadenomas (>1 cm) are more frequently associated with pituitary hormone deficiencies, with at least one hormone deficiency present in all patients with tumours >4 cm in diameter. Most

Table 40.1 Causes of hypothalamo–pituitary axis dysfunction

Tumours involving HP Axis
Hypothalamic tumour: Hamartoma, glioma, ganglioglioma, choristoma
Pituitary adenoma/carcinoma
Parasellar tumours
Developmental cyst: Rathke cleft cyst, arachnoid, epidermoid, pineal cyst
Nonadenomatous neoplasm: Craniopharyngioma, meningioma, chordoma, germinoma
Metastasis
Inflammatory/infiltrative lesion: Sarcoid, lymphoma, amyloid, vasculitis, tuberculosis
Surgery for Sellar tumours
Radiotherapy
HP axis tumours
Brain tumours
Head and neck cancer
Acute lymphoblastic leukaemia
Drugs
Anti-PD1 monoclonal antibody: Nivolumab, pembrolizumab
Anti-CTLA4 monoclonal antibody (ipilimumab)

Table 40.2 Endocrine manifestations of hypothalamo–pituitary axis damage

Clinical syndrome	Hormone affected
Short stature in children Adult GH deficiency	Growth hormone/growth hormone-releasing hormone (GHRH)
Adrenal insufficiency	Corticotropin-releasing hormone (CRH)/ACTH deficiency
Hypothyroidism	Thyrotropin-releasing hormone (TRH)/TSH deficiency
Hypogonadism	Gonadotropin-releasing hormone (GnRH)/LH/FSH deficiency
Lactation failure	Prolactin deficiency
Galactorrhea	Prolactin excess due to decreased dopamine
Central precocious puberty	Premature GnRH activation
Diabetes insipidus	ADH deficiency
Essential hypernatremia	Damage to osmoreceptors
Obesity	Damage to hypothalamic nuclei
Syndrome of inappropriate ADH secretion	ADH excess

commonly encountered nonadenomatous lesions include Rathke's cleft cyst, craniophargyioma, and meningioma. Suprasellar lesions like craniopharyngiomas and hypothalamic tumours are associated with higher rates of hypopituitarism especially diabetes insipidus. New onset hypopituitarism can develop following surgery because of manipulation or damage to the normal pituitary gland or stalk. Transient DI or SIADH occurs in 20–30% of patients in the immediate postoperative period. All patients undergoing pituitary surgery can have possible postoperative deterioration in pituitary function, and assessment of pituitary function should be performed promptly after surgery. Patients undergoing transsphenoidal surgery are less likely to develop hypopituitarism than patients requiring a transcranial surgical approach. About 50% of patients recover in at least one pituitary axis function after transsphenoidal surgery.

40.2.2 Effect of Radiotherapy on Hypothalamo–Pituitary Function

HP axis dysfunction develops following conventional or stereotactic radiotherapy for a pituitary adenoma, but also in patients who receive radiation therapy for skull base, parasellar and primary brain tumours. Children who undergo prophylactic cranial or whole body irradiation are also at risk of hormone deficiency. Damage to both hypothalamus and pituitary can occur with radiation; however, hypothalamus in more susceptible than pituitary. Somatotropes are the most sensitive, and thyrotropes are the most resistant pituitary cell type to radiation damage. Somatotropes alone are affected by radiation doses below 20 Gy, whereas other pituitary

Table 40.3 Effects of cranial irradiation on the hypothalamic–pituitary axis in children and adolescents

Mean dose delivered (Gy)	Time to develop GH deficiency (months)	Effect on LH and FSH	ACTH and TSH deficiency
10–15	>60	Nil	Nil
15–30	36–60	Precocious puberty	Nil
30–50	12–36	Hypogonadism likely	Possible
>50	12	Very likely	Very likely

hormone deficiencies occur at doses between 30 and 50 Gy (Table 40.3). The development of hormonal abnormalities correlates strongly with biological effective dose delivered (total dose, fraction size and radiotherapy schedule) and the time interval after completion of radiotherapy [2]. Up to two-thirds of patients ultimately develop hormone insufficiency after a median dose of 50 Gy (5000 rad) directed at the skull base. Endocrine disorders may not develop for decades after completion of cancer treatment, underscoring the importance of lifelong surveillance for those at risk.

40.2.3 Effect of Drugs on Hypothalamo–Pituitary Function

Conventional chemotherapeutic agents are not associated with pituitary hormone deficiencies. Glucocorticoids are often used to treat cancers; long-term use can result in suppression of the hypothalamic–pituitary–adrenal axis. Adrenal insufficiency may persist even if steroids are gradually withdrawn and recovery of adrenal function can take months to years. Immunomodulatory antibodies (immune checkpoint inhibitors) that can be used to disrupt the immune system to generate an immune reaction against the tumour are associated with autoimmune endocrinopathies, mainly hypophysitis leading to multiple pituitary hormone deficiency [3]. These drugs include anti CTLA4 antibody (ipilimumab) and anti-PD1 antibody (nivolumab, pembrolizumab). Management includes stopping the offending drug, replacing affected hormones, and treatment with glucocorticoids for recovery for pituitary function.

40.3 Assessment of Pituitary Function

The presentation of hypopituitarism can be nonspecific, and a high index of suspicion is required to make the diagnosis. Therefore, all patients at risk of HP axis dysfunction (Table 40.1) should undergo periodic assessment for hormone deficiencies. Biochemical diagnosis of hormone insufficiency is made by demonstrating low levels of respective pituitary trophic hormones in the setting of low levels of target hormones (Table 40.4). Provocative test may be used to assess pituitary reserve in case of GH and ACTH deficiency.

Table 40.4 Assessment of pituitary function

Test	Interpretation
GH deficiency	
IGF1, IGFBP3 Dynamic GH tests Insulin tolerance test: Regular insulin (0.05–0.15 U/kg IV) GHRH test: 1 μg/kg IV Arginine test: 30 g IV over 30 min Dopa test: 500 mg PO Clonidine: 100 mg/m^2	Decreased GH levels >10 μg/L, normal response 5–10 μg/L, equivocal <5 μg/L, GH deficiency
ACTH deficiency	
ACTH 8 am serum cortisol ACTH stimulation test (synacthen) 250 μg IM or IV	Low Less than 5 μg/dL is deficiency Peak cortisol ≥21 μg/dL or ↑ ≥7 μg/dL, normal response 5–15 μg/dL, equivocal >15 μg/dL, normal
TSH deficiency	
TSH Free T4	Low, normal or slightly high Low
Prolactin	
Hypogonadism	
LH, FSH Testosterone in men Oestradiol in women	Low Less than 300 ng/ml Low
Central precocious puberty	
GnRH stimulation test Leuprolide 20 μg/kg Triptorelin 0.1 mg/m^2	LH levels at 1–3 h >5 IU/L Oestradiol at 24 h >50 pg/ml
ADH deficiency	
Serum sodium Urine osmolarity Water deprivation test if needed	Serum sodium >145 meq/L Urine osmolarity <300 mosm/kg

40.3.1 GH Deficiency

GH deficiency (GHD) is the most common endocrine disorder observed in survivors of CNS tumours, even after the exclusion of HP tumours. GH-deficient children and adolescents usually present with decreased linear growth velocity and short stature. Other clinical features in children include increased body fat, micropenis, high-pitched voice and a propensity to hypoglycaemia. In adults, GH deficiency manifest with changes in body composition, lipid metabolism, impaired quality of life and cardiovascular dysfunction. Insulin-like growth factor 1 (IGF1) and IGF-binding protein 3 (IGFBP-3) levels may not be accurate surrogate markers in cancer survivors. Dynamic test to determine GH status should be considered in children with poor growth velocity and in adults who are willing for GH replacement. GH replacement is indicated in children and adults with GH deficiency whose cancer is in remission. GH therapy improves final height in children and also helps in improving bone health, body composition, quality of life and cardiovascular risk factors both in adults and children. Children may not attain their adult height potential because of other factors such as skeletal/spinal sequelae to radiation, abnormal pubertal timing, chronic illness and primary disease burden. GH therapy is started 1 year after cancer remission in children and 2 years after in adults. Long-term follow-up data do not support increased risks of mortality or cancer recurrence in cancer survivors treated with GH. Treatment with hGH was associated with a higher risk of second neoplasms, mostly meningioma in two reports from a large multicentre childhood Cancer Survivor Study [1]. The increase in meningioma has been attributed to radiotherapy in other studies. Further long-term follow-up studies are needed to establish whether GH therapy affects the risk of secondary cancer.

40.3.2 Central Precocious Puberty

Central precocious puberty (CPP) is defined by the onset of pubertal development before the age of 8 years in girls and 9 years in boys as a result of the premature activation of the HP gonadal axis. CPP is commonly seen in tumours arising near the hypothalamus or optic pathways (glioma, hamartoma). The prevalence of CPP in patients with these tumours is between 26% and 39% [2]. Exposure to HP axis radiation at a wide range of doses (20–50 Gy) has also been associated with CPP. Diagnosis of CPP requires clinical assessment of pubertal status (tanner staging) and GnRH stimulation test. The treatment of CPP relies on suppression of HP gonadal axis with gonadotropin-releasing hormone agonist depot preparations (leuprolide or triptorelin). Patients with a history of CPP may experience gonadotropin deficiency as an effect of CNS radiotherapy in later years.

40.3.3 Adrenocorticotropic Hormone Deficiency

ACTH deficiency is a potentially life-threatening component of hypopituitarism. Patients with chronic ACTH deficiency usually present with progressive symptoms of fatigue, anorexia, postural hypotension and weight loss. Children are at risk of hypoglycaemia. In contrast to primary adrenal failure, hypocortisolism associated with pituitary failure is not accompanied by hyperpigmentation or mineralocorticoid deficiency. Diagnosis is by measurement of 8 am plasma cortisol, ACTH and ACTH stimulation test. Glucocorticoid replacement therapy improves most symptoms of adrenal insufficiency, oral hydrocortisone 10–15 mg/m^2 in three divided doses or prednisolone 2.5–5 mg once daily can be used. Doses should be increased several fold during periods of acute illness or stress to prevent adrenal crisis.

40.3.4 Thyroid-Stimulating Hormone Deficiency

TSH deficiency occurs late in most pituitary disorders. Symptoms are often milder than in primary hypothyroidism. Levothyroxine is the treatment as in primary hypothyroidism; starting dose in most patients is ~1.6 μg/kg/day. As thyroxine increases the metabolism of cortisol and can precipitate an adrenal crisis, ACTH deficiency must always be excluded or treated before initiation of thyroxine therapy. Levothyroxine dose can be titrated to increase serum-free T4 into the upper-normal range. Measurement of TSH is not helpful in the monitoring of T4 replacement therapy unlike primary hypothyroidism.

40.3.5 Gonadotropin Deficiency

Hypogonadism may present with pubertal delay, arrested puberty, primary or secondary amenorrhea or sex hormone deprivation symptoms (decreased libido, loss of muscle mass and decreased facial and body hair growth in males, breast atrophy and vaginal dryness in females) depending on attained pubertal stage at the time of onset of LH/FSH deficiency. The laboratory diagnosis is based on the measurement of LH, FSH and oestradiol (females) or testosterone (males). The treatment relies on sex hormone replacement therapy (oestrogen with or without progesterone in females and testosterone in males). Fertility can be achieved in both male and female patients with GnRH or gonadotropin therapy (HCG with or without recombinant HMG/FSH).

40.3.6 ADH Deficiency (Vasopressin)

Central diabetes insipidus (DI) commonly occurs following pituitary surgery; up to 30% of patients develop transient DI within 24 h of surgery. After a few days, it may transition to a 2- to 3-week period of inappropriate antidiuresis (SIADH), after which the DI may or may not recur permanently. DI is uncommon with pituitary adenoma, more commonly seen in hypothalamic tumours and craniopharyngioma. Patients present with polydipsia and polyuria (>250 mL/h) which is hypotonic (specific gravity <1.005 or urine osmolality <200 mOsm/kg H_2O). An increase in serum osmolality and hypernatremia develops if fluid intake is restricted. The investigation for confirmation of DI is water deprivation test. ADH deficiency can then be distinguished from nephrogenic DI by the response to vasopressin. Desmopressin (synthetic analogue of vasopressin) is the treatment for central DI; either oral or intranasal desmopressin can be used.

40.4 Pituitary Adenoma and Hypersecretion Syndromes

Pituitary adenomas are the most common cause of sellar mass, accounting for about 15% of all intracranial neoplasms. Approximately half of pituitary adenomas are functional; they secrete distinct pituitary hormones leading to various clinical syndromes [4]. Secreted hormones include prolactin (40–50%) growth hormone (10–15%) or adrenocorticotropic hormone (3–6%) [4]. Thyroid-stimulating hormone (TSH)-secreting and gonadotropin-secreting adenomas are rare (<1% of pituitary adenomas). Tumours are currently classified into five main IHC types—PRL,

Table 40.5 Clinical and pathological classification of pituitary tumours

Adenoma type	Immunohistochemistry	Clinical syndrome
Lactotroph (prolactin secreting) Densely granulated Sparsely granulated	PRL PRL (a-SU)	Galactorrhea/hypogonadism
Somatotroph (GH secreting) Monohormonal Densely granulated Sparsely granulated Plurihormonal Mixed GH-PRL Mammosomatotroph Acidophilic stem cell	 GH GH GH, PRL (a-SU, b-TSH) GH, PRL (a-SU, b-TSH) PRL, GH	Acromegaly/gigantism Acromegaly/galactorrhea/ hypogonadism
Thyrotropinoma (TSH secreting) Monohormonal Plurihormonal	b-TSH, a-SU b-TSH, a-SU, GH, PRL	Hyperthyroidism
Corticotroph (ACTH secreting)	ACTH, b-endorphin, b-LPH	Cushing's syndrome
Gonadotrophin(FSH-LH secreting)	b-FSH, b-LH, (a-SU)	Silent

a-SU alpha subunit of TSH, FSH, LH, *b-LPH-beta* lipotropin

GH, ACTH, TSH and FSH–LH—which can be monohormonal or plurihormonal, with or without (silent) clinical features of hormone excess (Table 40.5). While these tumours are commonly benign, they have potent endocrine effects that lead to significant morbidity. Hormone-secreting pituitary adenomas require defined management protocols that can include relief of mass effect and control of hormone hypersecretion combining medical therapy, surgery and radiotherapy.

40.4.1 Prolactinoma [5]

Prolactin hypersecretion is the most prevalent hypothalamic–pituitary dysfunction, with prolactin secreting pituitary adenoma or prolactinoma being the most common cause of hyperprolactinemia. Prolactinoma is the most common pituitary adenoma and contributes to 40–45% of all pituitary adenomas. About 60% of the prolactinomas are microadenomas (diameter less than 10 mm).

40.4.1.1 Clinical Features

Prolactinomas are more common in females; female to male ratio for micro adenoma is 20:1, while for macroadenoma gender ratio is almost equal. Clinical manifestations in women are menstrual irregularities, galactorrhea and infertility, while men present with symptoms of hypogonadism like decreased shaving frequency, reduced libido, erectile dysfunction and infertility. Galactorrhea is uncommon in postmenopausal women and men due to lack of glandular tissue. Men usually harbour larger adenomas and present with tumour mass effects rather than hypogonadism.

40.4.1.2 Diagnosis

All patients with pituitary tumours should have serum prolactin levels measured. An elevated level of serum prolactin, above the gender-specific reference range, is called hyperprolactinemia (women >25 ng/ml and men >20 ng/ml). Prolactinomas are usually associated with a serum prolactin >100 ng/ml and a level >250 ng/ml is virtually diagnostic of a prolactinoma. Prolactin levels correlate with tumour size, and microadenomas tend to have a lower prolactin levels, commonly less than 200 ng/ml. Non-tumoral causes of hyperprolactinemia should be considered with serum prolactin <100 ng/ml despite the presence of microadenoma. Drugs are the most common cause of non-tumoral hyperprolactinemia, mainly antipsychotics, D2 antagonists, antidepressants and H2 antihistaminic. Other causes include primary hypothyrodisim, renal or hepatic insufficiency, polycystic ovary disease and chest/spine lesions. Occasionally, a patient with a very high serum PRL level might have a normal or slightly high result reported (false negative), called the "high-dose hook effect". The "hook effect", also called as prozone phenomenon, is a false low estimation of high levels of prolactin due to saturation of antibodies used in immunometric assays. To rectify "hook effect", prolactin should be estimated in dilution in all suspected prolactin secreting macro adenomas. In other cases with macroadenomas and mildly elevated serum prolactin (<100 ng/ml), elevation of prolactin is due to the stalk effect rather than the tumour secreting prolactin (compression of the pituitary stalk resulting in decreased tonic inhibition of prolactin by dopamine from the hypothalamus).

40.4.1.3 Treatment

Medical therapy with dopamine agonists (D2 receptor agonist, DA) is the first-line treatment for prolactinomas. Cabergoline and bromcriptine are the two commonly used dopamine agonists. Cabergoline due to its greater efficacy, longer duration of action and fewer adverse events has surpassed bromocriptine as the first-line therapeutic choice for most patients except in pregnancy. The major adverse effects of DA are nausea, vomiting and postural hypotension. Valvulopathy reported with higher dose of dopamine agonists are rarely seen in prolactinoma management. Cabergoline is initiated at a lower dose of 0.25–0.5 mg per week, dose has to be titrated up monthly in response to prolactin levels, till it normalizes. Patients with macroprolactinoma require clinical monitoring, visual field analysis and serum prolactin estimation monthly. In addition, MR imaging should be performed at an interval of 6–12 months to look for tumour size reduction. Follow-up MRI and visual field analysis are not required in microadenoma since the chance of tumour progression is very rare (5%). Control of hyperprolactinemia requires doses of cabergoline ranging from 0.25 to 3 mg/week; however, rarely patients may require doses up to 11 mg per week. After 2 years of medical treatment, dopamine-agonist therapy may be tapered to cessation if no adenoma is visible on MR imaging and serum prolactin levels are normalized. However, medical therapy may need to be reinstituted if hyperprolactinemia recurs. Responses to dopamine agonists are variable. Failure to normalize prolactin or to reduce tumour size by >50% after 1 year of treatment with optimal doses of DA is termed as dopamine agonist resistance.

Ten percent of patients with microadenomas and 18% of patients with macroadenoma show cabergoline resistance. Patients with cabergoline resistance should be offered surgery and/or radiotherapy. The indications of surgery in prolactinoma include DA resistance, drug intolerance and invasive macroadenoma compromising vision which fails to improve after drug treatment. Other indications for surgery are pituitary apoplexy and concurrent CSF rhinorrhoea which cannot be managed conservatively.

40.4.2 Growth Hormone-Secreting Pituitary Adenoma [6]

Growth hormone excess state produces gigantism and acromegaly before and after closure of epiphysis respectively. The most common cause of acrogigantism is pituitary adenoma (99%), commonly a benign macroadenoma (80%). Other rare causes of GH excess include GH-producing islet cell tumour/lymphoma or GHRH-producing tumours (hypothalamic hamartoma, choristoma, bronchial carcinoid, pheochromocytoma, etc.).

40.4.2.1 Clinical Features

Acromegaly is a Greek word meaning "akros" (extremity) and "megalos" (enlargement). Manifestations of acromegaly are caused by either central pressure effects of the pituitary mass or peripheral actions of excess GH and IGF1. When GH excess occurs in childhood; it leads to acrogigantism (defined as the height >97th percentile or 3SD above normal mean height for age). The somatic and metabolic effects of GH-IGF1 excess occur insidiously over several years, often resulting in delayed diagnosis, ranging from 6 to 10 years. Characteristic features of acromegaly include large fleshy lips and nose, spade-like hands, frontal skull bossing, prominent supraorbital ridges, macroglossia, prognathism, hyperhidrosis, oily skin and coarsened facial features (Fig. 40.1). Acromegaly is associated with an increased risk for diabetes mellitus, systemic hypertension, dyslipidaemia, osteoarthritis, cardiomyopathy, obstructive sleep apnoea, carpel tunnel syndrome, proximal myopathy, goitre and increased cancer mortality. In a meta-analysis of 16 studies, overall mortality rate was found to be increased in acromegaly with a standardized mortality ratio of 1.72.

40.4.2.2 Diagnosis

The best screening test for the diagnosis of acromegaly is elevated serum insulin-like growth factor 1 (IGF1), with a sensitivity of 97% and specificity of 90%. GH suppression test after 75 g oral glucose load is considered as the "gold standard" for the diagnosis of GH excess. The inability to suppress serum GH to 0.4 ng/mL using ultrasensitive GH assays or GH to <1.0 ng/mL using standard assays after oral glucose load is diagnostic of active acromegaly. Single random GH measurements are not usually recommended due to inherent episodic GH secretion from both normal and adenomatous pituitary. Contrast-enhanced MRI sella should be done to localize the source of GH excess. All patients should be screened for associated

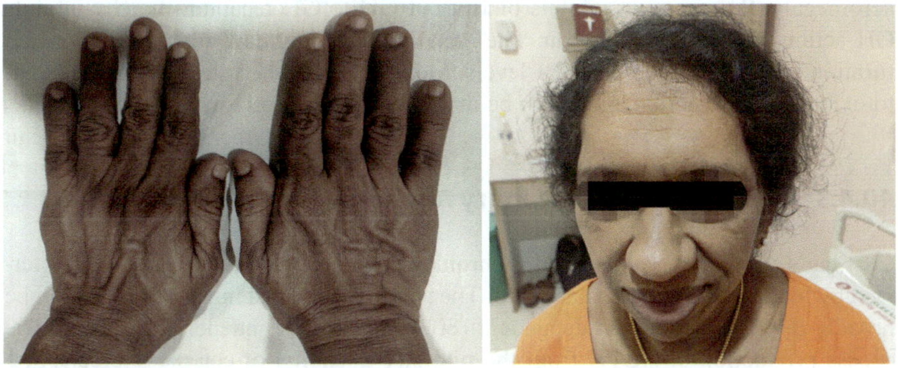

Fig. 40.1 Features of acromegaly

comorbidities. Colonoscopy at diagnosis is recommended due to the increased incidence of colonic polyps in acromegaly.

40.4.2.3 Treatment

Transsphenoidal surgery is the primary modality of treatment in all patients with acromegaly, irrespective of tumour size. The surgical cure rate is 80–90% in patients with microadenoma and 50% in macroadenoma. Patients with large and invasive tumours especially cavernous sinus invasion, initial serum GH >40 ng/ml and younger age often do not go into remission after surgery. Generally patients are reassessed after 12 weeks of surgery for possible remission, a glucose supressed GH level < 1 ng/ml and normal IGF1 levels are considered as cure. The options available to treat patients who are not in remission after surgery include repeat surgery, medical therapy and radiotherapy. Second surgery can be considered when the residual tumour is intrasellar as they are likely to be cured. Medical therapy for GH adenoma includes somatostatin receptor ligands (SRLs), dopamine agonists (DA) and GH receptor antagonist (pegvisomant) either alone or in combination. Medical therapy is lifelong and expensive. Radiotherapy can be considered in patients who cannot afford or tolerate medical therapy. However, a long lag time for normalization of GH–IGF1, need for medical management during interim period and progressive developments of hypopituitarism are the limitations of radiotherapy. SRLs, which bind on somatostatin receptor subtype 2 and subtype 5, thereby inhibiting GH release from the adenoma, are the preferred drugs in medical management. Octreotide long-acting release (20–40 mg intramuscular once a month) and Lanreotide Autogel (subcutaneous 60–120 mg once a month) are the commonly used SRLs. Pasireotide another SRL which acts predominantly on somatostatin receptor subtype 5 can be used in patients resistant to octreotide or lanreotide. Pegvisomant or cabergoline can be added in a patient with inadequate response to an SRL. Dopamine agonists (cabergoline and bromocriptine) can be considered alone in patients with mild symptoms and modest elevation of IGF1 levels (less than two times ULN). Pegvisomant is used in SRL therapy-resistant or intolerant acromegaly. Pegvisomant inhibits binding of GH to GH receptor, thereby inhibits GH action and IGF1 generation. The dose is 10–40 mg/day subcutaneously, and titration of dose is based on IGF1 levels. Major side effect of pegvisomant is hepatotoxicity; liver function test has to be monitored periodically while on treatment.

40.4.3 ACTH-Producing Pituitary Adenoma [7]

Cushing's syndrome is a disorder of chronic glucocorticoid excess and is characterized by features of protein catabolism. The aetiology of Cushing's syndrome can be broadly divided into ACTH dependent (80%) and ACTH independent (20%). The excess production of ACTH from the pituitary adenoma or ectopic source (commonly carcinoma lung, bronchial carcinoid, thymic tumour) causes glucocorticoid excess in ACTH-dependent Cushing's syndrome. Adrenal tumours or hyperplasia cause ACTH-independent Cushing's syndrome. Cushing's disease (pituitary cause) is the most common cause of endogenous Cushing's syndrome. Approximately 90% of patients with Cushing's disease have microadenoma, while macroadenoma contributes to the rest.

40.4.3.1 Clinical Features

Cushing's disease occurs more commonly in females, female: male ratio is 3:1. Easy bruising, facial plethora, purple striae >1 cm in width, cuticular atrophy and proximal myopathy are some of the features that best discriminate Cushing's syndrome (Fig. 40.2). Other clinical features include moon facies, truncal obesity, excess dorsocervical pad of fat, pedal oedema, diabetes mellitus, systemic hypertension, unexplained osteoporosis, cataract, depression, menstrual irregularities and hirsutism. Hyperpigmentation of the skin and mucosal areas occur in ACTH-dependent Cushing's syndrome. Insidious onset of disease, female gender, young age and features of androgen excess suggest a diagnosis of pituitary Cushing's syndrome.

40.4.3.2 Diagnosis

In a patient with suspected Cushing's syndrome endogenous hypercortisolaemia has to be confirmed with any two of the following tests:

- Urine free cortisol (at least two measurements).
- Late-night salivary cortisol (two measurements).
- Midnight serum cortisol.
- 1-mg overnight dexamethasone suppression test—serum cortisol value >1.8 μg/dl is considered positive.
- Low-dose dexamethasone suppression test (2 mg/day)—serum cortisol value >1.8 μg/dl is considered positive.

Once the diagnosis of Cushing's syndrome is confirmed, elevated ACTH can rule out an adrenal cause. High-dose dexamethasone suppression and corticotropin-releasing hormone (CRH) stimulation tests may be used to differentiate between

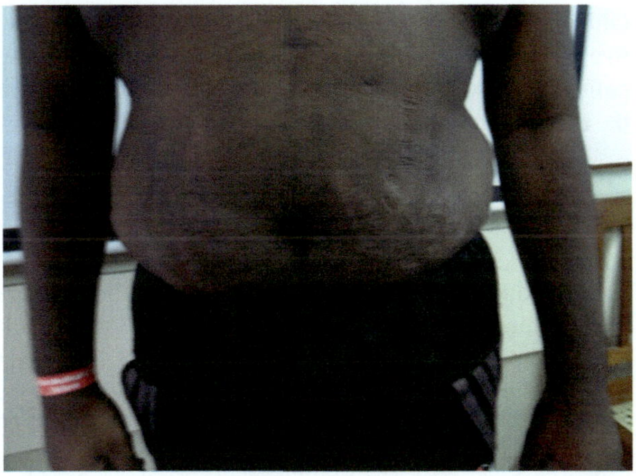

Fig. 40.2 Striae in ectopic Cushing's syndrome

pituitary and ectopic sources of ACTH hypersecretion, diagnostic accuracy of these tests are suboptimal: 65–100% sensitivity and 60–100% specificity for the high-dose dexamethasone test and 70–93% sensitivity and 95–100% specificity for the CRH test. High-resolution, gadolinium-enhanced dynamic MRI of the pituitary with specified protocols has to be done in patients with suspected pituitary adenoma. Because the adenomas are often small (as small as 2 mm), visible tumour may not be apparent in 25–40% of cases. Inferior petrosal sinus sampling (IPSS) should be done to localize ACTH hypersecretion (pituitary vs. ectopic) in cases with negative or equivocal imaging (adenoma size <6 mm) when the clinical and biochemical investigations were suggestive of a pituitary source.

40.4.3.3 Treatment [8]

Transphenoidal resection of adenoma is the treatment of choice for ACTH-secreting pituitary adenoma. Remission rate is around 70–90% for microadenoma and 50% for macroadenoma. Postoperative biochemical remission may be assessed by morning serum cortisol, serum ACTH and urine-free cortisol within the first week after surgery. Hypocortisolaemia develops in patients who go into remission; glucocorticoid replacement should be given until recovery of normal corticotroph function occurs (typically 3–18 months after surgery). Total or partial hypophysectomy may be performed when adenoma cannot be identified during surgery which results in 50–70% remission rates. Management of recurrent or persistent disease includes repeat surgery, radiotherapy or medical management. Medical therapies include steroidogenesis inhibitors, corticotroph-directed agents and glucocorticoid receptor blockers. Medical therapy is used in patients with failure to localize the source of ACTH excess, as preoperative preparation, during interim period after radiotherapy, presence of concurrent comorbidities which renders patient at high risk for surgery. Commonly used adrenal steroidogenesis inhibitors include ketoconazole, osilodrostat, metyrapone, mitotane and etomide. Somatostatin receptor ligand pasireotide acts at the pituitary level, reducing ACTH production and tumour growth.

Glucocorticoid receptor antagonist mifepristone can also be used, especially effective in hyperglycaemia. Bilateral adrenalectomy can be considered as an option in patient with severe refractory disease when all the others modalities fail to control the disease.

40.4.4 TSH-Producing Pituitary Adenoma

TSH-producing adenomas are rare (1–2% of the pituitary adenomas). These tumours are commonly invasive and large, but for the most part benign. Patient can present with symptoms of hyperthyroidism (weight loss, palpitation, tremors, insomnia, heat intolerance) and goitre. Biochemical evaluation shows elevated thyroid hormones (T4 and T3) along with normal or elevated TSH. Alpha subunit levels and TRH stimulation can differentiate between TSH overproduction by a TSH-secreting tumour and thyroid hormone insensitivity. About 30% of TSHomas can co-secrete growth hormone or prolactin. Consequently, features of acromegaly or hyperprolactinemia may also be present. Surgical resection is typically the first-line therapy. Remission rates are low since the tumour is commonly large and invasive. Radiation therapy or medical therapy with SRLs (octreotide) can be used in patients who do not achieve biochemical remission after surgery.

References

1. Chemaitilly W, Cohen LE. Diagnosis of endocrine disease: endocrine late-effects of childhood cancer and its treatments. Eur J Endocrinol. 2017;176:R183–203.
2. Sklar CA, Antal Z, Chemaitilly W, et al. Hypothalamic–pituitary and growth disorders in survivors of childhood cancer: an endocrine society- clinical practice guideline. J Clin Endocrinol Metabol. 2018;103:2761–84.
3. Crowne E, Gleeson H, Benghiat H, et al. Effect of cancer treatment on hypothalamic–pituitary function. Lancet Diabet Endocrinol. 2015;3:568–76.
4. Mehta GU, Lonser RR. Management of hormone-secreting pituitary adenomas. Neuro-Oncology. 2016;19(6):762–73.
5. Melmed S, Casanueva FF, Hoffman AR, et al. Diagnosis and treatment of hyperprolactinemia: an endocrine society clinical practice guideline. J Clin Endocrinol Metabol. 2011;96:273–88.
6. Katznelson L, Laws ER, Melmed S, et al. Acromegaly: an endocrine society clinical practice guideline. J Clin Endocrinol Metabol. 2014;99:3933–51.
7. Nieman LK, Biller BMK, Findling JW, et al. The diagnosis of Cushing's syndrome: an endocrine society clinical practice guideline. J Clin Endocrinol Metabol. 2008;93:1526–40.
8. Nieman LK, Biller BMK, Findling JW, et al. Treatment of Cushing's syndrome: an endocrine society clinical practice guideline. J Clin Endocrinol Metabol. 2015;100:2807–31.

GPSR Compliance

The European Union's (EU) General Product Safety Regulation (GPSR) is a set of rules that requires consumer products to be safe and our obligations to ensure this.

If you have any concerns about our products, you can contact us on ProductSafety@springernature.com

In case Publisher is established outside the EU, the EU authorized representative is:

Springer Nature Customer Service Center GmbH
Europaplatz 3
69115 Heidelberg, Germany

Batch number: 10726058

Printed by Printforce, the Netherlands